Molecular Pathogenesis of Cardiac Arrhythmia

Molecular Pathogenesis of Cardiac Arrhythmia

Editors

Yosuke Okamoto
Kyoichi Ono

MDPI • Basel • Beijing • Wuhan • Barcelona • Belgrade • Manchester • Tokyo • Cluj • Tianjin

Editors

Yosuke Okamoto
Department of Cell Physiology
Akita University Graduate School of Medicine
Akita
Japan

Kyoichi Ono
Executive Director for Research, Regional Revitalization and Public Relations
Akita University
Akita
Japan

Editorial Office
MDPI
St. Alban-Anlage 66
4052 Basel, Switzerland

This is a reprint of articles from the Special Issue published online in the open access journal *Biomolecules* (ISSN 2218-273X) (available at: www.mdpi.com/journal/biomolecules/special_issues/pathogenesis_cardiac_arrhythmia).

For citation purposes, cite each article independently as indicated on the article page online and as indicated below:

LastName, A.A.; LastName, B.B.; LastName, C.C. Article Title. *Journal Name* **Year**, *Volume Number*, Page Range.

ISBN 978-3-0365-5614-7 (Hbk)
ISBN 978-3-0365-5613-0 (PDF)

Contents

biomolecules

Editorial

Molecular Pathogenesis of Cardiac Arrhythmia

Yosuke Okamoto * and Kyoichi Ono

Department of Cell Physiology, Akita Graduate School of Medicine, Hondo 010-8543, Japan
* Correspondence: okamoto@med.akita-u.ac.jp

The heart is a necessary organ for sustaining life in mammals, and it is the first organ to function during early development. Once this primary engine begins to operate, it continues to pump blood throughout the body until the individual animal dies. The miraculous homeostasis of the heartbeat, which involves tremendous energy consumption, is naturally synchronized with systemic homeostasis. The minimum unit of energy in the body is called the ATP, which is produced in large quantities in organelles called mitochondria. Recently, it has been shown that the quality maintenance of cardiac mitochondria is achieved by mobilizing the immune system [1]. In other words, cardiac mitochondrial function is being addressed as a systemic challenge. In this Special Issue, Dr. Takeuchi and Prof. Matsuoka provide a review focusing on Ca^{2+} dynamics in cardiac mitochondria [2]. Mitochondrial biophysical variables inside the cell are indirectly measured by selective drugs, selective fluorescent reagents, and selective genetic modification. Sophisticated studies have shown that mitochondrial ATP production and mitochondrial Ca^{2+} concentration are closely linked. During heart contraction, the mitochondrial Ca^{2+} concentration and NADH oxidation rate increase synchronously. Without a mechanism to excrete increased intramitochondrial Ca^{2+}, the Ca^{2+} overload causes excessive oxidative phosphorylation, which produces reactive oxygen species (ROS); this, in turn, induces ventricular arrhythmias. The $Na^+/Ca^{2+}/Li^+$ exchanger (NCLX) is responsible for mitochondrial Ca^{2+} excretion. While NCLX is thought to contribute to mitochondrial homeostasis, chronic activation of NCLX in heart failure conditions, where ATP production has shifted from being fatty acid dependent to being carbohydrate dependent, is expected to result in mitochondrial Ca^{2+} deprivation. Thus, the evaluation of disease status using mitochondrial function is critical and, at the same time, largely complicated. In order to understand such complex biological phenomena, Dr. Takeuchi and Prof. Matsuoka emphasize the importance of mathematical model analysis using computers.

Dr. Tsumoto and Prof. Kurata reviewed the bifurcation analysis on the cardiac action potentials, a type of mathematical model analysis [3]. In the bifurcation analysis, a set of differential equations (dynamic equations) is considered a system, and a sudden change from one stable status to another stable status is referred to as a bifurcation. The highly abstract approach of describing the action potential as a system enables us to identify specific single molecules that are essential for the stability of the system. Dr. Tsumoto and Prof. Kurata present various bifurcation analyses of cardiomyocyte excitation phenomena. While their review is effectively illustrative of the power of bifurcation analysis, they also identify a problem with bifurcation analysis, namely, that the interpretation of the bifurcations is not simple. For example, in analyzing early after-depolarization (EAD) by reducing the conductance of I_{Ks} and I_{Kr}, more than four bifurcations have been observed. Each bifurcation is an arrhythmia with a different molecular mechanism. In addition, different underlying models naturally lead to different results; in some cases, EAD can be explained by action potential reactivation, while in others, it is accompanied by intracellular Ca^{2+} release. Eventually, the accumulation of precise experimental data and trial and error in the analysis seem to be indispensable.

On the other hand, to elucidate the mechanism of cardiac function, Dr. Namekata et al. and Dr. Okada et al. employed a strategy of comparing different animal species [4,5]. To

Citation: Okamoto, Y.; Ono, K. Molecular Pathogenesis of Cardiac Arrhythmia. *Biomolecules* **2022**, *12*, 1393. https://doi.org/10.3390/biom12101393

Received: 27 September 2022
Accepted: 28 September 2022
Published: 29 September 2022

our knowledge, the normal beating heart is driven by a coupled-clock system in which Ca^{2+} release preceding the action potential is reflected in depolarization via a Na^+–Ca^{2+} exchanger (NCX). Surprisingly, Dr. Namekata et al. pharmacologically excluded the need for an NCX in guinea pig heartbeat. Indeed, similar reports have previously been reported in cellular experiments [6]. Furthermore, the authors of the coupled-clock system themselves have recently reported in vivo observations demonstrating that the action potentials of pacemaker cells are heterogeneous, with Ca^{2+} release preceding the action potential, and the action potential preceding Ca^{2+} release [7]. It seems that the coupled clock is simply a phenomenological observation. Similarly, Dr. Okada et al. dismissed the high-scoring report stating that Tbx18 is the most critical transcription factor that characterizes pacemaker cells [8]. They reported that SHOX2 is the essential transcription factor for pacemaker cells when animal species, including humans, are considered [5].

The study of heart rhythm in this way has been a passion, and there are a wide variety of experimental and analytical methods. The most straightforward approach is to inhibit specific ion channels in the study of arrhythmias. For example, inhibition of human ether-a-go-go-related gene (hERG) channels is known to produce ventricular arrhythmias, and Dr. Tamura et al. reported that synthetic estrogen has a protective effect against hERG channel inhibition [9]. Dr. Shimizu et al. reported that the intravenous anesthetic propofol suppresses hyperpolarization-activated, cyclic nucleotide-gated (HCN) channels and is associated with bradyarrhythmia [10]. Furthermore, Dr. Flowler and Dr. Zissimopoulos reported a review of ventricular arrhythmias associated with ryanodine receptor 2 (RyR2) mutations [11]. RyR2 is a giant ion channel with ~560 kDa subunits forming a tetramer, and more than 350 mutations have been reported. Mutations, thus, need to be classified as indicated in this review.

By the way, even though the molecular mechanisms of cardiac excitation and contraction are extremely well regulated, there is one heart region that is innately susceptible to arrhythmias: the pulmonary veins. The pulmonary veins are the source of the most frequent arrhythmia: atrial fibrillation. Bredeloux et al. have provided a review in this Special Issue on the history and latest findings in pulmonary vein research [12], and Okamoto et al. contributed new insights into the molecular mechanisms of pulmonary vein arrhythmias [13].

Finally, Dr. Omatsu-Kanbe discovered atypical cardiomyocytes (AMCs), which she reviewed in this Special Issue [14]. AMC is a new cell type, discovered in the cardiac myoblast removal fraction that was previously discarded in the experimental process. Because of their cell fusion ability, AMCs have the potential to play an active role in the healing process after myocardial injury, but their complete characteristics remain a mystery. It is deeply satisfying to see that the review manuscript on AMCs has finally been accepted for publication. New discoveries always have to fight against traditional standards. Still, researchers cannot stop fighting. We are proud to have produced a Special Issue that encourages new discoveries, such as that of AMCs.

Acknowledgments: We would like to acknowledge all the authors for this Special Issue.

Conflicts of Interest: The authors declare no conflict of interest.

References

1. Nicolás-Ávila, J.A.; Lechuga-Vieco, A.V.; Esteban-Martínez, L.; Sánchez-Díaz, M.; Díaz-García, E.; Santiago, D.J.; Rubio-Ponce, A.; Li, J.L.Y.; Balachander, A.; Quintana, J.A.; et al. A Network of Macrophages Supports Mitochondrial Homeostasis in the Heart. *Cell* **2020**, *183*, 94–109.e23. [CrossRef] [PubMed]
2. Takeuchi, A.; Matsuoka, S. Physiological and Pathophysiological Roles of Mitochondrial Na^+-Ca^{2+} Exchanger, NCLX, in Hearts. *Biomolecules* **2021**, *11*, 1876. [CrossRef] [PubMed]
3. Tsumoto, K.; Kurata, Y. Bifurcations and Proarrhythmic Behaviors in Cardiac Electrical Excitations. *Biomolecules* **2022**, *12*, 459. [CrossRef] [PubMed]
4. Namekata, I.; Jitsukata, K.; Fukuda, A.; Odaka, R.; Hamaguchi, S.; Tanaka, H. Intracellular Ca^{2+}-Mediated Mechanisms for the Pacemaker Depolarization of the Mouse and Guinea Pig Sinus Node Tissue. *Biomolecules* **2022**, *12*, 377. [CrossRef] [PubMed]

5. Okada, D.; Okamoto, Y.; Io, T.; Oka, M.; Kobayashi, D.; Ito, S.; Yamada, R.; Ishii, K.; Ono, K. Comparative Study of Transcriptome in the Hearts Isolated from Mice, Rats, and Humans. *Biomolecules* **2022**, *12*, 859. [CrossRef] [PubMed]
6. Himeno, Y.; Toyoda, F.; Satoh, H.; Amano, A.; Cha, C.Y.; Matsuura, H.; Noma, A. Minor contribution of cytosolic Ca^{2+} transients to the pacemaker rhythm in guinea pig sinoatrial node cells. *Am. J. Physiol. Heart Circ. Physiol.* **2011**, *300*, 251–261. [CrossRef] [PubMed]
7. Bychkov, R.; Juhaszova, M.; Tsutsui, K.; Coletta, C.; Stern, M.D.; Maltsev, V.A.; Lakatta, E.G. Synchronized Cardiac Impulses Emerge From Heterogeneous Local Calcium Signals Within and Among Cells of Pacemaker Tissue. *JACC Clin. Electrophysiol.* **2020**, *6*, 907–931. [CrossRef]
8. Kapoor, N.; Liang, W.; Marbán, E.; Cho, H.C. Direct conversion of quiescent cardiomyocytes to pacemaker cells by expression of Tbx18. *Nat. Biotechnol.* **2013**, *31*, 54–62. [CrossRef] [PubMed]
9. Tamura, F.; Sugimoto, S.; Sugimoto, M.; Sakamoto, K.; Yamaguchi, M.; Suzuki, T.; Fukuda, K.; Ieda, M.; Kurokawa, J. The Effect of a Synthetic Estrogen, Ethinylestradiol, on the hERG Block by E-4031. *Biomolecules* **2021**, *11*, 1385. [CrossRef] [PubMed]
10. Shimizu, M.; Mi, X.; Toyoda, F.; Kojima, A.; Ding, W.-G.; Fukushima, Y.; Omatsu-Kanbe, M.; Kitagawa, H.; Matsuura, H. Propofol, an Anesthetic Agent, Inhibits HCN Channels through the Allosteric Modulation of the cAMP-Dependent Gating Mechanism. *Biomolecules* **2022**, *12*, 570. [CrossRef] [PubMed]
11. Fowler, E.D.; Zissimopoulos, S. Molecular, Subcellular, and Arrhythmogenic Mechanisms in Genetic RyR2 Disease. *Biomolecules* **2022**, *12*, 1030. [CrossRef] [PubMed]
12. Bredeloux, P.; Pasqualin, C.; Bordy, R.; Maupoil, V.; Findlay, I. Automatic Activity Arising in Cardiac Muscle Sleeves of the Pulmonary Vein. *Biomolecules* **2021**, *12*, 23. [CrossRef] [PubMed]
13. Okamoto, Y.; Aung, N.Y.; Tanaka, M.; Takeda, Y.; Takagi, D.; Igarashi, W.; Ishii, K.; Yamakawa, M.; Ono, K. Preferential Expression of Ca^{2+}-Stimulable Adenylyl Cyclase III in the Supraventricular Area, including Arrhythmogenic Pulmonary Vein of the Rat Heart. *Biomolecules* **2022**, *12*, 724. [CrossRef] [PubMed]
14. Omatsu-Kanbe, M.; Fukunaga, R.; Mi, X.; Matsuura, H. Atypically Shaped Cardiomyocytes (ACMs): The Identification, Characterization and New Insights into a Subpopulation of Cardiomyocytes. *Biomolecules* **2022**, *12*, 896. [CrossRef] [PubMed]

biomolecules

Review

Physiological and Pathophysiological Roles of Mitochondrial Na^+-Ca^{2+} Exchanger, NCLX, in Hearts

Ayako Takeuchi [1,2,*] **and Satoshi Matsuoka** [1,2]

1 Department of Integrative and Systems Physiology, Faculty of Medical Sciences, University of Fukui, Fukui 910-1193, Japan; smatsuok@u-fukui.ac.jp
2 Life Science Innovation Center, University of Fukui, Fukui 910-1193, Japan
* Correspondence: atakeuti@u-fukui.ac.jp; Tel.: +81-776-61-8311

Abstract: It has been over 10 years since *SLC24A6/SLC8B1*, coding the $Na^+/Ca^{2+}/Li^+$ exchanger (NCLX), was identified as the gene responsible for mitochondrial Na^+-Ca^{2+} exchange, a major Ca^{2+} efflux system in cardiac mitochondria. This molecular identification enabled us to determine structure–function relationships, as well as physiological/pathophysiological contributions, and our understandings have dramatically increased. In this review, we provide an overview of the recent achievements in relation to NCLX, focusing especially on its heart-specific characteristics, biophysical properties, and spatial distribution in cardiomyocytes, as well as in cardiac mitochondria. In addition, we discuss the roles of NCLX in cardiac functions under physiological and pathophysiological conditions—the generation of rhythmicity, the energy metabolism, the production of reactive oxygen species, and the opening of mitochondrial permeability transition pores.

Keywords: mitochondria; heart; mitochondrial Na^+-Ca^{2+} exchanger; NCLX; metabolism; Ca^{2+} signaling

Citation: Takeuchi, A.; Matsuoka, S. Physiological and Pathophysiological Roles of Mitochondrial Na^+-Ca^{2+} Exchanger, NCLX, in Hearts. *Biomolecules* **2021**, *11*, 1876. https://doi.org/10.3390/biom11121876

Academic Editors: Yosuke Okamoto and Kyoichi Ono

Received: 30 November 2021
Accepted: 10 December 2021
Published: 14 December 2021

1. Introduction

It has been almost a half century since a mitochondrial Na^+-Ca^{2+} exchange (NCX_{mit}) system was discovered in the heart [1], and it has been more than 10 years since *SLC24A6/SLC8B1*, coding the $Na^+/Ca^{2+}/Li^+$ exchanger (NCLX), was identified as the gene responsible for the system [2]. Owing to this molecular identification, our understanding of physiological and pathophysiological roles of NCX_{mit} has dramatically increased in various types of cells, including cardiomyocytes, neurons, astrocytes, B lymphocytes, pancreatic β cells, and brown adipocytes [3–10]; see also reviews [11–13]. In the heart, NCX_{mit} comprises the major Ca^{2+} efflux mechanism to balance against Ca^{2+} influx via mitochondrial Ca^{2+} uniport (CU_{mit}) activity. For the physiological functions of cardiomyocytes, it is vital for mitochondrial matrix Ca^{2+} to be maintained within an appropriate range because several metabolic enzymes are activated by Ca^{2+} to supply ATP, meeting cellular ATP demand, but excess Ca^{2+} causes mitochondrial dysfunction via opening of mitochondrial permeability transition pores (mPTP) (see other reviews for details [14,15]). In this review, we provide an overview of recent findings regarding NCX_{mit}, especially focusing on its biophysical properties, distributions, and physiological and pathophysiological roles in the heart.

2. Tissue-Specific Characteristics of Mitochondrial Ca^{2+} Dynamics

The mitochondrial Ca^{2+} handling system varies from tissue to tissue, which may contribute to tissue-specific tuning of mitochondrial as well as cellular functions. It is well accepted that CU_{mit} activity is lower in the heart than in other tissues [16]. This lower CU_{mit} activity may be due to differences in stoichiometry among a pore-forming protein (MCU), a dominant negative subunit (MCUb), and/or an EF-hand containing Ca^{2+}-sensitive regulator (MICU1)—an MCUb:MCU ratio that is higher and an MICU1:MCU ratio that is lower in the heart [17–19]. In addition, Wescott et al. [20] reported that "gatekeeping" of CU_{mit} via a certain cytosolic Ca^{2+} threshold was not observed in heart, though it has been well described functionally and structurally in other tissues or cell types (see review [21]). These heart-specific characteristics of the mitochondrial Ca^{2+} influx system may contribute to preventing mitochondrial Ca^{2+} overload in the heart, where cytosolic Ca^{2+} periodically rises.

In order to balance against the mitochondrial Ca^{2+} influx, NCX_{mit} and H^+-Ca^{2+} exchange (HCX_{mit}) extrude Ca^{2+} from mitochondria, with the former accounting for the major component in excitable tissues such as the heart and brain, and the latter being dominant in non-excitable tissues such as the liver and kidney (see review [22]). Rysted et al. [23] quantitatively compared the NCX_{mit} activity in mitochondria isolated from mouse brains, livers, and hearts. By evaluating extra-mitochondrial Ca^{2+} using Calcium Green-5N, they demonstrated that the rate of Na^+-dependent Ca^{2+} efflux from mitochondria was ~3-fold larger in the brain than in the heart. This well agrees with the lower CU_{mit} activity in the heart compared with other tissues [16,18]. Interestingly, the NCX_{mit} activity in the liver was negligible, despite the fact that it has the highest mRNA expression level of NCLX. The authors attributed this to extra-mitochondrial expression of NCLX protein in the liver.

In the heart, the fraction of Na^+-dependent Ca^{2+} efflux to total Ca^{2+} efflux is 60–100%, depending on species and experimental conditions [8,23–25]. The remaining fraction should be mediated by HCX_{mit}, though its contribution in the heart has been controversial. Leucine-zipper-EF hand-containing transmembrane (Letm1), which was initially shown to mediate H^+-dependent Ca^{2+} influx into mitochondria [26,27], was proposed as the gene responsible for HCX_{mit}. Natarajan et al. [28] detected H^+-induced Ca^{2+} efflux from rat cardiac mitochondria, which were dependent on the free matrix Ca^{2+} concentration. Furthermore, they confirmed Letm1-mediated Ca^{2+} efflux from mitochondria by demonstrating a diminished Ca^{2+} efflux rate in permeabilized H9c2 cells due to Letm1 knockdown. Interestingly, they found that the expression level of the Letm1 protein in mitochondria was higher in the heart than in the liver, though the functional contribution of HCX_{mit} was much higher in the liver than in the heart. Post-translational modifications or extra-mitochondrial localization of Letm1 protein in the heart, just as reported for NCLX [23], may explain the disparity between the expression level and function.

3. Biophysical Properties of NCX_{mit}

The electrogenicity of NCX_{mit} had been controversial [29–32]. These controversies were raised largely because mitochondrial membrane potential ($\Delta\Psi$) is affected by H^+ movements across the mitochondrial inner membrane via the electron transport chain, F_1F_o-ATP synthase, and so on. Therefore, detecting an NCX_{mit}-mediated $\Delta\Psi$ change with the exclusion of the $\Delta\Psi$ change via the fluctuation of energy metabolism was challenging. In order to overcome this problem, it is necessary to clamp $\Delta\Psi$. Recently, our group succeeded in recording membrane currents through NCX_{mit} in mouse cardiac mitochondria using whole-mitoplast patch clamp methods, thus settling the controversy [8].

The characteristics of the NCX_{mit} current in forward mode, i.e., an extra-mitochondrial Na^+-induced inward current with Ca^{2+} in the pipette, corresponded well to those in previous reports—the currents were diminished by the NCX_{mit} inhibitor CGP-37157 [32,33], the Hill coefficient for cytosolic Na^+ was around 3–4 [29,32], and Li^+ can be substitutable for Na^+ with ~70% lower efficacy [23,24]. One deviation was that the half-maximum concentration for Na^+, 35.6 mM, was higher than the reported value of 1–8 mM in the heart (8 mM in [29]; 1 mM in [32]), possibly because the Na^+-permeable background current could exist in the mitoplast preparations. The lower efficacy of Li^+ compared to Na^+ in exchanging for Ca^{2+} was also reported in brain mitochondria, to a similar extent as in the heart [10,23]. Through molecular modeling analysis based on *Metanococcus jannaschii* NCX_Mj and *Archaeoglobus fulgidus* CAX_Af, combined with functional analysis of human NCLX mutants, distinct amino acid residues in NCLX were identified as determining Na^+ or Li^+ binding [34]. That is, N149, P152, D153, N467, S468, and G494 were proposed to render Li^+ selectivity, whereas D471 was proposed to render Na^+ selectivity [34]. Giladi et al. [35] independently analyzed NCX_Mj-derived mutant NCLX_Mj, with nine substituted resides causing a NCLX-like phenotype, and found that peptides 248–255 were sensitive only to Li^+ binding, but not to Na^+ nor Ca^{2+} binding. Therefore, it is reasonable that the efficacy of exchanging for Ca^{2+} was different between Na^+ and Li^+. Although the three-dimensional (3D) structure of NCLX has not been solved yet, recent advances in artificial intelligence-based structure prediction methods makes it possible to easily visualize a putative 3D structure of NCLX. Figure 1 shows a putative 3D structure of human NCLX (Q6J4K2), predicted using AlphaFold [36], with specific residues highlighted that are suggested to be functionally important.

In the whole-mitoplast patch clamp experiments, the NCX_{mit} current in reverse mode—an extra-mitochondrial Ca^{2+}-induced outward current with Na^+ in the pipette—could not be recorded [8]. This was rather surprising to us because the reverse mode of NCX_{mit} activity was previously reported to exist in mitochondria of rat cardiomyocytes [32,37]. Further evaluation of intra-mitochondrial Ca^{2+} using Fluo-8 in isolated mitochondria revealed that the reverse mode of NCX_{mit} activity did exist in the heart. That is, CGP-37157-sensitive and intra-mitochondrial Na^+-dependent Ca^{2+} influx was detected, but the rate was too slow to be recorded electrophysiologically [8]. What is the mechanism underlying the slow NCX_{mit} activity in reverse mode? One possible explanation may be an allosteric regulation of NCLX by $\Delta\Psi$, as reported in SH-SY5Y neuronal cells and in HEK-293T cells [38]. The authors showed that mild $\Delta\Psi$ depolarization inhibited NCX_{mit} via two clusters of positively charged residues, which are putatively located in the regulatory loop around the inner membrane (yellow sticks in Figure 1). They also showed that phosphorylation of S258 in human NCLX, known to be a protein kinase A (PKA) target site [39] (blue sticks in Figure 1), could override the regulation. Since mitoplasts and isolated mitochondria were free of cytosolic ingredients, it could be possible that phosphorylation at the residue was not sufficient to override the depolarization-mediated inhibition under the experimental conditions of [8]. The unfavorable reversal of NCX_{mit} was also reported in leukotriene C_4-stimulated mast cells with depolarized mitochondria [40]. Interestingly, however, mitochondrial fusion protein mitofusin (MFN) 2 knockdown caused repetitive reversal of NCX_{mit} even under depolarized conditions, resulting in mitochondrial and cytosolic Ca^{2+} oscillation. It is worth examining phosphorylation status at NCLX S258 in MFN2-knockdown cells.

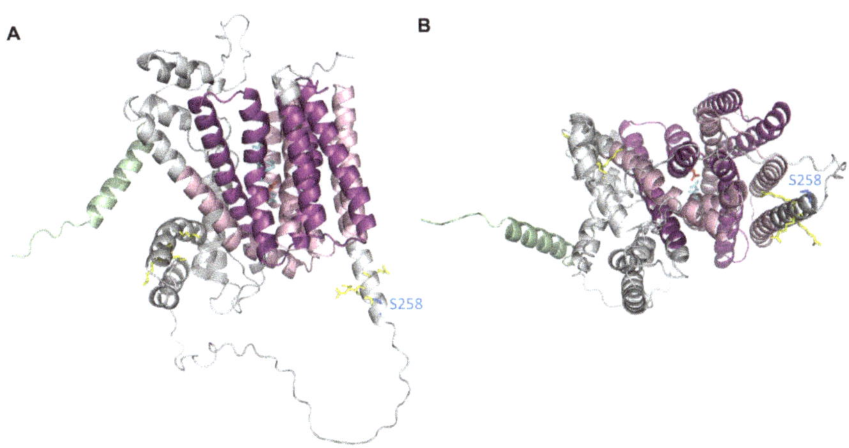

Figure 1. Putative three-dimensional structure of human NCLX (UniProtKB accession number Q6J4K2) predicted using AlfaFold [36]. The pdb file (AF-Q6J4K2-F1-model_v1) was downloaded from the AlphaFold Protein Structure Database (https://alphafold.ebi.ac.uk/ accessed on 05 November 2021) and graphics were prepared using PyMOL v.2.1.0. (**A**) Side view, (**B**) bottom view. Putative mitochondria transit peptide and two sodium/calcium exchanger membrane regions are shown in green and pale and dark pink, respectively. Putative protein kinase A (PKA) phosphorylation site, S258 [39], is shown as blue sticks. Putative amino acids rendering Li^+ selectivity, Na^+ selectivity [34], and those sensitive to $\Delta\Psi$ depolarization [38] are shown as light blue, red, and yellow sticks, respectively.

In dopaminergic neurons, it was demonstrated that PTEN-induced putative kinase 1 (PINK1) at mitochondria activated PKA, thereby phosphorylating S258 of NCLX [39]. PINK1 deficiency is closely associated with mitochondrial abnormalities and the progression of early-onset familial Parkinson's disease [41,42]. In addition, recent studies demonstrated deficiencies of PINK1 and mitochondrial function in failing hearts, such as in hearts with late stages of dystrophic cardiomyopathy and sepsis [43,44]. Abnormal NCX_{mit} activity via PINK1 deficiency may be associated with mitochondrial dysfunction in these failing hearts.

4. Spatial Distribution of NCX_{mit} in Cardiomyocytes

Mitochondria are physically and electrically connected with each other via intermitochondrial junctions and form "mitochondrial reticulum" throughout the cell in the skeletal and cardiac muscles [45,46]. However, there is still functional heterogeneity depending on their spatial distributions—one just beneath the sarcolemmal membrane (subsarcolemmal mitochondria, SSM), one between myofibrils (interfibrillar mitochondria, IFN), and one near the nucleus (perinuclear mitochondria, PNM). This heterogeneity may contribute to dealing with region-specific energy demands, sensitivity to oxidative stress, Ca^{2+} handling, and so on [47,48].

Using mitochondrial Ca^{2+} indicator Myticam-expressing rabbit cardiomyocytes, Lu et al. [49] demonstrated that 1 Hz electrical stimulation induced faster Ca^{2+} uptake in IFM than in PNM, whereas the post-stimulation Ca^{2+} efflux was comparable. The higher uptake but comparable efflux of mitochondrial Ca^{2+} in IFM than in PNM resulted in the higher sensitivity to phenylarsine oxide for the opening of mPTP, shown as faster $\Delta\Psi$ depolarization. In addition, the spatial difference of CU_{mit} activity was preserved in permeabilized cardiomyocytes under the conditions of a clamped cytosolic Ca^{2+} concentration with a disabled sarcoplasmic reticulum (SR) Ca^{2+} pump (SERCA), but the efficacy was lower than that observed in intact cardiomyocytes. Therefore, an intrinsic difference in CU_{mit} activity may be further amplified by local SR Ca^{2+} release-associated excitation–contraction coupling in intact cardiomyocytes.

While Ca^{2+} efflux activity was comparable in IFM and PNM [49], it was shown to be larger in SSM than in IFM [50]. Immunofluorescence analyses using stochastic optical reconstruction microscopy revealed that NCLX localized mainly in SSM near (<20 nm) voltage-dependent Na^+ channel $Na_v1.5$ clusters, and hardly existed in IFM of rabbit ventricular myocytes. Functional analyses showed that the treatment of cells with a Na^+ channel blocker, tetrodotoxin, increased intensity of Rhod-2, an indicator of mitochondrial Ca^{2+}, in SSM but not in IFM. Since NCX_{mit} inhibition by CGP-37157 produced similar results as those obtained using tetrodotoxin, it was suggested that Na^+ accumulation at the subsarcolemmal space via $Na_v1.5$ potentiated NCLX-mediated Ca^{2+} efflux from mitochondria, preventing mitochondrial Ca^{2+} accumulation. The authors further examined the contribution of the physical and functional coupling of $Na_v1.5$-NCLX to reactive oxygen species (ROS) production in mitochondria, which is closely associated with mitochondrial Ca^{2+}, as will be described in Section 5.3. However, they failed to detect spatial differences in mitochondrial ROS production, possibly due to the propagation of ROS via the mitochondrial reticulum [45,46,51]. Further analyses are necessary to prove the physiological and pathophysiological roles of this $Na_v1.5$–NCLX coupling. Quantitative analysis of spatial Na^+ regulation via the Na^+-K^+ ATPase and Na^+ channel, as performed by Skogestad et al. [52] would help in understanding this subject.

In addition to the heterogeneities of mitochondrial Ca^{2+} dynamics at differentially localized cardiac mitochondria, intra-mitochondrial heterogeneities were also reported. Lu et al. [53] evaluated mitochondrial Ca^{2+} transients using Myticam-expressing rat ventricular myocytes. They analyzed 0.2 Hz stimulation-induced small mitochondrial Ca^{2+} transients, which increased from ~150 nM by ~30 nM and found that the upstroke was faster at position near the Z-line than near the M-line but the decay was comparable. Although MCU immunofluorescence showed a uniform distribution over the mitochondrion, it was suggested that CU_{mit} activity was higher at mitochondria facing junctional SR (jSR) than at those facing bulk cytosol, whereas Ca^{2+} efflux activity was comparable. This is reasonable because mitochondria–jSR association creates high Ca^{2+} microdomains near the dyadic space, which enables them to meet the low affinity of CU_{mit} for Ca^{2+} uptake (see review [54]).

De La Fuente et al. [55,56] further explored the spatial heterogeneities of mitochondrial Ca^{2+} dynamics. Using conventional and super-resolution immunofluorescence analyses of isolated cardiac mitochondria and isolated cardiomyocytes, they demonstrated that about 50% of MCU were closely co-localized with the SR Ca^{2+} release channel ryanodine receptor (RyR) 2 [55]. The authors explained that the divergence of this biased MCU distribution from the previously reported uniform distribution [53] was attributable to the antibodies chosen, since one used in [53] gave non-specific signals in MCU knockout mouse-derived cardiomyocytes. Supporting the idea of MCU-RyR2 colocalization, MCU and EMRE, which are essential CU_{mit} regulator proteins, were more abundant in crude mitochondria than in Percoll-purified mitochondria, and were also found in jSR [55]. On the other hand, the NCLX protein was more abundant in pure mitochondria than in crude mitochondria, and was not found in jSR [56]. Moreover, the authors strengthened their findings on distinct distributions of MCU and NCLX by means of functional assays. CU_{mit} activity—CU_{mit} inhibitor Ru360-sensitive $^{45}Ca^{2+}$ uptake corrected with citrate synthase activity—was much higher in isolated jSR than that in isolated mitochondria. On the contrary, $^{45}Ca^{2+}$ retention assays revealed that Na^+- and CGP-37157-sensitive mitochondrial Ca^{2+} efflux activity was much higher in pure mitochondria than that in jSR. This $^{45}Ca^{2+}$ efflux activity became larger and smaller in heart-specific NCLX overexpressing and knockout mice, respectively. The authors proposed that the spatially separated distribution of MCU-RyR2 and NCLX contributes to minimizing the energy cost for maintaining $\Delta\Psi$. In other word, if MCU-RyR2 were near NCLX, $\Delta\Psi$ would depolarize both due to Ca^{2+} influx via CU_{mit} and due to Ca^{2+} efflux via NCX_{mit}. Accordingly, the spatial separation of MCU-RyR2 and NCLX should be necessary for optimizing mitochondrial Ca^{2+} signals and energy cost. Interestingly, it was demonstrated that NCLX efficiently supplies Ca^{2+} from mitochondria to the SR/endoplasmic reticulum (ER) via SERCA, thereby regulating the automaticity of HL-1 cardiomyocytes, as well as antigen receptor-mediated Ca^{2+} signaling of B lymphocytes [3,5]. It is worth evaluating the physical coupling of NCLX and SERCA in cardiomyocytes, which would fill in the last piece in our understanding of the efficient Ca^{2+} cycling between SR and mitochondria.

5. Role of NCX_{mit} in Cardiomyocyte Functions

5.1. Role of NCX_{mit} in Cardiac Rhythmicity

Since CU_{mit} comprises only 1–2% of total Ca^{2+} removal from cardiomyocytes [53,57,58], the contribution of NCX_{mit} to cytosolic Ca^{2+} transients and to action potential generation has been considered negligible. However, in a spontaneously beating cell line, HL-1, originating from mouse atrial myocytes, NCLX knockdown by siRNA significantly decelerated the upstrokes of action potentials and Ca^{2+} transients, and prolonged the cycle lengths [5]. Consistently with the results of other studies [53,57,58], NCLX knockdown did not change the rest and peak fluorescence ratios of Indo-1, an indicator of the cytosolic Ca^{2+} level. Rather, it was demonstrated that NCLX knockdown decreased caffeine-responsive SR Ca^{2+} content and slowed subsequent SR Ca^{2+} reuptake rate, evaluated using a FRET protein Cameleon D1ER. Further analyses of a mathematical model combined with experimental validation revealed that (1) the automaticity of HL-1 cells was driven by the so-called "Ca^{2+} clock" mechanism, in which a Ca^{2+} leak from SR potentiates the inward current via sarcolemmal Na^+-Ca^{2+} exchange to facilitate membrane depolarization, (2) NCLX knockdown decreased the Ca^{2+} supply from mitochondria to SR, thereby decelerating SR Ca^{2+} leak, delaying the activation of the inward current through sarcolemmal Na^+-Ca^{2+} exchange, and thus delaying the activations of voltage-dependent Na^+ and Ca^{2+} currents, causing a cycle length prolongation.

Considering that HL-1 cells are derived from atrial myocytes, which are quiescent under physiological conditions, NCX_{mit} may be involved in abnormal automaticity of atria, such as atrial flutter and atrial ectopic tachycardia. In addition, it may also be plausible that abnormal NCX_{mit} function causes ventricular arrhythmias. In fact, the involvement of abnormal NCX_{mit} activity in altered rhythmicity was suggested in mouse embryonic stem cell-derived as well as in human induced pluripotent stem cell-derived ventricular myocytes, where the "Ca^{2+} clock" drives the automaticity [59]. In addition, arrhythmic evens with QRS interval widening were observed in tamoxifen-induced heart-specific conditional NCLX-knockout mice, though the events only occurred immediately before death [25].

The question of whether NCX_{mit} participates in the automaticity of normal pacemaker cells, i.e., sinoatrial (SA) node cells, is still a big issue. The automaticity of SA node cells has been proposed to be driven by a "coupled-clock" pacemaking system, which is composed of a sarcolemmal ion channel/transporter-derived rhythm ("membrane clock") and subsarcolemmal Ca^{2+} release (LCR)-related rhythm ("Ca^{2+} clock") [60–62]. In the former, pacemaker channels such as the hyperpolarization-activated cation channel and various other inward membrane currents at the plasma membrane drive diastolic depolarization. In the latter, LCR from SR activates the inward current via sarcolemmal Na^+-Ca^{2+} exchange to drive diastolic depolarization. NCX_{mit} may modulate the "Ca^{2+} clock" part in SA node cells, as observed in HL-1 cells [5]. In fact, application of an NCX_{mit} inhibitor, CGP-37157, slowed the firing rate of rabbit as well as mouse SA node cells [63,64]. However, recent imaging studies of mouse SA node preparations revealed marked heterogeneity of LCR and action potential-induced Ca^{2+} transients within and among SA node cells [65]. That is, some SA node cells generated only LCR and did not fire; some only generated action potential-induced Ca^{2+} transients and did not generate LCR; and some generated LCR during the diastolic phase before an occurrence of action potential-induced Ca^{2+} transients. These data suggest that the coupling degree of the "coupled-clock" system may differ among SA node cells in vivo. Our model analyses suggested that NCX_{mit} reduction in an SA node cell which is solely driven by the "membrane clock" accelerates, instead of decelerating, the firing rate [11]. NCX_{mit} reduction-mediated slowing of automaticity in "Ca^{2+} clock"-driven cells may be compromised by NCX_{mit}-mediated acceleration of automaticity in "membrane clock" cells in the SA node region. In fact, tamoxifen-induced NCLX deletion in the adult mouse heart, with a 70% reduction of NCLX protein 3 days after tamoxifen treatment, did not show altered sinus rhythms except for on the date of death, 8–10 days after tamoxifen treatment [25]. In vivo imaging of the SA node of NCLX-knockout mice would clarify the quantitative roles of NCX_{mit} in pacemaking activity.

5.2. Role of NCX_{mit} in Cardiac Energetics

The heart is continuously pumping blood around the body, which is energetically driven by ATP hydrolysis. In the healthy adult heart, ATP synthesis is mainly dependent on mitochondrial oxidative phosphorylation, and the process is strictly regulated to balance the large, dynamically changing energy demands [66]. One candidate factor for the regulation is Ca^{2+}, which activates three mitochondrial dehydrogenases—pyruvate dehydrogenase complex (PDHC), isocitrate dehydrogenase, and 2-oxoglutarate dehydrogenase (OGDH) [67] (Figure 2). The product NADH is oxidized in the electron transport chain, causing the proton motive force to be utilized for ATP synthesis via F_1F_o ATP synthase. Therefore, the contribution of NCX_{mit}, as one of the determinants of the mitochondrial Ca^{2+} level, to cardiac energetics has been an issue to be clarified.

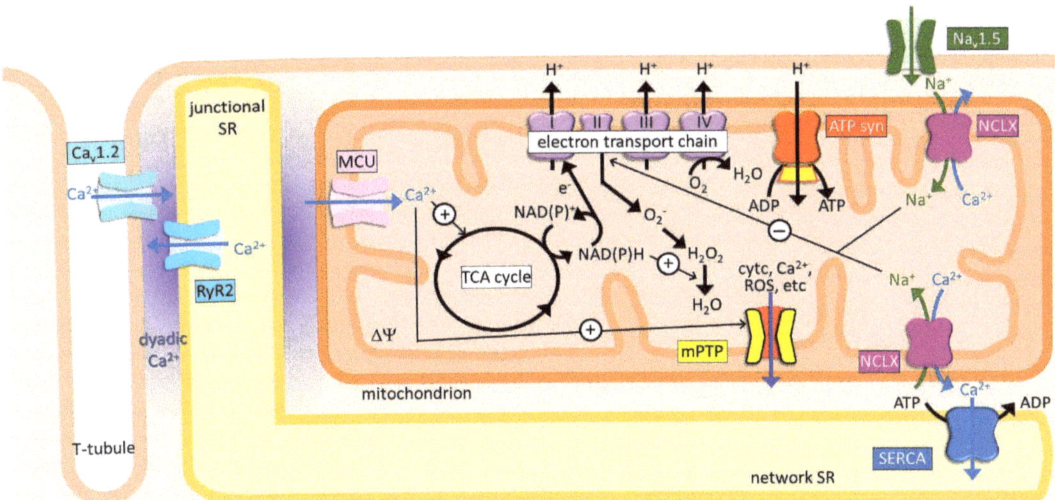

Figure 2. An overview of the NCLX-mediated physiological and pathophysiological functions in a cardiomyocyte. ATPsyn, F_1F_o-ATP synthase; $Ca_V1.2$, L-type Ca^{2+} channel; cytc, cytochrome c; $\Delta\Psi$, mitochondrial membrane potential; MCU, mitochondrial Ca^{2+} uniporter complex; mPTP, mitochondrial permeability transition pores; $Na_V1.5$, voltage-dependent Na^+ channel; ROS, reactive oxygen species; RyR2, ryanodine receptor 2; SERCA, sarcoplasmic reticulum Ca^{2+} pump; SR, sarcoplasmic reticulum; TCA, tricarboxylic acid.

This was first shown in whole-cell patch clamp experiments using guinea pig ventricular myocytes loaded with Rhod-2 for evaluating mitochondrial Ca^{2+} changes [68]. It was demonstrated that when NCX_{mit} became more active with 15 mM compared with 5 mM Na^+ in the pipette, the mitochondrial Ca^{2+} increase induced by an abrupt workload increase (3–4 Hz pacing in the presence of isoproterenol) was diminished. At the same time, under the condition of 15 mM Na^+ in the pipette, NADH autofluorescence decreased upon the workload increase, indicating that mitochondrial Ca^{2+} was not sufficient enough to activate NADH production by mitochondrial dehydrogenases. An NCX_{mit} inhibitor, CGP-37157, restored the workload-induced Ca^{2+} accumulation in mitochondria and attenuated the NADH decrease [69]. Since a cytosolic Na^+ increase and energy starvation are characteristic properties of failing heart [70], the authors further studied a guinea pig model of heart failure which was induced by aortic constriction with/without β-adrenergic receptor stimulation [69,71]. In ventricular myocytes from failing hearts, where cytosolic Na^+ evaluated from SBFI ratio image was ~15 mM compared to ~5 mM in sham myocytes, the abrupt workload increase caused essentially the same responses of mitochondrial Ca^{2+} (Rhod-2 or Myticam) and NADH as those reported with 15 mM Na^+ in the pipette [68,69]—the diminished increase of mitochondrial Ca^{2+} and the subsequent NADH starvation upon the workload increase. More importantly, the changes were restored in the presence of an NCX_{mit} inhibitor, CGP-37157, to levels similar to those observed in sham myocytes. These results suggested a causative role of NCX_{mit} in the energy starvation of the failing heart. In addition, as will be explained in Section 5.3, chronic treatment of the animals with CGP-37157 partially prevented cardiac dysfunctions. Accordingly, the authors proposed that blocking of NCX_{mit} is a novel strategy for treating heart failure [71].

However, the contribution of NCX_{mit} to cardiac energetics in the failing heart may not be as large as that expected from experiments using cardiomyocytes, where an extreme workload change was applied—rapid 3–4 Hz pacing from a quiescent state, which hearts in situ never experience [68,69,71]. Recently, the effects of chronic and acute myocardial Na^+ loads on cardiac energetics were extensively studied in Langendorff-perfused mouse hearts with ^{23}Na, ^{31}P, ^{13}C NMR, and ^{1}H-NMR metabolomic profiling [72]. Chronic (phospholemman PLM^{3SA} mouse) and acute (treatment with ouabain and blebbistatin) inhibition of Na^+-K^+ ATPase, as well as pressure-overload-induced cardiac hypertrophy caused a cytosolic Na^+ increase, and switched the substrate preference from fatty acid to carbohydrate oxidation, which are characteristic features frequently observed in failing hearts [70,73]. The acute Na^+ elevation resulted in the most severe metabolic alterations, such as decreased metabolite levels of tricarboxylic acid (TCA) cycle intermediates downstream from OGDH (succinate, fumarate, and malate), suggesting the reduced Ca^{2+}-dependent activation of TCA cycle dehydrogenases. However, regardless of the strategy for cytosolic Na^+ elevations, the energy supply was maintained, as is evident from the preserved ATP, phosphocreatine (PCr), PCr/ATP ratio, NADH, and pH. Metabolome profiles obtained with NMR, as well as in silico predictions using CardiNet, revealed that they were achieved at the cost of extensive metabolic flux remodeling. Therefore, the impact of impaired cytosolic Na^+ homeostasis on mitochondrial ATP production should be mechanistically more complex than what has been suggested in isolated cardiomyocytes. In all three sets of hearts with elevated cytosolic Na^+, treatment with CGP-37157 reversed the substrate preference from carbohydrate to fatty acid oxidation with normalized levels of the depleted metabolites. This suggests a therapeutic potential for CGP-37157 in the treatment of the metabolic reprograming that occurs before energetic impairments.

To the contrary, detrimental contributions of NCX_{mit} to cardiac energetics were not suggested in heart-specific NCLX-overexpression mice [25]. There were no apparent differences between control and NCLX-overexpression mice's ventricular myocytes in terms of the NAD^+/NADH ratio, oxidative phosphorylation evaluated by seahorse analyses with either pyruvate or palmitate as energy substrates (basal, ATP-linked and maximum respirations, spare capacity, and proton leak), nor in the phosphorylation level of mitochondrial Ca^{2+}-responsive PDHC. These findings suggested that NCLX overexpression had marginal effects on cardiac energetics. Rather, as will be explained in Section 5.3, NCLX overexpression prevented the cardiac dysfunctions of ischemia-reperfusion injury and ischemic heart failure. It should be noted that the basal mitochondrial Ca^{2+} level, evaluated as carbonyl cyanide 4-(trifluoromethoxy)phenylhydrazone)-responsive Fura-2 intensity, was comparable between control and NCLX-overexpressing cardiomyocytes, indicating that NCLX overexpression did not cause excessive deprivation of mitochondrial Ca^{2+}, in spite of an increase in mitochondrial Ca^{2+} efflux rate by 88%. Our model analyses suggested that cytosolic Ca^{2+} within its physiological range, 100 nM–2 μM, does not largely affect steady-state levels of energy substrates, though a lower cytosolic Ca^{2+} level collapsed the system because of mitochondrial Ca^{2+} deprivation [74,75]. Therefore, some compensation or backup mechanisms may work to prevent mitochondrial Ca^{2+} deprivation via NCLX overexpression. It would be informative to further evaluate the mitochondrial Ca^{2+} level, cytosolic Na^+ level, and metabolome profiles in failing hearts with or without NCLX overexpression.

5.3. Role of NCX_{mit} in ROS Production and mPTP Opening

Mitochondria are a major source of ROS, the production of which is tightly coupled with ATP synthesis—Complex I and Complex III in the electron transport chain produce O_2^- from O_2 oxidation. Then, manganese-dependent superoxide dismutase converts O_2^- to H_2O_2, which is eliminated by antioxidant scavenge systems, such as glutathione peroxidase and peroxiredoxin [76] (Figure 2). Excessive amounts of ROS, either via overproduction or via reduced scavenging pathways, exert detrimental effects on mitochondrial function, such as uncoupling of the electron transport chain to reduce ATP production, and triggering mPTP opening by sensitizing mPTP to mitochondrial Ca^{2+}. mPTP opening causes a burst of ROS released from mitochondria, in a process named ROS-induced ROS release, which impairs excitation–contraction coupling via modulating multiple ion channels and transporters, as well as via chronic remodeling [77,78] (Figure 2). Based on the fact that mitochondrial Ca^{2+} activates three dehydrogenases and the product NADH promotes ROS by-production via the electron transport chain, and that mitochondrial Ca^{2+} overload is a key trigger for mPTP opening, followed by a burst ROS release, altered mitochondrial

Ca^{2+} dynamics have been implicated to be closely associated with ROS dynamics in failing and injured hearts [79,80].

Hamilton et al. [81] demonstrated the involvement of NCX_{mit} in ROS production, SR Ca^{2+} handling, and arrhythmogenesis in rat ventricular myocytes. They monitored mitochondrial Ca^{2+} using a biosensor mtRCamp1h, and showed that NCX_{mit} inhibition by CGP-37157 decelerated mitochondrial Ca^{2+} decay, thereby enhancing mitochondrial Ca^{2+} accumulation triggered by 2 Hz electrical stimulation in the presence of isoproterenol. This resulted in larger $\Delta\Psi$ depolarization monitored by TMRM, increased ROS in the mitochondria-SR microdomain evaluated using ER-tuned redox sensor ER-roGFP_iE, increased RyR oxidation as evident from increased immunoprecipitation with anti-dinitrophenyl-antibody, and increased proarrhythmic Ca^{2+} waves. The authors also showed that this cascade further exacerbated proarrhythmic-triggered activity in hypertrophied hearts, which were induced by thoracic aortic banding.

The detrimental consequences of NCX_{mit} inhibition were more prominent in NCLX-knockout mice. The germline NCLX knockout was unsuccessful, and adult acute heart-specific NCLX knockout, in which NCLX protein expression was reduced by ~70%, caused ~87% lethality within 2 weeks due to severe myocardial dysfunction accompanying increased ROS, evaluated with dihydroethidium and MitoSox red, and mitochondrial swelling [25]. This lethality was attributable to mitochondrial Ca^{2+} overload-mediated mPTP opening, because the depletion of the mPTP component cyclophilin D on the NCLX conditional knockout background rescued the myocardial dysfunction and lethality following tamoxifen-induced NCLX ablation. Those authors suggested that NCLX-mediated Ca^{2+} efflux was necessary to maintain an appropriate mitochondrial Ca^{2+} level, which was vital for preventing mPTP opening and excessive ROS production, and for survival. The idea was further confirmed in heart-specific NCLX-overexpression mice subjected to ischemia-reperfusion [25]. Accordingly, NCLX overexpression reduced the ROS level evaluated using dihydroethidium in hearts with 40 min-left coronary artery ligation followed by 24 h reperfusion, and tended to decrease it 4 weeks after permanent occlusion of the left coronary artery. In addition, cardiac dysfunctions characterized by TUNEL-positive interstitial cells, fibrosis, and contractile dysfunction were all improved by NCLX overexpression. The above findings clearly indicated beneficial contributions of NCX_{mit} in ischemia-induced failing hearts.

However, a contradictory mechanism was proposed by Liu et al. [71]. As explained in Section 5.2, an abrupt workload increase resulted in a diminished increase in mitochondrial Ca^{2+}, followed by NADH starvation, in failing ventricular myocytes, possibly because elevated cytosolic Na^{+} excessively extruded Ca^{2+} from mitochondria via NCX_{mit} [71]. Interestingly, dichlorodihydrofluorescein diacetate oxidation, an index of the ROS level, was dramatically increased upon an abrupt workload increase in the failing cardiomyocytes but not in the sham cardiomyocytes and this ROS production was completely diminished in the presence of an NCX_{mit} inhibitor, CGP-37157. Moreover, chronic treatment of the animals with CGP-37157 using an osmotic pump partially prevented the animals from developing heart failure, as evident from improved hypertrophic remodeling, interstitial fibrosis, contractile dysfunction, and occurrence of arrhythmia. The authors attributed the mechanism to reduced ROS scavenging capacity due to the reduced NAD(P)H levels in failing cardiomyocytes. Accordingly, these findings indicated a detrimental contribution of NCX_{mit} in failing hearts.

The abovementioned contradictory roles of NCX_{mit} in failing hearts suggested that mitochondrial Ca^{2+} did not simply correlate with ROS production. Recently, a brand-new mechanism underlying hypoxia-induced ROS production via NCX_{mit} was proposed—Na^{+}–phospholipid interaction-mediated ROS regulation [82] (Figure 2). The authors first confirmed that NCX_{mit} was involved in hypoxia-induced ROS production in primary bovine aortic endothelial cells and mouse embryonic fibroblasts. Pharmacological inhibition with CGP-37157 or genetical reduction (siRNA or knockout) of NCLX diminished the cytosolic Ca^{2+} increase and cytosolic Na^{+} decrease, attenuated the reduction of the inner mitochondrial membrane fluidity and the mitochondrial ROS production caused by exposure of the cells to hypoxia (exposure of the cells to 1% O_2). Then, the authors showed that hypoxia-induced matrix acidification via Complex I inhibition caused Ca^{2+} solubilization from calcium phosphate precipitation in the matrix, as evident from morphological (electron microscopy images) as well as from functional assays (measurements of free mitochondrial Ca^{2+} in isolated mitochondria as well as in cells). Since a mitochondrial free Ca^{2+} increase enables NCX_{mit} to extrude Ca^{2+} in exchange for Na^+, the authors then focused on the roles of matrix Na^+ on electron transport chains and found that only Complex II-dependent respirations were decreased by Na^+, which was NCX_{mit}-dependent, resulting in increased ROS production. The authors filled in the final piece by showing that Na^+ directly bound to the phospholipid bilayer, as evident from infrared spectroscopy, which reduced the fluidity of the inner mitochondrial membrane for ubiquinone diffusion in the inner mitochondrial membrane, increasing the ROS production. Taken together, these findings clarified a distinct scheme of ROS production—regulation by matrix Na^+ via NCX_{mit}—from those proposed in previous reports.

6. Future Perspectives

As has been described so far, knowledge on the biophysical properties, distributions, and the physiological and pathophysiological significance of NCX_{mit} in the heart is rapidly increasing. The more knowledge is accumulated, the more complicated systems are elucidated, sometimes introducing difficulties into our understanding as a whole. Taking ROS dynamics under pathological conditions as an example, some experimental evidence supports the roles of NCX_{mit} in increasing ROS production [82], whereas others support its preventive roles in relation to ROS increases [25,81]. NCX_{mit} directly modulates and is affected by cytosolic and mitochondrial concentrations of Na^+ and Ca^{2+} ions, and $\Delta\Psi$, which are associated with ROS balance regulation via different pathways (Figure 2). Therefore, differences in ionic conditions and mitochondrial viability under different experimental conditions or diseased states would result in different contributions of NCX_{mit}.

In order to understand these complicated networks, the integration of NCX_{mit} activity and cellular/mitochondrial functions with mathematical modeling could be a powerful tool. Very recently, Cortassa et al. [83] succeeded in reconciling the apparently paradoxical roles of NCX_{mit} in ROS dynamics (see details [83]). In brief, they built two scenarios, the "Na^+-driven oxidized scenario" and the "Ca^{2+}-driven reduced scenario", and demonstrated that variations in redox status, cytoplasmic Na^+ concentrations and energetic capacity resulted in different mitochondrial Ca^{2+} levels and bioenergetic responses driving ATP supply and oxidative stress. The former scenario could be represented by heart failure with a reduced ejection fraction (HFrEF) in which considerable cytosolic Na^+ overload occurs, and the latter by heart failure with a preserved or moderate ejection fraction (HFpEF, HFmEF) in which only a modest Na^+ increase is expected. Integrating the model of matrix Ca^{2+} solubilization and precipitation from and to calcium phosphate [84] into Cortassa's model [83] would further facilitate our understandings in this area.

The discrepancies in experimental findings obtained from isolated mitochondria, isolated cardiomyocytes, and whole hearts are other issues that remain to be solved. Recent advances in imaging techniques used to evaluate electrophysiological and metabolic properties of single cells and even organelles in tissue are promising [65,85]. By utilizing these techniques, it is expected that our understandings of the roles of NCX_{mit} in healthy as well as in failing hearts will be further deepened.

Author Contributions: Conceptualization, A.T. and S.M.; writing—original draft preparation, A.T.; writing—review and editing, S.M.; visualization, A.T. and S.M.; funding acquisition, A.T. and S.M. All authors have read and agreed to the published version of the manuscript.

Funding: This research was funded by JSPS KAKENHI (grant number 18K06869 (A.T.) and 19H03400 (S.M.)) and by Research Grant from University of Fukui (grant number LSI21205 (S.M.)).

Institutional Review Board Statement: Not applicable.

Informed Consent Statement: Not applicable.

Data Availability Statement: Not applicable.

Conflicts of Interest: The authors declare no conflict of interest.

References

1. Carafoli, E.; Tiozzo, R.; Lugli, G.; Crovetti, F.; Kratzing, C. The release of calcium from heart mitochondria by sodium. *J. Mol. Cell Cardiol.* **1974**, *6*, 361–371. [CrossRef]
2. Palty, R.; Silverman, W.F.; Hershfinkel, M.; Caporale, T.; Sensi, S.L.; Parnis, J.; Nolte, C.; Fishman, D.; Shoshan-Barmatz, V.; Herrmann, S.; et al. NCLX is an essential component of mitochondrial Na^+/Ca^{2+} exchange. *Proc. Natl. Acad. Sci. USA* **2010**, *107*, 436–441. [CrossRef] [PubMed]
3. Kim, B.; Takeuchi, A.; Koga, O.; Hikida, M.; Matsuoka, S. Pivotal role of mitochondrial Na^+-Ca^{2+} exchange in antigen receptor mediated Ca^{2+} signalling in DT40 and A20 B lymphocytes. *J. Physiol.* **2012**, *590*, 459–474. [CrossRef] [PubMed]
4. Nita, I.I.; Hershfinkel, M.; Fishman, D.; Ozeri, E.; Rutter, G.A.; Sensi, S.L.; Khananshvili, D.; Lewis, E.C.; Sekler, I. The mitochondrial Na^+/Ca^{2+} exchanger upregulates glucose dependent Ca^{2+} signalling linked to insulin secretion. *PLoS ONE* **2012**, *7*, e46649. [CrossRef] [PubMed]
5. Takeuchi, A.; Kim, B.; Matsuoka, S. The mitochondrial Na^+-Ca^{2+} exchanger, NCLX, regulates automaticity of HL-1 cardiomyocytes. *Sci. Rep.* **2013**, *3*, 2766. [CrossRef] [PubMed]
6. Parnis, J.; Montana, V.; Delgado-Martinez, I.; Matyash, V.; Parpura, V.; Kettenmann, H.; Sekler, I.; Nolte, C. Mitochondrial exchanger NCLX plays a major role in the intracellular Ca^{2+} signaling, gliotransmission, and proliferation of astrocytes. *J. Neurosci.* **2013**, *33*, 7206–7219. [CrossRef]
7. Kim, B.; Takeuchi, A.; Hikida, M.; Matsuoka, S. Roles of the mitochondrial Na^+-Ca^{2+} exchanger, NCLX, in B lymphocyte chemotaxis. *Sci. Rep.* **2016**, *6*, 28378. [CrossRef] [PubMed]
8. Islam, M.M.; Takeuchi, A.; Matsuoka, S. Membrane current evoked by mitochondrial Na^+-Ca^{2+} exchange in mouse heart. *J. Physiol. Sci.* **2020**, *70*, 24. [CrossRef]
9. Assali, E.A.; Jones, A.E.; Veliova, M.; Acin-Perez, R.; Taha, M.; Miller, N.; Shum, M.; Oliveira, M.F.; Las, G.; Liesa, M.; et al. NCLX prevents cell death during adrenergic activation of the brown adipose tissue. *Nat. Commun.* **2020**, *11*, 3347. [CrossRef] [PubMed]
10. Takeuchi, A.; Matsuoka, S. Minor contribution of NCX to Na^+-Ca^{2+} exchange activity in brain mitochondria. *Cell Calcium.* **2021**, *96*, 102386. [CrossRef] [PubMed]
11. Takeuchi, A.; Kim, B.; Matsuoka, S. The destiny of Ca^{2+} released by mitochondria. *J. Physiol. Sci.* **2015**, *65*, 11–24. [CrossRef] [PubMed]
12. Takeuchi, A.; Kim, B.; Matsuoka, S. Physiological functions of mitochondrial Na^+-Ca^{2+} exchanger, NCLX, in lymphocytes. *Cell Calcium* **2020**, *85*, 102114. [CrossRef]
13. Katoshevski, T.; Ben-Kasus Nissim, T.; Sekler, I. Recent studies on NCLX in health and diseases. *Cell Calcium* **2021**, *94*, 102345. [CrossRef]
14. Brown, D.A.; Perry, J.B.; Allen, M.E.; Sabbah, H.N.; Stauffer, B.L.; Shaikh, S.R.; Cleland, J.G.; Colucci, W.S.; Butler, J.; Voors, A.A.; et al. Expert consensus document: Mitochondrial function as a therapeutic target in heart failure. *Nat. Rev. Cardiol.* **2017**, *14*, 238–250. [CrossRef] [PubMed]
15. O'Rourke, B.; Ashok, D.; Liu, T. Mitochondrial Ca^{2+} in heart failure: Not enough or too much? *J. Mol. Cell Cardiol.* **2021**, *151*, 126–134. [CrossRef]
16. Fieni, F.; Lee, S.B.; Jan, Y.N.; Kirichok, Y. Activity of the mitochondrial calcium uniporter varies greatly between tissues. *Nat. Commun.* **2012**, *3*, 1317. [CrossRef] [PubMed]
17. Raffaello, A.; De Stefani, D.; Sabbadin, D.; Teardo, E.; Merli, G.; Picard, A.; Checchetto, V.; Moro, S.; Szabò, I.; Rizzuto, R. The mitochondrial calcium uniporter is a multimer that can include a dominant-negative pore-forming subunit. *EMBO J.* **2013**, *32*, 2362–2376. [CrossRef]
18. Paillard, M.; Csordás, G.; Szanda, G.; Golenár, T.; Debattisti, V.; Bartok, A.; Wang, N.; Moffat, C.; Seifert, E.L.; Spät, A.; et al. Tissue-specific mitochondrial decoding of cytoplasmic Ca^{2+} signals is controlled by the stoichiometry of MICU1/2 and MCU. *Cell Rep.* **2017**, *18*, 2291–2300. [CrossRef] [PubMed]
19. Huo, J.; Lu, S.; Kwong, J.Q.; Bround, M.J.; Grimes, K.M.; Sargent, M.A.; Brown, M.E.; Davis, M.E.; Bers, D.M.; Molkentin, J.D. MCUb induction protects the heart from postischemic remodeling. *Circ. Res.* **2020**, *127*, 379–390. [CrossRef]
20. Wescott, A.P.; Kao, J.P.Y.; Lederer, W.J.; Boyman, L. Voltage-energized calcium-sensitive ATP production by mitochondria. *Nat. Metab.* **2019**, *1*, 975–984. [CrossRef] [PubMed]
21. Pallafacchina, G.; Zanin, S.; Rizzuto, R. From the identification to the dissection of the physiological role of the mitochondrial calcium uniporter: An ongoing story. *Biomolecules* **2021**, *11*. [CrossRef] [PubMed]
22. Bernardi, P. Mitochondrial transport of cations: Channels, exchangers, and permeability transition. *Physiol. Rev.* **1999**, *79*, 1127–1155. [CrossRef] [PubMed]
23. Rysted, J.E.; Lin, Z.; Walters, G.C.; Rauckhorst, A.J.; Noterman, M.; Liu, G.; Taylor, E.B.; Strack, S.; Usachev, Y.M. Distinct properties of Ca^{2+} efflux from brain, heart and liver mitochondria: The effects of Na^+, Li^+ and the mitochondrial Na^+/Ca^{2+} exchange inhibitor CGP37157. *Cell Calcium* **2021**, *96*, 102382. [CrossRef]

24. Crompton, M.; Künzi, M.; Carafoli, E. The calcium-induced and sodium-induced effluxes of calcium from heart mitochondria. Evidence for a sodium-calcium carrier. *Eur. J. Biochem.* **1977**, *79*, 549–558. [CrossRef] [PubMed]
25. Luongo, T.S.; Lambert, J.P.; Gross, P.; Nwokedi, M.; Lombardi, A.A.; Shanmughapriya, S.; Carpenter, A.C.; Kolmetzky, D.; Gao, E.; van Berlo, J.H.; et al. The mitochondrial Na^{+}/Ca^{2+} exchanger is essential for Ca^{2+} homeostasis and viability. *Nature* **2017**, *545*, 93–97. [CrossRef] [PubMed]
26. Jiang, D.; Zhao, L.; Clapham, D.E. Genome-wide RNAi screen identifies Letm1 as a mitochondrial Ca^{2+}/H^{+} antiporter. *Science* **2009**, *326*, 144–147. [CrossRef]
27. Jiang, D.; Zhao, L.; Clish, C.B.; Clapham, D.E. Letm1, the mitochondrial Ca^{2+}/H^{+} antiporter, is essential for normal glucose metabolism and alters brain function in Wolf-Hirschhorn syndrome. *Proc. Natl. Acad. Sci. USA* **2013**, *110*, E2249–E2254. [CrossRef]
28. Natarajan, G.K.; Glait, L.; Mishra, J.; Stowe, D.F.; Camara, A.K.S.; Kwok, W.M. Total matrix Ca^{2+} modulates Ca^{2+} efflux via the Ca^{2+}/H^{+} exchanger in cardiac mitochondria. *Front. Physiol.* **2020**, *11*, 510600. [CrossRef] [PubMed]
29. Crompton, M.; Capano, M.; Carafoli, E. The sodium-induced efflux of calcium from heart mitochondria. A possible mechanism for the regulation of mitochondrial calcium. *Eur. J. Biochem.* **1976**, *69*, 453–462. [CrossRef]
30. Affolter, H.; Carafoli, E. The Ca^{2+}-Na^{+} antiporter of heart mitochondria operates electroneutrally. *Biochem. Biophys. Res. Commun.* **1980**, *95*, 193–196. [CrossRef]
31. Jung, D.W.; Baysal, K.; Brierley, G.P. The sodium-calcium antiport of heart mitochondria is not electroneutral. *J. Biol. Chem.* **1995**, *270*, 672–678. [CrossRef] [PubMed]
32. Kim, B.; Matsuoka, S. Cytoplasmic Na^{+}-dependent modulation of mitochondrial Ca^{2+} via electrogenic mitochondrial Na^{+}-Ca^{2+} exchange. *J. Physiol.* **2008**, *586*, 1683–1697. [CrossRef]
33. Cox, D.A.; Conforti, L.; Sperelakis, N.; Matlib, M.A. Selectivity of inhibition of Na^{+}-Ca^{2+} exchange of heart mitochondria by benzothiazepine CGP-37157. *J. Cardiovasc. Pharm.* **1993**, *21*, 595–599. [CrossRef] [PubMed]
34. Roy, S.; Dey, K.; Hershfinkel, M.; Ohana, E.; Sekler, I. Identification of residues that control Li^{+} versus Na^{+} dependent Ca^{2+} exchange at the transport site of the mitochondrial NCLX. *Biochim Biophys. Acta Mol. Cell Res.* **2017**, *1864*, 997–1008. [CrossRef]
35. Giladi, M.; Lee, S.Y.; Refaeli, B.; Hiller, R.; Chung, K.Y.; Khananshvili, D. Structure-dynamic and functional relationships in a Li^{+}-transporting sodium-calcium exchanger mutant. *Biochim Biophys. Acta Bioenerg.* **2019**, *1860*, 189–200. [CrossRef]
36. Jumper, J.; Evans, R.; Pritzel, A.; Green, T.; Figurnov, M.; Ronneberger, O.; Tunyasuvunakool, K.; Bates, R.; Žídek, A.; Potapenko, A.; et al. Highly accurate protein structure prediction with alphafold. *Nature* **2021**, *596*, 583–589. [CrossRef] [PubMed]
37. Griffiths, E.J. Reversal of mitochondrial Na/Ca exchange during metabolic inhibition in rat cardiomyocytes. *FEBS Lett.* **1999**, *453*, 400–404. [CrossRef]
38. Kostic, M.; Katoshevski, T.; Sekler, I. Allosteric regulation of NCLX by mitochondrial membrane potential links the metabolic state and Ca^{2+} signaling in mitochondria. *Cell Rep.* **2018**, *25*, 3465–3475.e3464. [CrossRef]
39. Kostic, M.; Ludtmann, M.H.; Bading, H.; Hershfinkel, M.; Steer, E.; Chu, C.T.; Abramov, A.Y.; Sekler, I. PKA phosphorylation of NCLX reverses mitochondrial calcium overload and depolarization, promoting survival of PINK1-deficient dopaminergic neurons. *Cell Rep.* **2015**, *13*, 376–386. [CrossRef]
40. Samanta, K.; Mirams, G.R.; Parekh, A.B. Sequential forward and reverse transport of the Na^{+} Ca^{2+} exchanger generates Ca^{2+} oscillations within mitochondria. *Nat. Commun.* **2018**, *9*, 156. [CrossRef]
41. Gandhi, S.; Wood-Kaczmar, A.; Yao, Z.; Plun-Favreau, H.; Deas, E.; Klupsch, K.; Downward, J.; Latchman, D.S.; Tabrizi, S.J.; Wood, N.W.; et al. PINK1-associated Parkinson's disease is caused by neuronal vulnerability to calcium-induced cell death. *Mol. Cell* **2009**, *33*, 627–638. [CrossRef] [PubMed]
42. Pickrell, A.M.; Youle, R.J. The roles of PINK1, Parkin, and mitochondrial fidelity in Parkinson's disease. *Neuron* **2015**, *85*, 257–273. [CrossRef]
43. Kang, C.; Badr, M.A.; Kyrychenko, V.; Eskelinen, E.L.; Shirokova, N. Deficit in PINK1/PARKIN-mediated mitochondrial autophagy at late stages of dystrophic cardiomyopathy. *Cardiovasc. Res.* **2018**, *114*, 90–102. [CrossRef] [PubMed]
44. Zhou, Q.; Xie, M.; Zhu, J.; Yi, Q.; Tan, B.; Li, Y.; Ye, L.; Zhang, X.; Zhang, Y.; Tian, J.; et al. PINK1 contained in huMSC-derived exosomes prevents cardiomyocyte mitochondrial calcium overload in sepsis via recovery of mitochondrial Ca^{2+} efflux. *Stem Cell Res.* **2021**, *12*, 269. [CrossRef]
45. Glancy, B.; Hartnell, L.M.; Malide, D.; Yu, Z.X.; Combs, C.A.; Connelly, P.S.; Subramaniam, S.; Balaban, R.S. Mitochondrial reticulum for cellular energy distribution in muscle. *Nature* **2015**, *523*, 617–620. [CrossRef]
46. Glancy, B.; Hartnell, L.M.; Combs, C.A.; Femnou, A.; Sun, J.; Murphy, E.; Subramaniam, S.; Balaban, R.S. Power grid protection of the muscle mitochondrial reticulum. *Cell Rep.* **2017**, *19*, 487–496. [CrossRef] [PubMed]
47. Romashko, D.N.; Marban, E.; O'Rourke, B. Subcellular metabolic transients and mitochondrial redox waves in heart cells. *Proc. Natl. Acad. Sci. USA* **1998**, *95*, 1618–1623. [CrossRef] [PubMed]
48. Kuznetsov, A.V.; Usson, Y.; Leverve, X.; Margreiter, R. Subcellular heterogeneity of mitochondrial function and dysfunction: Evidence obtained by confocal imaging. *Mol. Cell Biochem.* **2004**, *256–257*, 359–365. [CrossRef] [PubMed]
49. Lu, X.; Thai, P.N.; Lu, S.; Pu, J.; Bers, D.M. Intrafibrillar and perinuclear mitochondrial heterogeneity in adult cardiac myocytes. *J. Mol. Cell Cardiol.* **2019**, *136*, 72–84. [CrossRef] [PubMed]
50. Perez-Hernández, M.; Leo-Macias, A.; Keegan, S.; Jouni, M.; Kim, J.C.; Agullo-Pascual, E.; Vermij, S.; Zhang, M.; Liang, F.X.; Burridge, P.; et al. Structural and functional characterization of a $Na_{V}1.5$-mitochondrial couplon. *Circ. Res.* **2021**, *128*, 419–432. [CrossRef]

51. Aon, M.A.; Cortassa, S.; Marban, E.; O'Rourke, B. Synchronized whole cell oscillations in mitochondrial metabolism triggered by a local release of reactive oxygen species in cardiac myocytes. *J. Biol. Chem.* **2003**, *278*, 44735–44744. [CrossRef] [PubMed]
52. Skogestad, J.; Lines, G.T.; Louch, W.E.; Sejersted, O.M.; Sjaastad, I.; Aronsen, J.M. Evidence for heterogeneous subsarcolemmal Na^+ levels in rat ventricular myocytes. *Am. J. Physiol. Heart Circ. Physiol.* **2019**, *316*, H941–H957. [CrossRef]
53. Lu, X.; Ginsburg, K.S.; Kettlewell, S.; Bossuyt, J.; Smith, G.L.; Bers, D.M. Measuring local gradients of intramitochondrial $[Ca^{2+}]$ in cardiac myocytes during sarcoplasmic reticulum Ca^{2+} release. *Circ. Res.* **2013**, *112*, 424–431. [CrossRef]
54. Rizzuto, R.; De Stefani, D.; Raffaello, A.; Mammucari, C. Mitochondria as sensors and regulators of calcium signalling. *Nat. Rev. Mol. Cell Biol.* **2012**, *13*, 566–578. [CrossRef]
55. De La Fuente, S.; Fernandez-Sanz, C.; Vail, C.; Agra, E.J.; Holmstrom, K.; Sun, J.; Mishra, J.; Williams, D.; Finkel, T.; Murphy, E.; et al. Strategic positioning and biased activity of the mitochondrial calcium uniporter in cardiac muscle. *J. Biol. Chem.* **2016**, *291*, 23343–23362. [CrossRef] [PubMed]
56. De La Fuente, S.; Lambert, J.P.; Nichtova, Z.; Fernandez-Sanz, C.; Elrod, J.W.; Sheu, S.S.; Csordás, G. Spatial separation of mitochondrial calcium uptake and extrusion for energy-efficient mitochondrial calcium signaling in the heart. *Cell Rep.* **2018**, *24*, 3099–3107.e3094. [CrossRef]
57. Bassani, R.A.; Bassani, J.W.; Bers, D.M. Relaxation in ferret ventricular myocytes: Unusual interplay among calcium transport systems. *J. Physiol.* **1994**, *476*, 295–308. [CrossRef]
58. Bassani, J.W.; Bassani, R.A.; Bers, D.M. Relaxation in rabbit and rat cardiac cells: Species-dependent differences in cellular mechanisms. *J. Physiol.* **1994**, *476*, 279–293. [CrossRef]
59. Xie, A.; Zhou, A.; Liu, H.; Shi, G.; Liu, M.; Boheler, K.R.; Dudley, S.C., Jr. Mitochondrial Ca^{2+} flux modulates spontaneous electrical activity in ventricular cardiomyocytes. *PLoS ONE* **2018**, *13*, e0200448. [CrossRef]
60. Mangoni, M.E.; Nargeot, J. Genesis and regulation of the heart automaticity. *Physiol. Rev.* **2008**, *88*, 919–982. [CrossRef]
61. Lakatta, E.G.; Maltsev, V.A.; Vinogradova, T.M. A coupled system of intracellular Ca^{2+} clocks and surface membrane voltage clocks controls the timekeeping mechanism of the heart's pacemaker. *Circ. Res.* **2010**, *106*, 659–673. [CrossRef]
62. Monfredi, O.; Maltsev, V.A.; Lakatta, E.G. Modern concepts concerning the origin of the heartbeat. *Physiology* **2013**, *28*, 74–92. [CrossRef]
63. Yaniv, Y.; Spurgeon, H.A.; Lyashkov, A.E.; Yang, D.; Ziman, B.D.; Maltsev, V.A.; Lakatta, E.G. Crosstalk between mitochondrial and sarcoplasmic reticulum Ca^{2+} cycling modulates cardiac pacemaker cell automaticity. *PLoS ONE* **2012**, *7*, e37582. [CrossRef]
64. Takeda, Y.; Matsuoka, S. Impact of mitochondria on local calcium release in murine sinoatrial nodal cells. *J. Mol. Cell Cardiol.* **2021**, *164*, 42–50. [CrossRef] [PubMed]
65. Bychkov, R.; Juhaszova, M.; Tsutsui, K.; Coletta, C.; Stern, M.D.; Maltsev, V.A.; Lakatta, E.G. Synchronized cardiac impulses emerge from heterogeneous local calcium signals within and among cells of pacemaker tissue. *JACC Clin. Electrophysiol.* **2020**, *6*, 907–931. [CrossRef]
66. Katz, A. *Physiology of the Heart*; Lippincott Williams & Wilkins: Philadelphia, PA, USA, 2010.
67. McCormack, J.G.; Halestrap, A.P.; Denton, R.M. Role of calcium ions in regulation of mammalian intramitochondrial metabolism. *Physiol. Rev.* **1990**, *70*, 391–425. [CrossRef] [PubMed]
68. Maack, C.; Cortassa, S.; Aon, M.A.; Ganesan, A.N.; Liu, T.; O'Rourke, B. Elevated cytosolic Na^+ decreases mitochondrial Ca^{2+} uptake during excitation-contraction coupling and impairs energetic adaptation in cardiac myocytes. *Circ. Res.* **2006**, *99*, 172–182. [CrossRef] [PubMed]
69. Liu, T.; O'Rourke, B. Enhancing mitochondrial Ca^{2+} uptake in myocytes from failing hearts restores energy supply and demand matching. *Circ. Res.* **2008**, *103*, 279–288. [CrossRef]
70. Ventura-Clapier, R.; Garnier, A.; Veksler, V.; Joubert, F. Bioenergetics of the failing heart. *Biochim Biophys. Acta* **2011**, *1813*, 1360–1372. [CrossRef]
71. Liu, T.; Takimoto, E.; Dimaano, V.L.; DeMazumder, D.; Kettlewell, S.; Smith, G.; Sidor, A.; Abraham, T.P.; O'Rourke, B. Inhibiting mitochondrial Na^+/Ca^{2+} exchange prevents sudden death in a guinea pig model of heart failure. *Circ. Res.* **2014**, *115*, 44–54. [CrossRef]
72. Aksentijević, D.; Karlstaedt, A.; Basalay, M.V.; O'Brien, B.A.; Sanchez-Tatay, D.; Eminaga, S.; Thakker, A.; Tennant, D.A.; Fuller, W.; Eykyn, T.R.; et al. Intracellular sodium elevation reprograms cardiac metabolism. *Nat. Commun.* **2020**, *11*, 4337. [CrossRef]
73. Carley, A.N.; Taegtmeyer, H.; Lewandowski, E.D. Matrix revisited: Mechanisms linking energy substrate metabolism to the function of the heart. *Circ. Res.* **2014**, *114*, 717–729. [CrossRef]
74. Saito, R.; Takeuchi, A.; Himeno, Y.; Inagaki, N.; Matsuoka, S. A simulation study on the constancy of cardiac energy metabolites during workload transition. *J. Physiol.* **2016**, *594*, 6929–6945. [CrossRef]
75. Takeuchi, A.; Matsuoka, S. Integration of mitochondrial energetics in heart with mathematical modelling. *J. Physiol.* **2020**, *598*, 1443–1457. [CrossRef]
76. Chen, Y.R.; Zweier, J.L. Cardiac mitochondria and reactive oxygen species generation. *Circ. Res.* **2014**, *114*, 524–537. [CrossRef]
77. Zorov, D.B.; Juhaszova, M.; Sollott, S.J. Mitochondrial reactive oxygen species (ROS) and ROS-induced ROS release. *Physiol. Rev.* **2014**, *94*, 909–950. [CrossRef]
78. Dey, S.; DeMazumder, D.; Sidor, A.; Foster, D.B.; O'Rourke, B. Mitochondrial ROS drive sudden cardiac death and chronic proteome remodeling in heart failure. *Circ. Res.* **2018**, *123*, 356–371. [CrossRef]

79. Santulli, G.; Xie, W.; Reiken, S.R.; Marks, A.R. Mitochondrial calcium overload is a key determinant in heart failure. *Proc. Natl. Acad. Sci. USA* **2015**, *112*, 11389–11394. [CrossRef]
80. Beretta, M.; Santos, C.X.; Molenaar, C.; Hafstad, A.D.; Miller, C.C.; Revazian, A.; Betteridge, K.; Schröder, K.; Streckfuss-Bömeke, K.; Doroshow, J.H.; et al. Nox4 regulates $InsP_3$ receptor-dependent Ca^{2+} release into mitochondria to promote cell survival. *EMBO J.* **2020**, *39*, e103530. [CrossRef]
81. Hamilton, S.; Terentyeva, R.; Kim, T.Y.; Bronk, P.; Clements, R.T.; O-Uchi, J.; Csordás, G.; Choi, B.R.; Terentyev, D. Pharmacological modulation of mitochondrial Ca^{2+} content regulates sarcoplasmic reticulum Ca^{2+} release via oxidation of the ryanodine receptor by mitochondria-derived reactive oxygen species. *Front. Physiol.* **2018**, *9*, 1831. [CrossRef]
82. Hernansanz-Agustín, P.; Choya-Foces, C.; Carregal-Romero, S.; Ramos, E.; Oliva, T.; Villa-Piña, T.; Moreno, L.; Izquierdo-Álvarez, A.; Cabrera-García, J.D.; Cortés, A.; et al. Na^+ controls hypoxic signalling by the mitochondrial respiratory chain. *Nature* **2020**, *586*, 287–291. [CrossRef]
83. Cortassa, S.; Juhaszova, M.; Aon, M.A.; Zorov, D.B.; Sollott, S.J. Mitochondrial Ca^{2+}, redox environment and ROS emission in heart failure: Two sides of the same coin? *J. Mol. Cell Cardiol.* **2021**, *151*, 113–125. [CrossRef]
84. Malyala, S.; Zhang, Y.; Strubbe, J.O.; Bazil, J.N. Calcium phosphate precipitation inhibits mitochondrial energy metabolism. *PLoS Comput. Biol.* **2019**, *15*, e1006719. [CrossRef]
85. Kosmach, A.; Roman, B.; Sun, J.; Femnou, A.; Zhang, F.; Liu, C.; Combs, C.A.; Balaban, R.S.; Murphy, E. Monitoring mitochondrial calcium and metabolism in the beating MCU-KO heart. *Cell Rep.* **2021**, *37*, 109846. [CrossRef]

biomolecules

MDPI

Review

Bifurcations and Proarrhythmic Behaviors in Cardiac Electrical Excitations

Kunichika Tsumoto and Yasutaka Kurata *

Department of Physiology II, Kanazawa Medical University, Uchinada 920-0293, Japan; tsumoto@kanazawa-med.ac.jp
* Correspondence: yasu@kanazawa-med.ac.jp; Tel.: +81-76-286-2211

Abstract: The heart is a hierarchical dynamic system consisting of molecules, cells, and tissues, and acts as a pump for blood circulation. The pumping function depends critically on the preceding electrical activity, and disturbances in the pattern of excitation propagation lead to cardiac arrhythmia and pump failure. Excitation phenomena in cardiomyocytes have been modeled as a nonlinear dynamical system. Because of the nonlinearity of excitation phenomena, the system dynamics could be complex, and various analyses have been performed to understand the complex dynamics. Understanding the mechanisms underlying proarrhythmic responses in the heart is crucial for developing new ways to prevent and control cardiac arrhythmias and resulting contractile dysfunction. When the heart changes to a pathological state over time, the action potential (AP) in cardiomyocytes may also change to a different state in shape and duration, often undergoing a qualitative change in behavior. Such a dynamic change is called bifurcation. In this review, we first summarize the contribution of ion channels and transporters to AP formation and our knowledge of ion-transport molecules, then briefly describe bifurcation theory for nonlinear dynamical systems, and finally detail its recent progress, focusing on the research that attempts to understand the developing mechanisms of abnormal excitations in cardiomyocytes from the perspective of bifurcation phenomena.

Keywords: action potential; nonlinear dynamical system; bifurcation theory; afterdepolarizations; cardiac arrhythmias

Citation: Tsumoto, K.; Kurata, Y. Bifurcations and Proarrhythmic Behaviors in Cardiac Electrical Excitations. *Biomolecules* **2022**, *12*, 459. https://doi.org/10.3390/biom12030459

Academic Editors: Yosuke Okamoto and Kyoichi Ono

Received: 17 February 2022
Accepted: 14 March 2022
Published: 16 March 2022

Publisher's Note: MDPI stays neutral with regard to jurisdictional claims in published maps and institutional affiliations.

1. Introduction

We can observe a wide variety of dynamics in life phenomena. Dynamics are the changes in states of a system over time. The system is defined as a set of objects with functions; if the states of the system change with time, then the system is called a dynamic system. A mathematical model is a description of the rule for state changes in the system, and such a model is called a dynamical system [1,2].

The heart is a highly hierarchical dynamic system consisting of molecules, cells, and tissues [3–5]. It is well known that cardiac muscle cells cause an electrical excitation. To reproduce the excitation phenomena observed in cardiac cells, many mathematical models have been proposed. Excitation is a nonlinear phenomenon; the excitatory dynamics of cardiomyocytes modeled as a nonlinear dynamical system could be complex. For understanding complicated dynamics, various methods based on nonlinear dynamical system theories have been utilized.

Electrical phenomena in the heart are initiated with the spontaneous excitation evoked in pacemaking cells in the sinoatrial (SA) node, followed by the excitation conduction to the atrium, atrioventricular (AV) node, His bundles, Purkinje fiber, and ventricle [6]. By regularly repeating the precise and coordinated excitation propagation, the heart plays an essential role as a circulating blood pump. Cardiac arrhythmias are disturbed heartbeats or uncoordinated excitation propagations.

In the 1980s–1990s, the idea of dynamical diseases was proposed by some researchers [7–9]. The idea assumes the existence of a certain kind of nonlinear dynamical system behind phenomena evoked in the heart and then links the normal function of the heart, i.e., normal heartbeat, to a stable state of the nonlinear dynamical system. This will allow us to consider the regular and coordinated excitation propagation that underlies the normal heartbeat as one of the desirable stable states for the cardiac system. Furthermore, the phenomenon that cardiac excitations and/or propagations are disturbed by ischemia, stress, drug action, etc., which is called *arrhythmia*, can be considered as the dynamical disease. In other words, cardiac arrhythmias can be captured as state changes in the cardiac system, with the desirable stable state being changed into another undesirable one by changing system parameters in the heart related to many factors.

Understanding how the proarrhythmic response occurs in the heart is extremely important for developing new ways to prevent and control cardiac arrhythmia and contractile dysfunction. Based on the concept of dynamical disease, a change in some of the system components may cause a breakdown in the electrical excitation of cardiomyocytes; this is a transition from a normal stable state to another stable state. Such a state change, a qualitative change in system behaviors, or a sudden change of phenomenon, is called a bifurcation (or bifurcation phenomenon) in nonlinear dynamical systems theory [10–13]. Bifurcation analysis is the search for system parameters whose changes abruptly alter system behaviors. This review summarizes recent progress in understanding the mechanisms of normal and abnormal excitations observed in cardiomyocytes from the perspective of bifurcation phenomena.

2. Ionic Mechanisms of Excitations in Cardiomyocytes

Working cardiac muscles such as atrial and ventricular myocytes are excitable cells (excitable membrane) and are quiescent at a constant potential called "resting membrane potential" (−80 to −90 mV) in the absence of an external stimulus. The membrane potential of atrial and ventricular myocytes changes transiently upon the input of an appropriate external stimulus that depolarizes the membrane potential above a threshold level. This membrane potential change is referred to as an action potential (AP). On the other hand, specialized cardiac muscles such as sinoatrial node and atrioventricular node cells do not have a constant resting membrane potential, but spontaneously and periodically generate APs. In both cases, membrane potential changes are caused by the movement of ions into and out of the cell through ion-transport molecules such as ion channels and transporters embedded in the cell membrane.

To understand how the proarrhythmic behavior of cardiomyocytes occurs, it is necessary to understand the mechanism of normal AP generation. To achieve this, the ionic contribution to APs should also be understood. In the following, we first summarize ion homeostasis in cardiomyocytes and the mechanisms of AP generation. The biophysical mechanisms of abnormal APs related to arrhythmogenesis will then be summarized.

2.1. *Mechanisms for Maintenance of Ion Homeostasis*

The electrical activity of the heart, e.g., AP generation, is regulated by a variety of ion-transport proteins such as ion channels and ion transporters/exchangers embedded in the cell membrane. In cardiomyocytes, the intra- and extracellular environments are separated by a lipid bilayer membrane, and there are differences in ion concentrations. The major ions that contribute to AP generations in cardiomyocytes are Na^+, K^+, and Ca^{2+}, and the extracellular concentrations of Na^+ and Ca^{2+} are higher than their intracellular concentrations. For example, when Na^+ or Ca^{2+} channels are opened within the range of physiological membrane potentials, the ions move from the outside to inside the cell by diffusion according to an electrochemical gradient, increasing their intracellular concentrations. On the other hand, the intracellular K^+ concentration is higher than the extracellular K^+ concentration. Thus, the opening of K^+ channels results in the outflow of K^+ ions from inside to outside the cell. When only K^+ channels remain open, the membrane potential will reach the reversal potential of the K^+ channel, i.e., the equilibrium potential for K^+. If all types of ion channels, Na^+, K^+, and Ca^{2+} channels, continue to open, then the concentration of each ion in the intracellular solution will reach that of the extracellular solution because of the thermodynamic principle. However, the intracellular ion concentrations are always maintained within certain ranges different from extracellular ones. In particular, the intracellular Ca^{2+} regulation (or Ca^{2+} handling) in cardiomyocytes is not only crucial for muscle contraction and relaxation but is also involved in the regulation of ion channel function through Ca^{2+}-dependent signaling as a second messenger [14–16]. Furthermore, intracellular Ca^{2+} handling is also involved in ion channel remodeling via changes in the activity of transcription factors. Ca^{2+}-handling abnormalities due to disruption of intracellular Ca^{2+} ion homeostasis, such as those in ischemia, hypertrophy, and heart failure, are a major cause of the development of lethal arrhythmias (for details, see reviews in [17,18]). In cardiomyocytes, three transporters, Na^+/K^+ ATPase (NKA) [19–23], Na^+-Ca^{2+} exchanger (NCX) [24–26], and Na^+-H^+ exchanger (NHE) [27], play a key role in maintaining ion homeostasis. NKA and NCX are electrogenic because they generate ionic currents, while NHE is electroneutral. Therefore, it is NKA and NCX that are directly involved in AP generation in cardiomyocytes.

NKA hydrolyzes ATP to generate chemical energy. The generated energy is used to exchange three intracellular Na^+ ions and two extracellular K^+ ions against each electrochemical gradient. When NKA works once, one extra cation is released from inside to outside the cell. Therefore, the NKA current (I_{NKA}) is an outward current, which can shorten the AP and hyperpolarize the resting membrane potential. So far, four isoforms of the α subunit, α1, α2, α3, and α4, are found to be expressed in mammals and three isoforms of the β subunit, β1, β2, and β3, have been identified [28,29]. The α1, α2, α3, α4, β1, β2, and β3-subunits are encoded by the genes *ATP1A1*, *ATP1A2*, *ATP1A3*, *ATP1A4*, *ATP1B1*, *ATP1B2*, and *ATP1B3*, respectively [30]. In rodents, α1 and α2 were the two major isoforms [31], while in dogs and macaques, α1 and α3 were mainly expressed [32]. In the human heart, three α isoforms, α1, α2, and α3, have been detected [32], and their expressions have been estimated to be 62%, 15%, and 23%, respectively [33,34]. In particular, the α1-subunit is widely distributed in the surface sarcolemma [24] and plays a major role in regulating Na^+ ion concentrations in the cytoplasm. The α2-subunit is mainly expressed in the dyadic cleft and may play a special role in regulating the transport of Ca^{2+} released from the sarcoplasmic reticulum via interaction with NCX [22]. The β-subunit is also essential for the pump function and plays a role in the α-subunit stabilization.

The main function of NCX is to exchange three extracellular Na^+ and one intracellular Ca^{2+} [35–37]. At this time, the electrochemical potential energy of Na^+ that is lost during its transport is used to transport Ca^{2+} against its electrochemical gradient. At the resting membrane potential in ventricular or atrial myocytes (~−80 mV), the ion Ca^{2+} is pumped out of myocytes by NCX. Furthermore, NCX generates an ionic current (I_{NCX}) inwardly for most of the time during AP generation. Thus, I_{NCX} can contribute to AP prolongations and depolarize the resting potential in working myocytes. Note that NCX can generate outward currents in the late phase of rapid AP depolarization where the membrane potential overshoots before an increase in the intracellular Ca^{2+} concentration. I_{NCX} turns outward in the late depolarization phase because the membrane potential becomes more depolarized than the reversal potential of NCX. This is called the "reverse mode" of NCX, causing Ca^{2+} influx into the cell. The reverse mode leads to increased intracellular Ca^{2+} concentrations, playing a role in the augmentation of the muscle contraction via the excitation–contraction coupling.

NCX is a member of the huge Ca^{2+}/cation antiporter superfamily [26]. Mammalian cells express three NCX isoforms, NCX1, NCX2, and NCX3, which are encoded by the genes *SLC8A1*, *SLC8A2*, and *SLC8A3*, respectively [38,39]. The representative of the cardiac NCX isoform is NCX1.1 [35].

2.2. Ionic Mechanisms of Cardiac Action Potentials

The transmembrane potential in cardiomyocytes changes by approximately 60 to 120 mV with AP generation. APs initiate from the resting membrane potential in working myocytes (~−80 mV) or the maximum diastolic potential in specialized cardiomyocytes (~−40 mV) that constitute the conduction system and exhibit automaticity. In working myocytes, the resting membrane potential is mainly influenced by the inwardly rectifying K^+ channel current (I_{K1}), which is mediated by the inward-rectifier K^+ channel (Kir channel) consisting of the Kir2.1 subunit encoded by *KCNJ2* [40,41]. The I_{K1} also functions to terminate the AP and to indirectly determine the excitability of cardiomyocytes. Furthermore, the NKA contributes to the maintenance of the resting membrane potential [23].

APs (time evolution of the membrane potential) of cardiomyocytes differ in shape in different parts of the heart, and also vary greatly between species. Here, we consider the AP of human ventricular myocytes (see Figure 1).

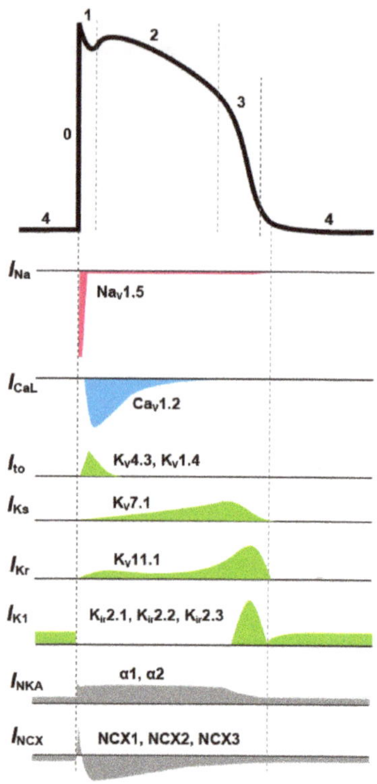

Figure 1. Schematic representation of a typical action potential (AP) in the ventricular myocyte and the respective ion channel currents that contribute to AP formation. It also indicates the major molecules that constitute ion channels and transporters (for the meanings of individual symbols, see text).

By applying appropriate stimulation to ventricular myocytes in a resting state, the cell membrane is depolarized. Depolarization of the membrane potential first activates Na^+ channels [42], which mainly consist of the $Na_V1.5\alpha$ subunit [43] encoded by *SCN5A*, causing Na^+ to flow into the cardiomyocyte. The influx of cations into the cell further depolarizes the membrane potential, leading to further activation of Na^+ channels. As a result, more Na^+ ions flow into the cell, and depolarization is accelerated. The influx of Na^+ into the myocyte in a positive-feedback manner causes a very large inward Na^+ channel current (I_{Na}) and rapidly depolarizes the membrane potential to form AP phase 0. However, the inactivation of Na^+ channels is also rapid, and I_{Na} will soon stop flowing.

Depolarization of the membrane potential also activates other ion channels such as K^+ and Ca^{2+} channels. Transient outward K^+ channels, which are mainly composed of

$K_v4.3$ and $K_v4.2$ α-subunits encoded by *KCND3* and *KCND2*, respectively, in human and canine ventricular myocytes [44,45], are activated following AP phase 0 and produce a large transient outward K^+ current (I_{to}) [46]. As a result, the membrane potential transiently repolarizes to form AP phase 1. In human cardiomyocytes, the $K_v1.4$ α-subunit is also expressed, making up 10–20% of the I_{to} density [47,48].

Following the AP overshoot, Ca^{2+} channels [49] and delayed rectifier K^+ channels [50] are activated to form AP phase 2. There are two major subtypes of Ca^{2+} channels expressed in cardiomyocytes: L-type Ca^{2+} channels (LTCCs) and T-type Ca^{2+} channels [51,52]. The most abundant type expressed in ventricular myocytes is the LTCC (CaV1.2), which carries Ca^{2+} currents through a pore-forming α-subunit (α1) encoded by the *CACNA1C* gene [46]. T-type Ca^{2+} channels (CaV3.1 and CaV3.2), encoded by the *CACNA1G* and *CACNA1H* genes, respectively [53,54], are believed to be involved in the automaticity of the sinoatrial node cell. However, their functional role in ventricular myocytes is not clear. Delayed rectifier K^+ channel currents, carrying K^+ out of the myocyte, can be separated into a rapid component (I_{Kr}) and a slow component (I_{Ks}) [50]. The I_{Kr} channel α-subunit is encoded by a human *ether-à-go-go-related gene* (hERG) and is also called the hERG channel and Kv11.1 pore-forming α-subunit [55–57]. In addition, I_{Ks} is thought to be carried by the Kv7.1 channel encoded by the *KCNQ1* gene [58–61]. Activation of the LTCC is sustained for a relatively long time and results in a slow inward current (I_{CaL}). On the other hand, I_{Kr} and I_{Ks} carry outward currents, though there is only a small current flow of I_{Kr} during AP phase 2 as hERG channels rapidly enter into the inactive state [62,63]. There is a certain balance between the inward Ca^{2+} and outward K^+ currents during this phase. This balance forms the plateau phase (AP phase 2) in AP of the ventricular myocyte.

Eventually, LTCCs also become inactive due to the two mechanisms, voltage- and Ca^{2+}-dependent inactivation, and I_{CaL} (inward current) becomes smaller. In the subsequent phase (AP phase 3), the outflow of cations by I_{Kr} and I_{Ks} causes the membrane potential to rapidly repolarize. Particularly for I_{Kr}, as the membrane potential starts to repolarize, hERG channels recover from inactivation and carry more I_{Kr}. These larger outward currents (I_{Kr} and I_{Ks}) contribute to the increased rate of repolarization. With the large flow of I_{K1} in the latter half of AP phase 3, the AP undergoes a transition from the repolarization phase to the resting state (AP phase 4). After that, I_{K1} channels continue to discharge K^+ out of the cardiomyocyte, consequently flowing the outward current and maintaining a deep resting membrane potential; NKA counteracts I_{K1} channels, returning K^+ to the intracellular space. The I_{K1} channels consist of the Kir2.1, Kir2.2, Kir2.3, and Kir2.4 pore-forming α-subunits, which are encoded by the *KCNJ2*, *KCNJ12*, *KCNJ4*, and *KCNJ14* genes, respectively [40,41,64,65].

2.3. *Mechanisms of Abnormal Action Potentials (Afterdepolarization)*

APD prolongation in ventricular myocytes increases the risk of the development of lethal ventricular arrhythmias such as torsades des pointes (TdP), often leading to cardiac sudden death in long QT syndrome (LQTS) [66–69]. At present, inherited LQTS, which is characterized by a congenital prolongation of QT interval, is subdivided into 17 different types depending on where genetic mutations occur (LQT1~LQT17) [66–69]. For instance, in patients with LQT1 and LQT2, the function of the ion channel responsible for I_{Ks} (LQT1) or I_{Kr} (LQT2) is reduced by loss-of-function mutations of the K^+ channels. In both cases, a decrease in repolarization currents delays AP repolarization and prolongs the QT interval. A genetic abnormality of $Na_V1.5$ (LQT3) that causes persistent inward Na^+ currents due to impaired Na^+ channel inactivation has also been observed; this can be called a gain-of-function mutation of the Na^+ channel. In the different genotypes, TdP may be precipitated by physical or emotional stress (LQT1), a startle (LQT2), or may occur at rest or during sleep (LQT3). These three types of LQTS account for more than 70% of all inherited LQTS (genetic abnormalities of LQT1, LQT2, and LQT3 account for 30~35%, 20~25%, and 5~10%, respectively) [68].

Normally, the AP of cardiomyocytes is completed by sufficient K^+ current flow during the repolarization phase (late AP phase 2 and phase 3) of the AP (Figure 1). The presence of multiple K^+ current components such as I_{Ks}, I_{Kr}, and I_{NKA}, all of which act to repolarize the AP, appears redundant as a system. However, the redundancy forms an available reserve for AP repolarization. This concept is referred to as the "repolarization reserve" [70,71]. If these outward K^+ channel currents, e.g., I_{Kr} and/or I_{Ks}, attenuate due to genetic abnormalities, drug actions, aging, etc., or if I_{Na} and/or I_{CaL} continue to flow during the plateau phase, then the repolarization is delayed, leading to excessive AP prolongation. In particular, the continuation of the plateau phase where the membrane potential is within the range of the I_{CaL} window current can lead to an excessive accumulation of Ca^{2+} in the cytoplasm. In this situation, abnormal behaviors called afterdepolarizations are likely to occur [70,72,73]. Afterdepolarizations include early afterdepolarization (EAD), which is the transient depolarization during the late AP phase 2 and phase 3, and delayed afterdepolarization (DAD) [74–76], which is the transient depolarization after AP completion, i.e., in AP phase 4. Many experimental studies have suggested that the reactivation of the LTCC current (I_{CaL}) during AP phase 2 and phase 3 in ventricular myocytes is a key mechanism of EAD formation [77–82]. Ca^{2+} is taken up into the cardiomyocyte by the Ca^{2+} influx through the LTCC. The LTCC activates at the membrane potential of approximately −30~+10 mV, resulting in a relatively large inward current. On the one hand, the LTCC exhibits a voltage-dependent inactivation (VDI) at the membrane potential of approximately −50~0 mV. The LTCC inactivates during AP phase 2 and Ca^{2+} influx terminates. Thus, during AP repolarization, I_{CaL} by LTCCs flows within a range of the membrane potential where the steady-state activation and inactivation curve of the LTCC overlap (see Figure 2), known as "the I_{CaL} window current region" [83,84]. AP repolarization results in recovery from the inactivation state of LTCCs. As a result, the LTCC slightly reactivates in the AP repolarization process and causes a small inward current. The repolarization of a normal AP proceeds rapidly, so the period of time during which the membrane potential is in the I_{CaL} window current range is short. The time for reactivation of LTCCs is very short, and sufficient current to cause EAD does not occur. However, when the repolarization reserve is reduced, i.e., I_{Kr} or I_{Ks} or both are reduced, the time at which the membrane potential is in the I_{CaL} window current range lengthens. For this reason, I_{CaL} can be augmented by LTCC reactivation. At this time, an increase in the inward current component during the AP repolarization phase causes the inward/outward current imbalance in the net membrane current (I_{net}). Since I_{K1} is not yet activated during the late AP phase 2 and early phase 3, the membrane resistance remains relatively high during these phases. This means that even a slight current causes a large change in the membrane potential. As a result, the first EAD is triggered in the late AP phase 2 to phase 3. As modulators of AP with EADs, at least, intracellular Ca^{2+}-handling (or SR Ca^{2+}-handling), intracellular Na^+-handling, and the heart rate can be considered [85] (Figure 3). For example, inhibition of NCX (e.g., by SEA-0400 administration [81]) primarily suppresses the Ca^{2+} efflux. As a result, the increase in Ca^{2+} concentrations in the intracellular subspace ($[Ca^{2+}]_{ss}$) and the cytoplasm ($[Ca^{2+}]_i$) promotes the Ca^{2+}-dependent inactivation(CDI) of LTCCs and secondarily decreases I_{CaL}, consequently suppressing EAD formation [86,87]. Conversely, enhanced NCX (e.g., by overexpression [82]) increases Ca^{2+} extrusion from the intracellular to extracellular space, resulting in a decrease in $[Ca^{2+}]_i$ (or $[Ca^{2+}]_{ss}$). The decrease in $[Ca^{2+}]_i$ (or $[Ca^{2+}]_{ss}$) suppresses the CDI of LTCCs and leads to an increase in the I_{CaL} window current, resulting in the promotion of EAD formation [88]. As exemplified by the administration of digitalis, NKA inhibition results in a decrease in outward current, i.e., repolarization reserve attenuation, and thus contributes to facilitating EAD generation. As suggested by Xie et al. [89], the membrane potential change via I_{CaL}, I_{NCX}, and I_{NKA}, and the interactions between intracellular Ca^{2+}- and Na^+-handling, influence the development of EADs. Another possible mechanism of EAD formation is spontaneous Ca^{2+} release from the sarcoplasmic reticulum (SR) [90], which will increase the Ca^{2+} concentration in the cytoplasm. The spontaneous SR Ca^{2+} release is also the recognized mechanism of DADs;

it has been reported that DADs are often accompanied by EADs [91–93]. As a result, the outflow of Ca^{2+} by NCX and inward I_{NCX} are enhanced and may cause transient depolarization of the membrane potential. For instance, the decrease in Na^+ efflux due to chronic inhibition of NKA causes an increase in the intracellular Na^+ concentration ($[Na^+]_i$). This adversely affects the activity of NCX, which operates by a loss of electrochemical energy during the transfer of Na^+ from outside to inside the cell. Consequently, increased $[Ca^{2+}]_i$ and $[Ca^{2+}]_{ss}$, and excessive Ca^{2+} accumulation in the SR ($[Ca^{2+}]_{SR}$) cause spontaneous Ca^{2+} release from the SR to the cytoplasm, which in turn facilitates EAD generation via NCX activation. Spontaneous SR Ca^{2+} releases have been suggested to induce EADs under certain conditions such as β-adrenergic stimulation via enhancing inward I_{NCX} [94,95]. This mechanism for EAD formation may be applicable to the EAD in the late AP phase 3 observed under β-adrenergic stimulation in an LQTS patient [92].

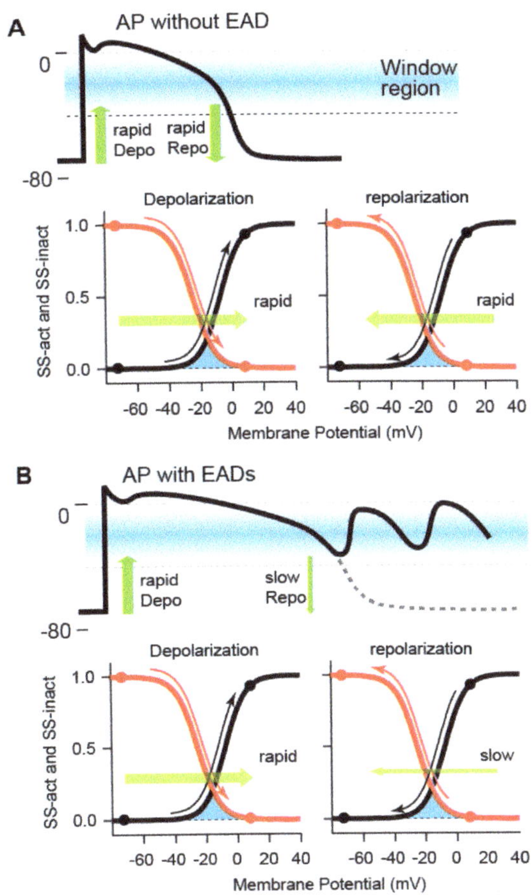

Figure 2. Relationships between membrane potential changes in response to depolarization and repolarization in a normal AP without EAD (**A**) and an abnormal AP with EADs (**B**) and voltage-dependent activation/inactivation of the L-type Ca^{2+} channel. Excessive APD prolongation due to slow repolarization leads to a long-lasting stay of the membrane potential in the L-type Ca^{2+} channel current (I_{CaL}) window current region (blue areas), resulting in large reactivation of I_{CaL}. SS-act: steady-state activation curve (black lines); SS-inact: steady-state inactivation curve (red lines); Depo: depolarization; Repo: repolarization.

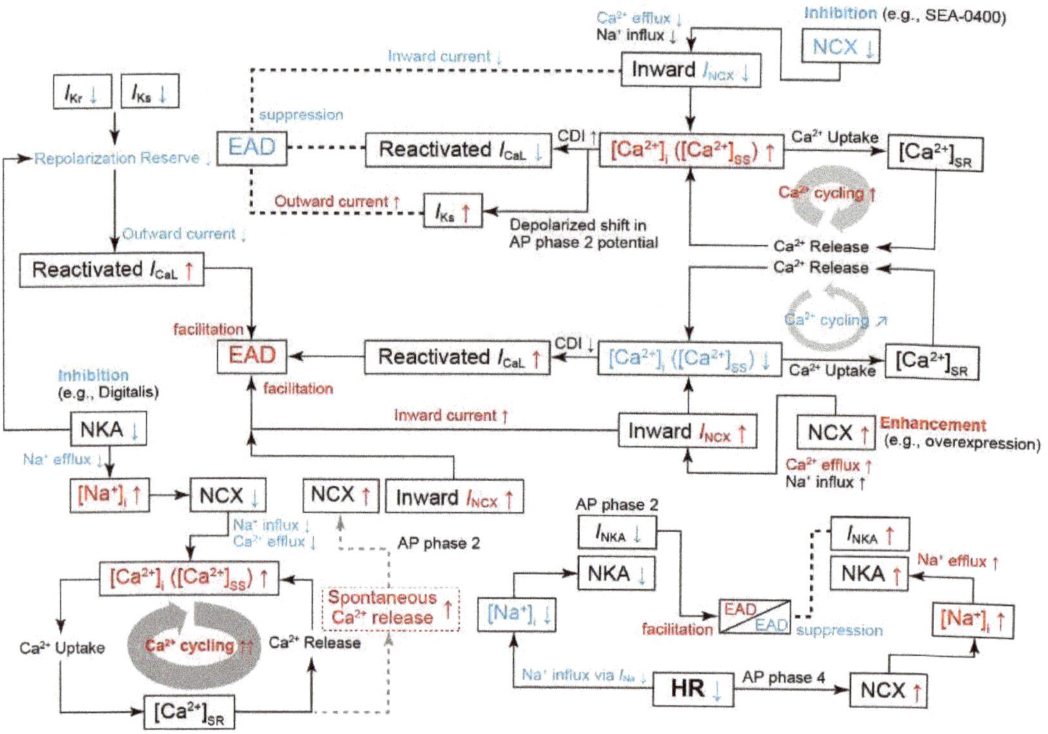

Figure 3. Schematic diagram showing how the dynamics of individual ion channels, transporters, $[Ca^{2+}]_I$, and $[Na^+]_i$ contribute to EAD generation and regulation. The diagram depicts the major functional components (I_{Ks}, I_{Kr}, I_{CaL}, I_{NCX}, and I_{NKA}), factors involved in Ca^{2+}- and Na^+-handling, heart rate (HR), and their interactions related to EAD generation and regulation. The upward and downward arrows represent an increase (or enhancements) and decrease (or attenuations) in each factor, respectively. A decrease in HR (bradycardia) decreases Na^+ influx via the Na^+ channel activation, resulting in a decrease in $[Na^+]_i$. This reduction in $[Na^+]_i$ facilitates EAD generation through a decrease in outward I_{NKA} during AP phase 2. On the other hand, a decrease in HR prolongs AP phase 4, i.e., diastolic interval (DI), and the prolongation of DI increases the amount of Na^+ influx through NCX and thus causes $[Na^+]_i$ elevation. This $[Na^+]_i$ elevation increases Na^+ efflux through NAK and outward I_{NAK} during AP phase 2. This may counteract the decrease in $[Na^+]_i$ due to reduced I_{Na} and the resulting reduction in I_{NKA} (modified from Figure 11 in [85]). For details of other depicted interactions that affect EAD formation, see text.

3. Mathematical Cardiomyocyte Models and Bifurcation Phenomena

Mathematical modeling of the electrophysiological properties of cardiac cells originated from the first AP model of squid giant axon based on experimental measurements by Hodgkin and Huxley in the 1950s [96]. Since a cardiac cell model in mammalian Purkinje fiber, employing this Hodgkin–Huxley (H–H) formalism, was first developed by Noble [97], mathematical models of heart muscle cells have been refined by incorporating diverse electrophysiological data. At present, cardiomyocyte models have reflected not only differences in cell types but also differences in experimental animal species, e.g., guinea pigs [98–101], rabbits [102,103], mice [104–108], dogs [109–111], and humans [112–118]. However, there are no substantial differences in the basic structures of mathematical models that reproduce electrophysiological properties of cardiomyocytes (see Figure 4A). For details on electrophysiological and mathematical models of heart muscle cells, refer to the original articles.

Furthermore, review articles written by Noble [119] and Amuzescu et al. [120] detail the history of changes in their models.

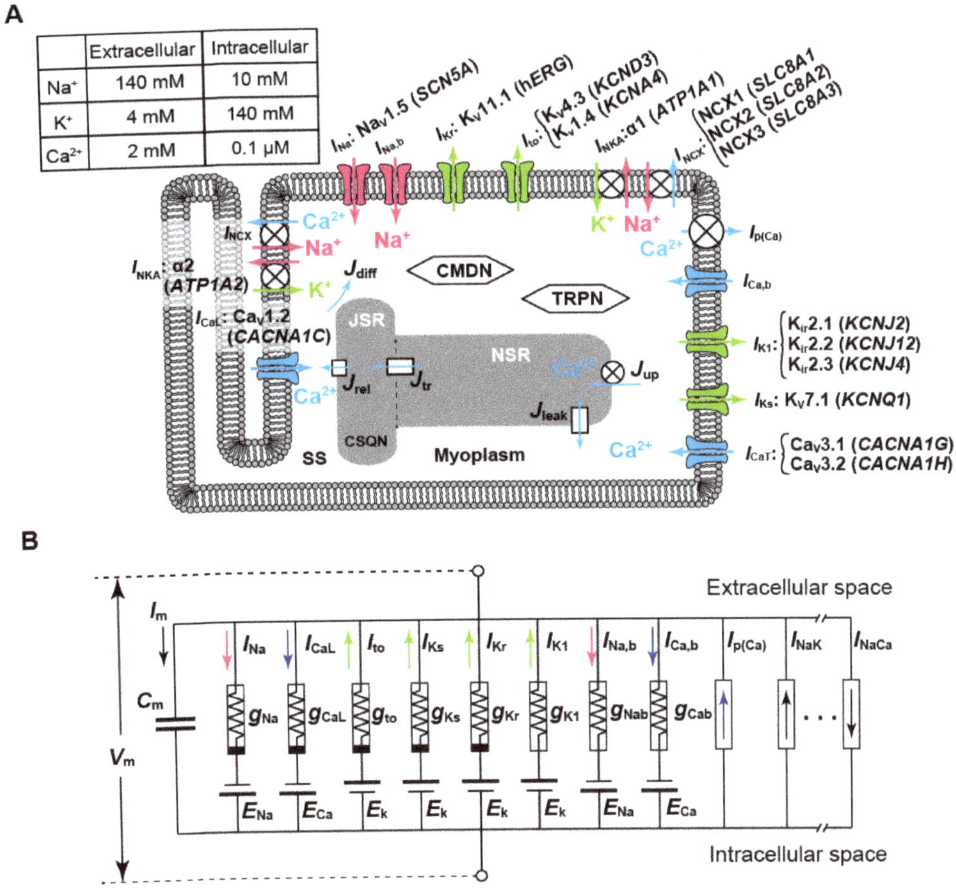

Figure 4. Schematic diagram of the structure and electrical properties of a ventricular myocyte. (**A**): Ion channel and transporter molecules embedded in the cell membrane, sarcoplasmic reticulum (Ca^{2+} cycling), and Ca^{2+}-binding molecules related to excitation and contraction. (**B**): An equivalent circuit model of the cell membrane, composed of the ion channel conductance (g), membrane capacitance (C_m), and electromotive force (E) representing ion concentration gradients. The current (I) flowing through each voltage-gated ion channel is represented by the product of the time-dependent variable conductance and driving force as the difference between the membrane potential (V_m) and individual reversal potential. g_x: maximum conductance of ion channels x, for x = Na^+, Ca^{2+}, K^+, etc.; I_{Na}: fast sodium channel current; I_{to}: transient outward K^+ channel current; I_{Kr} and I_{Ks}: fast and slow components, respectively, of delayed rectifier K^+ channel currents; I_{CaL}: L-type Ca^{2+} channel current; I_{CaT}: T-type Ca^{2+} channel current; I_{K1}: inward rectifier K^+ channel current; I_{NKA}: Na^+/K^+ ATPase current; I_{NCX}: Na^+/Ca^{2+} exchanger current; $I_{p(Ca)}$: Ca^{2+} pump current in the sarcolemmal membrane; J_{rel}: SR Ca^{2+} release flux by Ca^{2+}-induced Ca^{2+}-release; J_{up}: Ca^{2+} uptake flux via Ca^{2+} pump (SERCA) in the SR; J_{leak}: Ca^{2+} leakage flux from the SR; JSR: junctional SR; NSR: network SR; CMDN: calmodulin; TRPN: troponin; SS: a subspace of the myoplasm. For details of other symbols, refer to references [99,114–117].

3.1. Mathematical Models of Action Potential in Cardiomyocytes

The basic idea of the H–H formalism is just to consider the cell membrane as a simple electric circuit (Figure 4B). The capacitive property of the cell membrane is represented as the capacitor with a certain capacitance in the electric circuit. Na^+, K^+, and Ca^{2+} channels are modeled as the conductors that have conductances denoted as g_{Na}, g_K, and g_{Ca}, respectively, and batteries with the electromotive forces E_{Na}, E_K, and E_{Ca}, respectively. The electromotive forces of these batteries correspond to reversal potentials of each ion channel, representing ionic movements depending on ion concentration gradients across the cell membrane. In addition to these ion channels, ATP-driven ion pumps that carry ions against the concentration gradients, ion exchangers that exchange ions inside and outside the cell membrane, and other ion transporters also exist on the cell membrane of cardiomyocytes and generate currents. So, the membrane potential dynamics of cardiac cells are simply expressed by differential equations of the electric circuit shown in Figure 4B, generally described as a nonlinear ordinary differential equation as follows:

$$dV_m/dt = -(I_{ion} + I_{sitm})/C_m, \quad (1)$$

where V_m is the membrane potential, C_m is the membrane capacitance, and I_{stim} is the external stimulation current that can be a constant, a time-dependent function, or a single short pulse. I_{ion} is the net ionic current through the ion channels, transporters, and pumps, i.e., the sum of individual currents mediated by Na^+, K^+, and Ca^{2+} channels, NKA, NCX, the Ca^{2+} pump, etc., and it is denoted as follows:

$$I_{ion} = I_{Na} + I_K + I_{Ca} + I_{NKA} + I_{NCX} + \cdots. \quad (2)$$

For example, the fast Na^+ channel current I_{Na} in Equation (2) is denoted by the equation $I_{Na} = g_{Na} \times m^3 \times h \times j \times (V_m - E_{Na})$, which takes the form of (current) = (conductance) × (voltage), i.e., Ohm's law [99]. The voltage E_{Na} is the Nernst potential or the equilibrium potential of Na^+. The Nernst potential is the potential where the tendency of ions to move down their concentration gradient is exactly balanced with the force by the electric potential difference; no Na^+ current flows through the Na^+ channel when $V_m = E_{Na}$. Furthermore, the term $g_{Na} \times m^3 \times h \times j$ denotes the conductance of Na^+ channels, where the constant g_{Na} is the maximum conductance of the channel and $m^3 \times h \times j$ denotes the temporal change of the conductance, that is, the channel open probability. The gating (state) variables *m*, *h*, and *j* take (dimensionless) values between zero and unity and represent the open probabilities of the activation (*m*), fast inactivation (*h*), and slow inactivation (*j*) gates, respectively. Since the variable *m* is an increasing function of V_m and *h* (or *j*) is the decreasing one, *m* and *h* (or *j*) are also called the activation and inactivation variables, respectively, of the fast Na^+ channel. The dynamic opening and closing of the gates are described by the following equations:

$$dz/dt = \alpha_z(V_m)\cdot(1 - z) - \beta_z(V_m)\cdot z, \text{ or } dz/dt = (z^\infty(V_m) - z)/\tau_z(V_m), \quad (3)$$

$$z^\infty(V_m) = \alpha_z(V_m)\cdot\{\alpha_z(V_m) + \beta_z(V_m)\}^{-1}, \text{ and } \tau_z(V_m) = \{\alpha_z(V_m) + \beta_z(V_m)\}^{-1} \quad (4)$$

for $z = m, h$, and j, where $\alpha_z(V_m)$ is the rate constant for changing from a closed state to an open state, and $\beta_z(V_m)$ represents the rate constant for the transition from the open state to the closed state. In addition, $z^\infty(V_m)$ is the steady-state open probability to which the gating variable z asymptotically approaches at a given V_m, and $\tau_z(V_m)$ is the time constant of relaxation to a steady state. Note that the rate constants $\alpha_z(V_m)$ and $\beta_z(V_m)$, and the time constant $\tau_z(V_m)$ are not constant but dependent on V_m.

It is known that the specialized cells in the conduction system of the heart, represented by SA node cells, spontaneously produce APs, which is the behavior referred to as automaticity. This implies that AP dynamics can be repetitively produced without I_{stim} in Equation (1). On the other hand, atrial and ventricular myocytes known as working myocardium are driven by a sinus rhythm originating from the SA node, that is, I_{stim}

represents a periodic current input that repetitively stimulates the working myocardium at the sinus rhythm (regular rhythm). For instance, we assume that AP responses in a working myocardium model, such as that shown in Equation (1), are evoked by periodic external current stimuli (I_{stim}). The temporal variations in I_{stim} of the rectangular current pulse are expressed as follows:

$$\begin{aligned} I_{stim}(t) &= I_{stim,max} \qquad (0 \le t < T_1), \\ I_{stim}(t) &= 0 \qquad\qquad (T_1 \le t < T), \end{aligned} \tag{5}$$

where T is the pacing cycle length (i.e., period of the current stimuli) and T_1 is the duration for which the stimulus current is sustained at the maximum value ($I_{stim,max}$).

Mathematical models for cardiomyocytes, even now, continue to be refined by incorporating biophysical and biochemical processes. Elucidating the mechanism of proarrhythmic behavior in cardiomyocytes using mathematical models does not necessarily mean revealing unknown proarrhythmic factors. Instead, we aim to understand the time evolution of states, i.e., the dynamic phenomena of the cardiomyocyte. The AP evoked in cardiomyocytes is dynamic, and bifurcation analysis has been utilized to understand its dynamical behavior. Again, the purpose of such in silico studies is not only to model new molecular components involved in the development of proarrhythmic responses in cardiomyocytes. For that matter, such new components are often not the cause to be revealed through modeling. For example, ventricular myocytes may exhibit an AP with multiple transient depolarizations (EADs) during repolarization. The cause of EAD development is an excessive prolongation of the APD due to a decrease in repolarizing currents (or persistence of depolarizing currents). However, how much prolongation of the AP causes EAD and how many EADs occur during repolarization would not be able to be determined without analyzing the dynamics of model equations describing temporal changes of membrane potentials in the cardiomyocyte. This is also another motivation for such in silico studies.

Next, the bifurcation theory for nonlinear dynamical systems (bifurcation phenomena) will be briefly explained. One of the characteristics of nonlinear dynamical systems is that a slight change in a system parameter can cause a drastic change in system behavior. Such bifurcation phenomena observed in the cardiomyocyte and the heart will be summarized.

3.2. *Bifurcations in Nonlinear Dynamical Systems*

3.2.1. Definitions of Dynamical Systems

There are various models of dynamical systems. Here, we consider a system with time as an independent variable and with continuous state variables. Mathematical models that describe the system behavior are different depending on whether the time is continuous or discrete. In the former, the system is represented by a set of ordinary differential equations (ODEs). In the latter, it is a set of ordinary difference equations (i.e., recurrence formulas). Even if a mathematical model that we consider is continuous in terms of both states and time and is described by ODEs, it may be treated as a discrete-time system by sampling states at appropriate time intervals to analyze the behaviors of the system.

Now, we suppose that the nonlinear dynamical systems of interest are described by the following equations:

$$d\boldsymbol{x}/dt = \boldsymbol{f}(\boldsymbol{x}, \boldsymbol{\lambda}), \tag{6}$$

$$d\boldsymbol{x}/dt = \boldsymbol{f}(t, \boldsymbol{x}, \boldsymbol{\lambda}), \tag{7}$$

where t is the time, $\boldsymbol{x} = (x_1, x_2, \ldots, x_n)^{Tr}$, $\boldsymbol{\lambda} = (\lambda_1, \lambda_2, \ldots, \lambda_m)^{Tr}$, and $\boldsymbol{f} = (f_1, f_2, \ldots, f_n)^{Tr}$ are a state vector, parameter vector, and the vector field, respectively, and Tr represents transposition. As shown in Equation (7), a system that explicitly includes time is called a "non-autonomous system". On the other hand, the system of Equation (6), which does not explicitly include time, is called an "autonomous system". Behaviors of the system are different depending on whether it is an autonomous or a non-autonomous system, and thus, the methods for analyses are also different. The specialized cardiomyocytes that compose the conduction system in the heart possess automaticity (i.e., exhibit spontaneous and repetitive excitation), and do not require external stimuli for excitation. Therefore, specialized cardiomyocytes such as sinoatrial node and atrioventricular node cells can be modeled as autonomous systems. On the other hand, working myocytes such as atrial and ventricular muscles are typical examples of excitable systems and can be modeled as non-autonomous systems.

One of the characteristics of autonomous systems is that there can exist states where the right-hand-side vector of Equation (6), which defines the state velocity, becomes zero. The state where the velocity becomes zero, i.e., $\boldsymbol{f}(\boldsymbol{x}, \boldsymbol{\lambda}) = \boldsymbol{0}$, is called the "equilibrium point". The resting membrane potential corresponds to the equilibrium point in the myocardium system.

On the other hand, the non-autonomous system defined by Equation (7) includes time t in the state velocity, i.e., the right-hand side of Equation (7). Formally, let us consider the following autonomous system by considering time t as a new state variable τ ($t = \tau$) and rewriting Equation (7):

$$\mathrm{d}\boldsymbol{x}/\mathrm{d}\tau = \boldsymbol{f}(\boldsymbol{u}, \boldsymbol{\lambda}), \tag{8}$$

$$\mathrm{d}t/\mathrm{d}\tau = 1, \tag{9}$$

where $\boldsymbol{u} = (t, \boldsymbol{x})^{\mathrm{Tr}}$. In this case, since the right-hand side of Equation (9) is unity, then the right-hand side of the autonomous system of Equations (8) and (9) never becomes zero. In other words, there is no equilibrium point in the non-autonomous system of Equation (7). The states of a non-autonomous system flow at a constant speed in the direction of the time axis. For this reason, it is necessary to devise methods of analysis for non-autonomous systems, different from those for autonomous systems.

In this review, we will only consider the case where the vector field of a non-autonomous system is periodic with respect to time, i.e., $\boldsymbol{f}(t + \mathrm{CL}, \boldsymbol{x}, \boldsymbol{\lambda}) = \boldsymbol{f}(t, \boldsymbol{x}, \boldsymbol{\lambda})$, where CL is the period of external stimulus applied to the working myocardium, i.e., the pacing cycle length. Such a system is called a periodic non-autonomous system. Furthermore, states in such a system possess a periodic nature: $\boldsymbol{x}(t + k \times \mathrm{CL}) = \boldsymbol{x}(t)$, where $k = 1, 2, \ldots, m$, which is also referred to as a k-periodic solution. The periodic solution corresponds to the periodic oscillation observed in the original system.

Suppose that a cardiomyocyte is repeatedly stimulated at a given CL (Figure 5A). The CL has the following relationship: CL = APD + DI, where DI is the diastolic interval. Now, let us write a relationship between the APD of AP evoked by the nth stimulus and that of AP evoked by the next (n + 1)th stimulus as

$$\mathrm{APD}_{n+1} = P(\mathrm{APD}_n), \tag{10}$$

for $n = 0, 1, 2, \ldots$, where P is a mapping, which is also referred to as the "Poincaré map" [1,121–123]. This formula indicates that the APD of AP evoked by the (n + 1)th stimulus (APD_{n+1}) can be represented as a function of that of AP evoked by the nth stimulus (Figure 5B). In other words, APD_n is mapped to APD_{n+1} by the mapping P. If an AP train behavior evoked in the original non-autonomous system of Equation (7) is periodic and APDs of APs are equal for every stimulus, then APD_n satisfies the following relationship: $\mathrm{APD}_{n+1} - P(\mathrm{APD}_n) = 0$. This situation is exemplified in Figure 5B. Such a point in the (APD_n, APD_{n+1})-plane as shown in Figure 5B is called a "fixed point". Thus, the periodic response observed in Equation (7) can be in one-to-one correspondence with the fixed point of the Poincaré map P. Furthermore, if for some $\ell \neq 1$, $\mathrm{APD}_n = P^{\ell}(\mathrm{APD}_n)$, and if all $P^k(\mathrm{APD}_n)$, $k = 1, 2, \ldots, \ell - 1$, are different each other, the periodic behavior evoked in Equation (7) corresponds to an ℓ-periodic point. This case can also be studied simply by replacing P with P^{ℓ} or as the ℓth iterate of P. Usually, however, the explicit form of P cannot be obtained. Therefore, the Poincaré map, in general, must be obtained by acquiring the values of state variables for every CL in the numerical simulation.

In the following, we will briefly explain the bifurcation phenomena of equilibrium points and fixed (periodic) points.

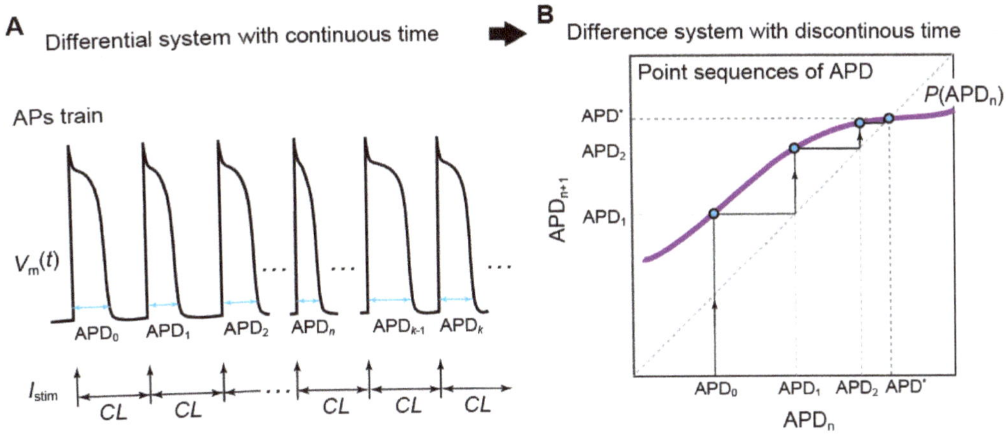

Figure 5. Examples of a dynamical system with continuous time and with discrete time. (**A**): An action potential (AP) train evoked in a cardiomyocyte with stimuli (I_{stim}) applied at a cycle length of *CL*, which is a typical example of the response of a dynamical system with continuous time. AP duration of the AP evoked by the *i*th stimulus is denoted as APD_i for $i = 0, 1, 2, \ldots, n, \ldots, k-1, k, \ldots$. (**B**): A point sequence consisting of APD values determined for the AP train and plotted on the (APD_n, APD_{n+1})-plane. If a mapping *P* (Poincaré map) is obtained to represent the relationship between a point (APD_n) and the next point (APD_{n+1}), i.e., the dynamics of the point sequence, then the differential system with continuous time is transformed into a difference system with discrete time. The dynamics of such the difference system with discrete time can then be studied by projecting the point sequence dynamics into the state space, e.g., (APD_n, APD_{n+1})-plane. Thereby, it is possible to examine the dynamics of the original differential system with continuous time more efficiently. In general, the Poincaré map is difficult to obtain analytically and is mostly obtained numerically. For examples of experimental and numerical methods for obtaining Poincaré maps, see [124,125].

3.2.2. Hopf Bifurcation

In autonomous systems, when the Hopf bifurcation is caused by changing the value of a parameter, an oscillatory response (or rhythmic dynamics), called the "limit cycle", emerges, as illustrated by the thick circle with arrowheads in Figure 6A,B. For instance, cells in the myocardial sleeve of the pulmonary vein do not normally exhibit automaticity [126,127]. However, for some reason, they suddenly acquire spontaneous excitations (automaticity), triggering paroxysmal atrial fibrillation [128]. The phenomenon such that the convergence behavior to a stationary state (an equilibrium point) suddenly switches to oscillatory motion (Figure 6C) is a typical example of the Hopf bifurcation [129].

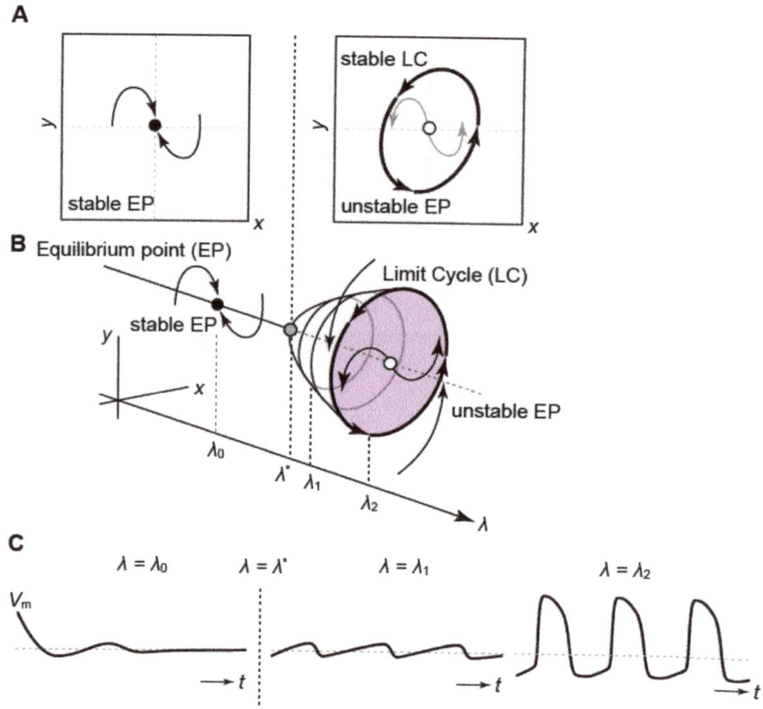

Figure 6. Hopf bifurcation. Examples of the dynamic responses to changes in the parameter λ in the 2-dimensional (x, y)-state space (**A**) and 3-dimensional (x, y, λ)-space (**B**). Schematic examples of membrane potential changes at the parameter λ_0, λ_1, λ_2 in the 3-dimensional (x, y, λ)-space of panel (**A**,**C**). Changing the system parameter λ, the stability of an equilibrium point (EP) changes via a Hopf bifurcation. Before and after the occurrence of a Hopf bifurcation (λ*), a stable EP becomes unstable with the emergence of a limit-cycle (LC) oscillation.

3.2.3. Saddle-Node Bifurcation

In the saddle-node bifurcation (or fold Bifurcation [123]) of equilibrium points, changes in parameters cause stable and unstable equilibrium points to coalesce and disappear. Thus, the number of equilibrium points can change (see Figure 7A). On the other hand, there are also saddle-node bifurcation phenomena of periodic solutions in non-autonomous systems (or of limit cycles in autonomous systems). As in the case of equilibrium points, this bifurcation causes a pair of stable and unstable points to disappear or emerge (Figure 7B). In numerical simulations, the states of equilibrium points and/or fixed (periodic) points may jump significantly when the parameter infinitesimally changes from this bifurcation point (this is referred to as a "jump phenomenon"). Moreover, a system that has reached a new state via the saddle-node bifurcation cannot be returned to its original state, even if the value of system parameters is returned to the bifurcation point (this phenomenon is called "hysteresis"). The hysteresis and jump phenomena are associated with the development of a bi- or multi-stable phenomenon of equilibrium points or periodically evoked APs [130,131].

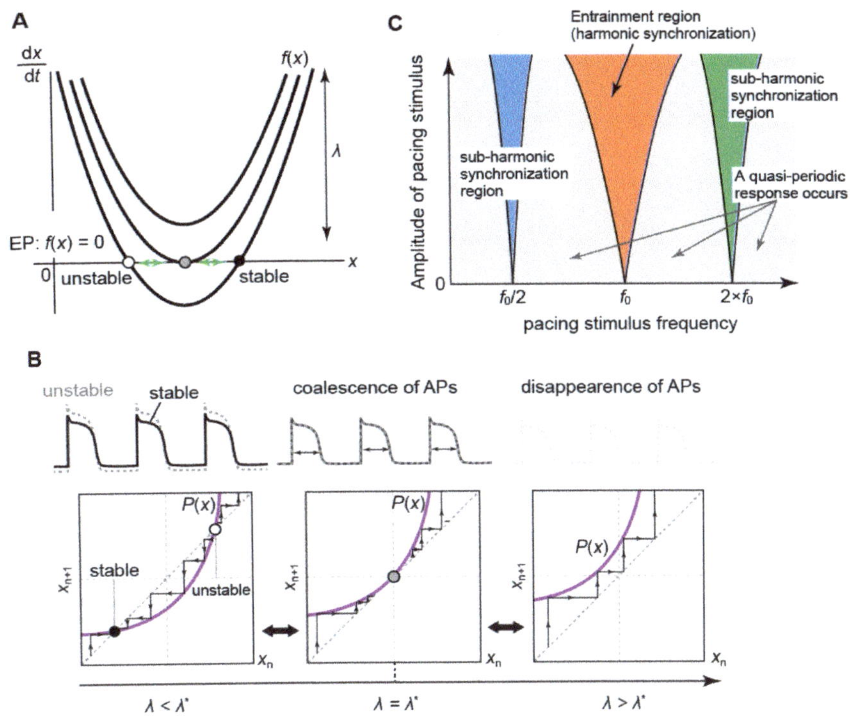

Figure 7. Saddle-node (SN) bifurcation. Schematic diagrams of the SN bifurcations of equilibrium (**A**) and fixed (**B**) points. As simple examples, in the dynamical systems $dx/dt = f(x, \lambda)$ (**A**) and $x_{n+1} = P(x_n, \lambda)$ (**B**), the functions $f(x)$ and $P(x_n)$ are varied up and/or down by changing system parameter λ. The SN bifurcation of equilibrium points occurs when the curve of $f(x)$ touches the x-axis ($dx/dt = 0$ axis). On the other hand, the SN bifurcation of fixed points occurs when the curve of $P(x_n)$ touches the diagonal line ($x_{n+1} = x_n$) at a value λ^* of parameters. After that, a node and a saddle point emerge or disappear. A schematic diagram of Arnold's tongue structure is also shown for a non-autonomous system with stimuli of various amplitudes and frequencies (**C**). Colored regions indicate parameter regions in which a periodic oscillation observed in an autonomous system can be entrained by external periodic stimuli, e.g., the orange region represents harmonic synchronized oscillation observed in the non-autonomous system. In general, the harmonic synchronized region and even the sub-harmonic synchronous region (e.g., blue and green regions) are divided in the parameter space by SN bifurcation sets. Oscillations that are asynchronous to the periodic stimuli, such as quasi-periodic oscillations, appear when set to parameters outside the synchronization region. f_0: an intrinsic frequency of the limit-cycle oscillation in the autonomous system.

Gadsby and Cranfield illustrated that there exist two stable resting (equilibrium) states in the cardiac Purkinje fiber cell and that the application of a stimulus causes the membrane potential to converge to different resting membrane potentials [132,133]. Jalife and Antzelevitch demonstrated that when a brief depolarizing pulse was applied during SA node pacemaking, spontaneous repetitive APs were annihilated and the membrane potential converged to another stable state; that is, the resting state [134]. These are typical examples that exhibited the bi-stability of two equilibrium states or a periodic response and an equilibrium state. To list other examples of the hysteretic phenomenon, previous experimental studies have investigated AP responses to changes in CL, i.e., rate-dependence and restitution properties of APD, using cardiomyocytes of guinea pigs, rabbits, and canines [124,135–137], and found hysteretic responses of APD to CL changes.

Finally, this bifurcation phenomenon is also involved in an entrainment phenomenon. The specialized cells in the conduction system of the heart, represented by SA node cells, spontaneously and periodically produce APs, being typical examples of self-excited oscillatory systems. When such a self-excited system is perturbed by an external periodic stimulus, if the ratio of the oscillation frequency (frequency of the limit cycle) of the self-excited system and the frequency of the external stimulus is close to a simple rational number, the self-excited oscillation may be entrained by the external periodic stimulus. This is a kind of synchronization phenomenon and is also called "entrainment". For example, a periodic AP generated in the SA node periodically stimulates the atrioventricular (AV) node (of a self-excited system) when it conducts from the atria to the ventricles. At this time, the self-excited oscillation in the AV node is entrained by the excitation frequency of the SA node cells, resulting in a synchronized periodic state. This frequency mismatch that allows the AV node cell to be entrained is limited to a certain frequency range, depending on the stimulus strength. The relationship between the frequency mismatch range, the so-called "phase lock range", and the stimulus strength is known to show a characteristic structure (called "Arnold's tongue structure" [123,138]), see Figure 7C. Then, the phase-locked state may disappear with a saddle-node bifurcation when the stimulus frequency or intensity changes. After that, asynchronous (quasi-periodic) rhythms may occur in which two rhythms with different frequencies proceed with each other without much interaction [9,131].

3.2.4. Homoclinic Bifurcation

The aforementioned saddle-node and Hopf bifurcations of the equilibrium point are classified as "local bifurcations" because they are attributed to the stability change of the dynamics near an equilibrium point. On the other hand, there also exist bifurcation phenomena such that the behavior of the solution is developed over a large region within the state space. These bifurcations are called "global bifurcations" because they are related to the global behavior of the solution [123].

Stable (resp. unstable) manifolds are hypersurfaces within the state space formed by sets of initial values approaching (resp. moving away from) an equilibrium point or a limit cycle over time (Figure 8A). When system parameters change, the stable and unstable manifolds of a given saddle equilibrium point may be connected and a closed orbit (also called a "homoclinic orbit") may emerge. This phenomenon is called "homoclinic bifurcation", see Figure 8B.

Miake et al. reported that in guinea pigs, the dominant-negative suppression of Kir channels by viral gene transfer could convert ventricular myocytes that exhibit a quiescent state into pacemaker-like cells that generate spontaneous and repetitive APs [139]. The study of mathematical modeling and bifurcation analyses on biological pacemaker activity of ventricular myocytes by Kurata et al. [112] illustrated that the disappearance of a resting state via a saddle-node bifurcation by suppressing I_{K1} evoked automaticity in the ventricular myocyte. This change in dynamics seems to involve the occurrence of a saddle-node homoclinic bifurcation [123] or a saddle-node on an invariant cycle (SNIC) bifurcation [140], in which a saddle-node bifurcation and a homoclinic bifurcation at an equilibrium point occur simultaneously (see Figure 8C, and Refs. [123,140] for details and complete definitions).

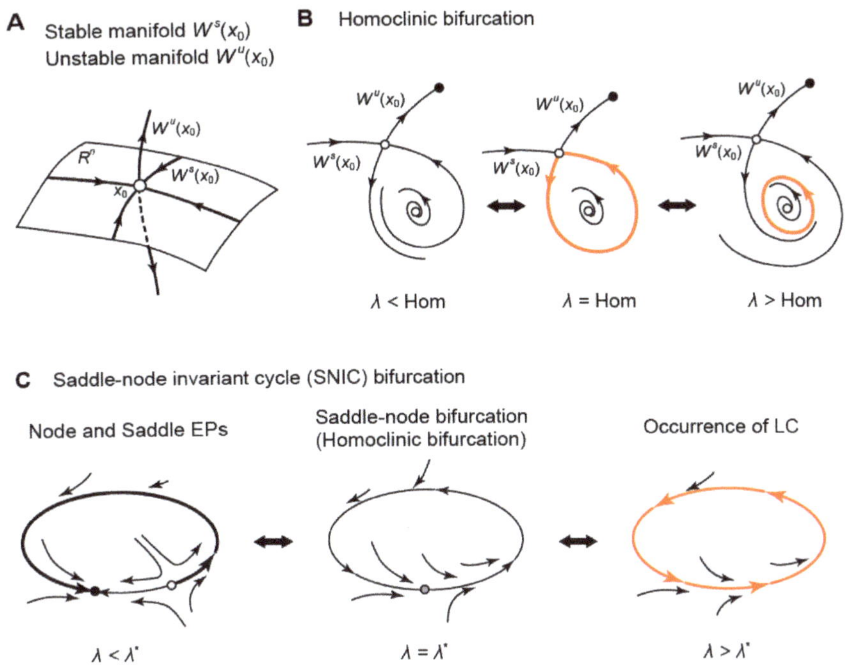

Figure 8. Bifurcation phenomena associated with stable and unstable manifolds of an equilibrium point. (**A**) A schema of stable and unstable manifolds of the equilibrium point x_0 in the state space. $W^s(x_0)$, stable manifold; $W^u(x_0)$, unstable manifold. (**B**) Homoclinic bifurcation. When the parameter λ becomes the value of Hom, the stable and unstable manifolds of the equilibrium point x_0 coincide, giving rise to a closed-loop orbit, which is referred to as "homoclinic orbit". This closed-loop orbit is vulnerable to changes in the parameters and breaks down immediately. (**C**) Saddle-node invariant cycle bifurcation. At the bifurcation parameter value λ^*, saddle-node and homoclinic bifurcations occur simultaneously.

3.2.5. Period-Doubling Bifurcation

The stability of a fixed point on the Poincaré section changes on the occurrence of this bifurcation and a periodic orbit with a double period emerges or disappears (see Figure 9A). The period-doubling bifurcation is also called a flip Bifurcation [123]. Typical examples of this bifurcation are the development of AP alternans [141–144]. When a cardiomyocyte is periodically stimulated at a given CL, the relation of APD and DI to CL is expressed as CL = APD + DI. In general, if the CL is shortened, then both APD and DI are also shortened. As the DI shortens, repolarization currents, especially I_{Ks} with relatively slow deactivation kinetics, accumulate with each stimulus; I_{Ks} does not have an inactivation mechanism. This implies that shortened DI causes I_{Ks} activation by the next AP generation before I_{Ks} completely returns to its original state by deactivation. As a result of the accumulation of activated I_{Ks} channels, the increased repolarization current accelerates AP repolarization, leading to APD shortening and DI prolongation. The DI prolongation leads to more progress in I_{Ks} deactivation and returns more I_{Ks} channels to a closed state. Thus, in the AP generated by the next stimulus, the repolarization current decreases, resulting in APD prolongation, i.e., DI shortening. This alternating change in APD, i.e., repeated prolongation and shortening of APD during stimuli, can also be described as follows:

$$\mathrm{APD}_n = P^2(\mathrm{APD}_n). \tag{11}$$

This is the phenomenon called "APD alternans". APD alternans associated with recovery from inactivation of Na^+ channels has also been reported, albeit in in silico studies [145,146]. Furthermore, the period-doubling bifurcation is known to successively occur, resulting in a chaotic response [142,147]. This implies that SA node pacemaking becomes irregular or arrhythmic [125], i.e., sinus arrhythmias occur via the sequence of period-doubling bifurcations.

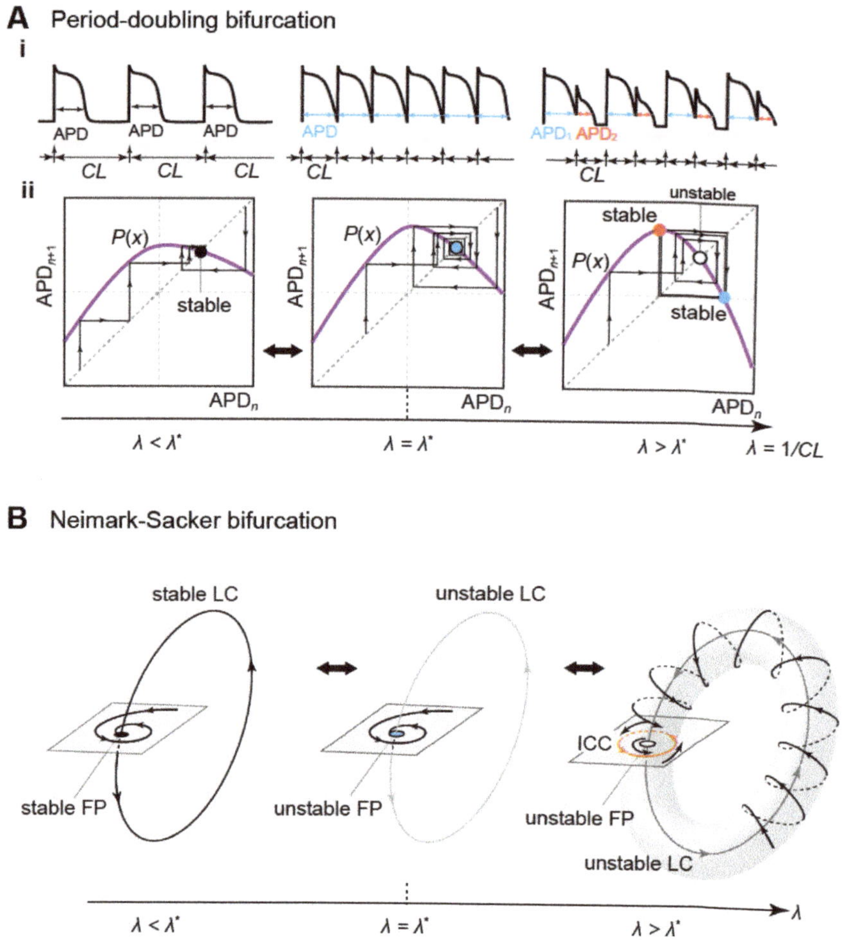

Figure 9. Schematic diagrams of period-doubling and Neimark–Sacker bifurcations. (**A**): A typical example of the period-doubling (PD) bifurcation in non-autonomous systems is AP alternans (**i**). As the cycle length (*CL*) becomes shorter, the stable fixed point on the (APD_n, APD_{n+1})-plane becomes unstable at the period-doubling bifurcation point (λ^*), and a pair of periodic points is generated (**ii**). (**B**): The change in a limit-cycle (LC) oscillation through the Neimark–Sacker (NS) bifurcation in autonomous systems. The NS bifurcation occurs at λ^* as the parameter λ changes. Then, the LC oscillation becomes unstable and a torus-like oscillatory response, so-called "quasi-periodic oscillation", occurs around the unstable LC oscillation. When the quasi-periodic oscillation is discretized as a point sequence via the Poincaré mapping, a closed curve, which is referred to as an "invariant closed curve (ICC)", appears around the unstable fixed point that reflects the unstable LC oscillation.

3.2.6. Neimark–Sacker Bifurcation

If this bifurcation occurs, quasi-periodic oscillations may emerge or disappear (see Figure 9B). In the state space, the stability of a fixed (or periodic) point on a Poincaré section changes, and a certain point sequence emerges around the fixed (periodic) points. The point sequence in a mapping constitutes a closed-loop such as a limit cycle in an autonomous system. Because of this, it is called an "invariant closed curve (ICC)". The ICC corresponds to a doubly periodic (quasi-periodic) oscillation of the original periodic non-autonomous system. The Neimark–Sacker bifurcation is also known as a torus bifurcation, or the Hopf bifurcation in discrete systems [123].

To our knowledge, there exists no experimental evidence that changes in dynamics via the Neimark–Sacker bifurcation were observed at a cellular level. On the other hand, an experimental study demonstrated that the reentrant wave became unstable with fluctuating circulatory time and APD, exhibiting a quasi-periodic pattern [148]. The excellent studies by Qu et al. [149,150] reported that the meandering reentrant excitation wave exhibits quasi-periodic dynamics. They focused on tip dynamics of spiral wave reentry. The reentrant excitation wave was periodically sampled and a return mapping of its tip dynamics was constructed; as the conductance of LTCCs was increased, point sequences generated by the return mapping changed from fixed point-like sequences to a closed-loop shape. In the context of bifurcation theory, this closed-loop is a certain type of ICC, and the emergence of ICC implies a dynamics change via the Neimark–Sacker bifurcation.

3.2.7. Bifurcation Analysis

Bifurcation analysis involves the investigation of bifurcation phenomena by obtaining a set of parameter values that cause bifurcation (bifurcation set) and a graph of these sets (bifurcation diagram). Even now, many researchers manually track their bifurcation sets by trial and error. These days, however, we can use powerful computational tools, for instance, AUTO [151,152], XPP-AUT [153], MATCONT [154], etc., to perform bifurcation analyses. Furthermore, a unique bifurcation analysis tool was developed by Kawakami in 1984 [155]. Since then, this tool has continued to be improved by him and his co-workers [156]. Parker and Chua showed a concrete procedure for the numerical calculation of bifurcation sets [157]. We will leave the details on the bifurcation phenomena of equilibrium points and their analyses to other references [2,10,13,152], and recommend relevant literature [1,10,158] as a good source for these topics. For readers who are interested in the bifurcation analysis of dynamical systems, we recommend an excellent monograph by Kuznetsov [123], which summarizes bifurcation phenomena from the standpoint of applications and would clarify the points that could not be explained in this review.

4. Bifurcation Analyses of Proarrhythmic Behaviors

Bifurcation analysis examines a sudden change in dynamic behaviors depending on changes in system parameters and enables us to determine the parameter ranges over which the AP response is stable and the transition mechanisms of the AP responses. One expectation for bifurcation analysis may be to know how to control dynamic responses such as APs from the viewpoint of prediction and/or prevention. When abnormal APs in cardiomyocytes occur, we will hope to control their dynamics and restore them to normal APs. However, in general, we have no way of knowing when, where, and how the parameters of a system should be modified to attain the best control and restoration. Bifurcation analysis is a way to gain insight into the fundamental properties of dynamical systems, and thus, may be able to provide helpful clues for controlling the AP response in cardiomyocytes.

4.1. Bifurcation Analyses of Cardiac Cell Models

In early studies, primitive mathematical models that assume only myocardial excitability and its refractoriness have been utilized to understand the complex dynamics of APs evoked in cardiomyocytes. Analyses of the AP dynamics evoked in these simple models helped to understand the dynamical mechanisms of the AV conduction block [144,159,160] and the modulation of the SA node pacemaker by the sympathetic and parasympathetic nervous systems [161–164]. Furthermore, the arrhythmia models of parasystole, in which the ventricle (or atria) is doubly governed by an ectopic pacemaker in addition to the pacemaker rhythm of the SA node, have been analyzed and the generative conditions for complex proarrhythmic dynamics of cardiac excitations, such as bigeminy and trigeminy, have been theoretically clarified [165,166].

As the electrophysiological refinement of cardiomyocyte models has progressed, studies targeting cardiac excitation dynamics of a higher degree of freedom have become mainstream. SA node cell models described as the autonomous system are compatible with existing computational tools such as AUTO, XPP-AUT, and MATCONT for bifurcation analysis [151–154]. Therefore, SA node cell models have been suitable for bifurcation analyses and have been utilized to elucidate how individual components consisting of ion-transport molecules contribute to normal pacemaking in SA node cells [167–176]. Bifurcation analyses were also utilized to examine the generation of abnormal oscillations or biological pacemaker activity in ventricular myocytes [177–180]. On the other hand, bifurcation analyses of periodic non-autonomous systems such as paced atrial and ventricular myocytes [181–183] require some kind of ingenuity because it is difficult to apply existing analysis tools to non-autonomous systems without modification. By artificially generating periodic stimuli corresponding to the SA node pacemaking and applying the periodic stimuli to atrial/ventricular myocytes while adjusting the pacing cycle length (PCL), the corresponding APD of the AP train evoked in the cardiomyocyte at every set PCL can be measured experimentally. From the experimentally obtained data, it is possible to derive a one-dimensional discrete dynamical system (e.g., Equation (10)) describing the time evolution of periodic AP trains based on the relationship between the PCL and APD. By investigating bifurcations in the one-dimensional discrete dynamical system, the dynamical mechanisms of abnormal APs observed in the original non-autonomous system (cardiomyocyte with periodic stimuli) have been elucidated. Furthermore, the guinea pig ventricular myocyte model developed by Luo and Rudy (LRd1 model) [184] has often been used for bifurcation analyses of AP dynamics observed in ventricular myocytes [141,185–189]. Although the LRd1 model is classified into a periodic non-autonomous system, investigations of bifurcation phenomena observed in the LRd1 model were first started from a perspective of an autonomous system, that is, analyzing bifurcations of equilibrium points when the amplitude of constant depolarized current stimuli was changed. As a result, various bifurcation phenomena, such as abnormal spontaneous oscillations evoked in ventricular myocytes, have been studied. As will be shown in the next section, the pioneering work of Tran et al., who explained the mechanism of EAD development in terms of bifurcation theory, has also used the LRd1 model [190].

Since the LRd1 model was published in 1991 [184], cardiomyocyte models have become more sophisticated and complex in the past 30 years [101,114–118,191–193]. Although intracellular ion concentration changes were not taken into account in the LRd1 model, by integrating intracellular and extracellular ion concentrations into some cardiomyocyte models as state variables, it is currently possible to examine the effects of intracellular Ca^{2+}-handling and Na^{+}-handling on AP dynamics [194–196]. Bifurcation analyses using such cardiomyocyte models with a higher degree of freedom were previously difficult due to low computational power and computing performance, but have recently become possible with improved computer performance [85,130,131,197–199]. In the following section, we summarize bifurcation analyses of the AP with EADs observed in more sophisticated cardiomyocyte models.

4.2. Bifurcation Analyses of Early Afterdepolarizations

To understand dynamic mechanisms underlying repolarization abnormality, i.e., the development of EAD, the effects of changes in the maximum conductance of various ion channels on APs with EAD observed in aforementioned various ventricular myocyte models have been investigated [81,112,114,117,184,200]. So far, many experimental [201–204] and theoretical [81,205–209] studies have shown that excessive AP prolongation results in the occurrence of EAD. Assuming decreases in I_{Ks} for LQT1 and I_{Kr} for LQT2, the LQT1 and LQT2 versions of human ventricular myocyte models, which can reproduce EADs during β-adrenergic stimulation and bradycardia, respectively, have been constructed [85,130,198]. Relationships between EAD occurrence and changes in parameters that are related to factors involved in the genesis and modulation of EAD such as repolarization currents (I_{Kr} and/or I_{Ks}), I_{CaL}, I_{NCX}, SR Ca^{2+} pump current (I_{up}), and heart rate have been comprehensively examined [85,130,198].

How will EAD emergence caused by the LTCC reactivation be explained from the perspective of dynamical system theory? As a strategy to elucidate the generative mechanisms of EADs [85,198], bifurcation analyses and AP simulations were combined (Figure 9a). First, the model of human ventricular myocytes was regarded as an autonomous system, and bifurcation phenomena of equilibrium points and limit cycles that occur in the system were investigated using MATCONT [154] (see Figure 10A, top). Next, AP simulations of the ventricular myocyte model, which is a periodic non-autonomous system, were performed with a fixed set of parameters, and then a phase diagram was constructed by mapping the information of steady-state AP responses (whether an EAD occurred or not) into a parameter plane (or space) (see Figure 10A, bottom). After that, the bifurcation diagram obtained by analyses of the autonomous system and the phase diagram for the non-autonomous system were merged to examine the relationship of bifurcations and dynamic changes in APs that occurred in the original system (Figure 10B). The moderately complex human ventricular myocyte model proposed by Kurata et al. [85,112] (K05) and the more complex models proposed by ten Tusscher et al. [114,115,198] (TP06) and O'Hara et al. [85,117] (ORd) had a common mechanism for EAD development, namely the reactivation of LTCC. Furthermore, EADs distinct in generation mechanism from LTCC reactivation-dependent ones, i.e., those caused by spontaneous Ca^{2+} releases from the SR were identified in the TP06 model [198]. In our previous study [130], whether the dynamic change of APs that causes EAD can be explained as the occurrence of local bifurcations was directly investigated for APs observed in the K05 model as a periodic non-autonomous system without modification (Figure 10C,D). Consequently, it was found that there were parameter regions in which multiple stable AP responses coexisted, and that hysteresis phenomena occurred for the change in the maximum conductance of I_{Kr}. However, EAD developments could not be simply explained as resulting from an occurrence of the local bifurcation.

A series of analyses advanced the understanding of the dynamical mechanisms of transitions between the different AP responses (APs with and without EAD) of ventricular myocytes due to functional changes in individual ion-transport molecules such as ion channels and transporters.

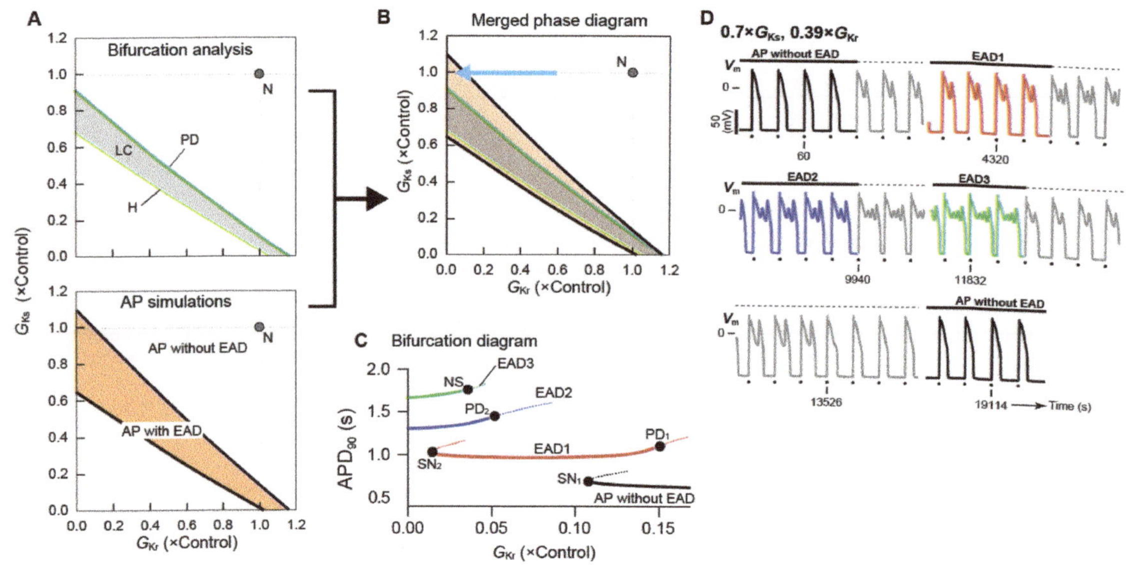

Figure 10. Bifurcation phenomena and action potential behaviors observed in the Kurata model [112] of a human ventricular myocyte. (**A**) A two-parameter bifurcation diagram on the (G_{Kr}, G_{Ks})-parameter plane (**top**) obtained using MATCONT [154] and a phase diagram (**bottom**) obtained by AP simulations. The maximum conductances of the rapid (G_{Kr}) and slow (G_{Ks}) components of delayed rectifier K^+ channel currents are expressed as normalized values, i.e., ratios to the control values. In the two-parameter bifurcation diagram (**top**), the symbols H and PD represent the loci of parameter sets that cause the Hopf bifurcation of an equilibrium point and period-doubling bifurcation of a limit cycle (LC), respectively. The gray region indicates the area of parameters in which a stable LC can be observed. On the other hand, in the phase diagram (**bottom**), the orange and white regions represent parameter regions in which an AP with and without EAD, respectively, can be observed in the Kurata model. The gray point with the symbol *N* indicates the control condition with the normal G_{Kr} and G_{Ks}. (**B**) A merged phase diagram. (**C**) A one-parameter bifurcation diagram of APD at 90% repolarization (APD_{90}) in each AP response as a function of G_{Kr}; see the blue arrow in panel (**B**), which indicates the change in G_{Kr} with the fixed normal G_{Ks} (1). The solid and dashed lines in (**C**) represent stable and unstable AP responses, respectively. SN: saddle-node bifurcation; EAD1–3: AP with 1–3 EAD(s); PD: period-doubling bifurcation. NS: Neimark–Sacker bifurcation. (**D**) An example of tetra-stable AP dynamics in the Kurata model at 0.39 × G_{Kr} with 0.70 × G_{Ks}. Colored and grey lines indicate the steady-state and transient responses, respectively. Dots indicate the application of current pulses. Pacing cycle length = 2 s. Each panel was modified from [85,130].

4.3. Slow–Fast Decomposition Analyses of Dynamic AP Responses with EADs

APs observed in ventricular myocytes are representative relaxation oscillations. The ventricular AP is composed of rapid and slow membrane potential changes. This means that the dynamical system describing the electrical behavior of ventricular myocytes is composed of a fast subsystem contributing to the rapid membrane potential changes and a slow subsystem contributing to the slow membrane potential changes [210]. Tran and co-workers examined the AP and EAD behavior observed in the LRd1 model [184] by decomposing the original (full) system into the fast and slow subsystems [190]. In their and many other theoretical studies [211–215], the activation gating variable of I_{Ks} was regarded as the slow subsystem or a parameter for the fast subsystem of the ventricular myocyte model. Kurata and co-workers [197] also employed the activation gating variable of I_{Ks} in the K05 model [112] for the slow subsystem, and in our study [198], the state variable of the intra-SR Ca^{2+} concentration was employed for the slow subsystem. As an example, I_{Ks} in [85,112,130,197] was formulated by the following equations:

$$I_{Ks} = g_{Ks} \cdot n^2 \cdot (V_m - E_{Ks}), \quad (12)$$

$$dn/dt = (n^{\infty} - n)/\tau_n, \quad (13)$$

where g_{Ks} and E_{Ks} are the I_{Ks} channel maximum conductance and reversal potential of I_{Ks}, respectively. n denotes a state variable representing the activation gate of I_{Ks} and n^2 corresponds to the open probability of I_{Ks} channels. In human ventricular myocyte models, when APs are generated, state variables such as V_m change rapidly, while n (n^2) changes slowly (Figure 11A,B, left). Considering the difference in response speed in the fast and slow subsystems, the slow dynamics of the state variable n can be regarded as a parameter change. Thus, bifurcations of dynamical responses observed in the fast subsystem were investigated as a function of the slow variable n (see Figure 11A,B, middle). Such a method is known as "slow-fast decomposition analysis" in the field of nonlinear dynamical system theory [140,210,216]. The slow–fast decomposition analysis can provide a definition of EAD and a clear reason why EADs occur in LQTS.

To explain the dynamical mechanism for the occurrence of EAD [190,197,198], dynamic behaviors (orbits) of the full system are superimposed on the bifurcation diagram of the fast subsystem (see Figure 11A,B, middle). The dynamical mechanism of EAD development in the K05 model was explained as follows [85,112,130,197]: The stimulus current (I_{stim}) input rapidly depolarized V_m in the full system and V_m overshot 0 mV. After that, the I_{Ks} channel open probability, n^2, which was the slow subsystem, slowly increased, consequently increasing I_{Ks} and repolarizing V_m (Figure 11A,B, right). With increasing n^2 as a parameter for the fast subsystem, an equilibrium point at depolarized V_m (qEQ_3) was destabilized via the Hopf bifurcation (see center panels in Figure 11A,B). When the I_{Kr} conductance decreased (as in LQT2; Figure 11B), the Hopf bifurcation point of qEQ_3 in the fast subsystem shifted significantly toward higher n^2 values (compare center panels in Figure 11A,B). This means that the region of stable qEQ_3 broadened in the direction of higher n^2, shifting the stable limit sets of the limit-cycle qLC to the right. Then, the trajectory of the full system was attracted to the stable limit sets for qEQ_3 and qLC in the fast subsystem during delayed repolarization. Such trapping constrained the orbit of the full system to move along the stable manifolds of the qEQ_3 and the qLC. Therefore, the full system showed oscillation-like behavior during late AP phase 2 to phase 3. Since the stable/unstable limit sets of qLCs were lost by crossing the SN bifurcation point, the increase in n^2 above the SN bifurcation point led to the release of the orbit in the full system trapped in the stable manifold of the qLC. Then, V_m quickly converged to a stationary state (resting state). This suggests that transient depolarization during AP repolarization, i.e., EAD can be explained as limit cycle-like oscillations occurring in the fast subsystem [197,198]. Thus, EADs can be defined as transient oscillations of a full system trajectory around the stable and unstable qEQ_3 close to Hopf bifurcation points or in the vicinity of the stable qLC during changes of the slow variable. The occurrence of EADs in LQTS can be attributable to the shift in a Hopf bifurcation point on qEQ_3 in bifurcation diagrams of the fast subsystem. In recent studies, slow–fast decomposition analyses have been performed by selecting various slow state variables, and attempts have been made to explain the dynamical mechanism underlying EADs [82,217–219]. We also recently reported another type of EAD formation distinct from the membrane-dependent mechanism described above, e.g., the development of EAD due to spontaneous Ca^{2+} releases from the SR [198].

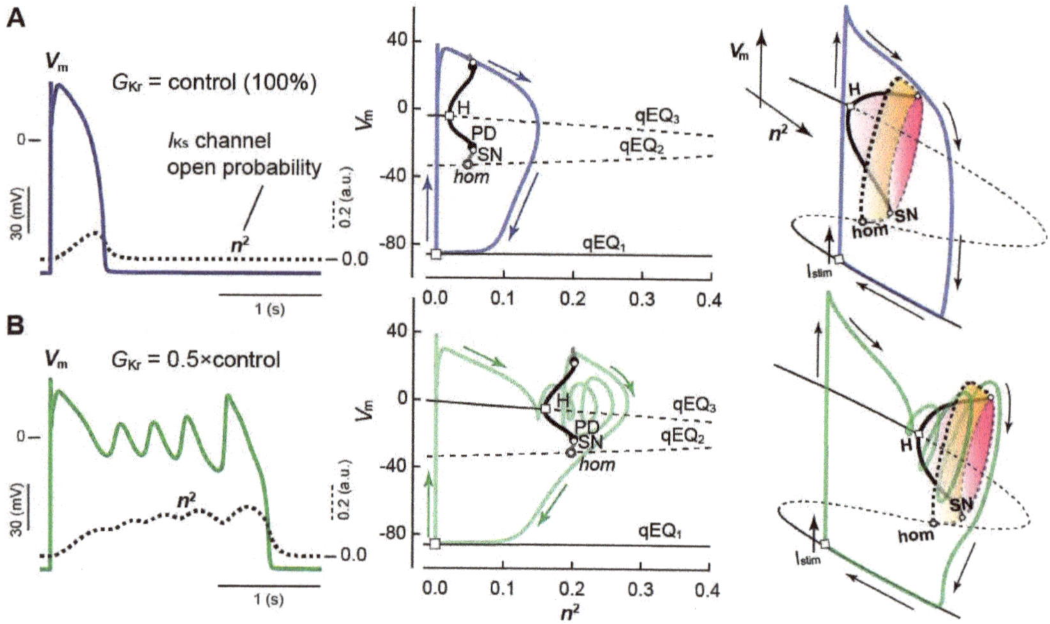

Figure 11. Examples of the slow–fast decomposition analysis. AP simulations and slow–fast decomposition analyses were performed using the Kurata model [112] with G_{Kr} = 100% (**A**) and G_{Kr} = 0.5 × control (**B**). (**Left**) The membrane potential (V_m) and the time course of I_{Ks} open probability (n^2) in the full system. (**Center**) One-parameter bifurcation diagrams of the fast subsystem, depicting quasi-equilibrium potentials ($qEQ_{1–3}$) and the potential extrema of quasi limit cycles (qLCs) as a function of the slow variable n^2. The trajectory of the full system projected onto the (V_m, n^2)-plane is superimposed on the one-parameter bifurcation diagram of the fast subsystem. Solid and dashed thin lines of $qEQ_{1–3}$ indicate stable and unstable equilibrium potentials, respectively. The black and gray thick lines are stable and unstable qLCs, respectively. H: Hopf bifurcation; SN: saddle-node bifurcation of limit cycles; PD: period-doubling bifurcation of limit cycles; hom: homoclinic bifurcation. (**Right**) Schematic diagrams representing the relationships between the dynamic behavior projected onto the 3-dimensional state space of the full system and the bifurcation structure of the fast subsystem. Modified from [197].

5. Future Directions

This review outlined fundamental points of dynamical systems and bifurcation phenomena and applications of bifurcation analyses to cardiac systems. The AP response of cardiomyocytes changes dynamically due to various factors. We employed bifurcation theory of the dynamical system to elucidate the mechanism of EAD, motivated by our expectation that the occurrence of EAD, which triggers the onset of fatal arrhythmias, can be associated with changes in the dynamical stability of APs (i.e., bifurcation phenomena).

We could analyze the bifurcation phenomena of AP responses observed in the cardiac cellular system without any approximation. Although the bifurcation analyses were able to make clear the dynamic transition mechanisms of AP responses depending on parameter changes, it was difficult to explain the mechanism of EAD development by local bifurcation alone. In this respect, slow–fast decomposition analysis was effective.

There will be no doubt that a theoretical approach based on bifurcation theory is effective in understanding the mechanisms of dynamic behaviors seen in cardiac systems. Although this review focused on understanding the EAD mechanism, the bifurcation analysis may be useful in understanding the developmental mechanism of DADs, which has been believed to be another triggering mechanism for lethal arrhythmias. However, current mathematical models of cardiac cells are still immature, and it cannot be said that actual complex dynamic responses seen in cardiomyocytes can be sufficiently reproduced by the model cells. In addition, cardiac arrhythmias do not denote abnormal excitation that occurs at the cellular level, but rather indicate abnormal electrical phenomena that occur in multicellular systems at the tissue and organ levels. Thus, we must consider bifurcations of excitation propagation (not AP responses). We previously reported several in silico studies [145,146] investigating the effects of system parameter changes on excitation propagations observed in simple models of myocardial fiber or strand using a brute force method, which is a kind of parameter study, and by constructing phase diagrams such as bifurcation diagrams. However, the existing bifurcation theory alone may be insufficient for analyzing the bifurcation phenomena with high degrees of freedom, such as changes of excitation propagation before and after the development of lethal arrhythmias. It will be important not only to improve model cells by further experimental studies but also to develop new analysis methods that can be applied to multicellular models. In the future, we would like to develop the control theory of cardiac arrhythmias and establish strategies for their treatments.

Author Contributions: Conceptualization, K.T. and Y.K.; writing—original draft preparation, K.T. and Y.K.; writing—review and editing, K.T. and Y.K.; visualization, K.T.; supervision, Y.K.; funding acquisition, K.T. and Y.K. All authors have read and agreed to the published version of the manuscript.

Funding: This work was funded by the Japan Society for the Promotion of Science (JSPS) KAKENHI, grant number 21K06776 for K.T. and 19K07290 for Y.K., and the Grant for Promoted Research from Kanazawa Medical University, grant number S2019-2, and Kisshokai Medical Research Grant 2020, for K.T.

Institutional Review Board Statement: Not applicable.

Informed Consent Statement: Not applicable.

Data Availability Statement: Not applicable.

Conflicts of Interest: The authors declare no conflict of interest.

References

1. Guckenheimer, J.M.; Holmes, P. *Nonlinear Oscillations, Dynamical Systems, and Bifurcations of Vector Fields*; Springer: New York, NY, USA, 1986; p. 453.
2. Nusse, H.E.; Yorke, J.A.; Kostelich, Y.J. *Dynamics: Numerical Explorations*; Springer: New York, NY, USA, 1994; p. 484.
3. Noble, D. *The Music of Life: Biology beyond the Genome*; Oxford University Press: Oxford, UK, 2006; 153p.
4. Noble, D. A theory of biological relativity: No privileged level of causation. *Interface Focus* **2012**, *2*, 55–64. [CrossRef] [PubMed]
5. Furutani, K.; Tsumoto, K.; Kurachi, Y. HD Physiology Project-Japanese efforts to promote multilevel integrative systems biology and physiome research. *NPJ Syst. Biol. Appl.* **2017**, *3*, 1. [CrossRef] [PubMed]
6. Sperelakis, N. *Cell Physiology Source Book: Essentials of Membrane Biophysics*, 4th ed.; Sperelakis, N., Ed.; Academic Press: San Diego, CA, USA, 2012; p. 996.
7. Belair, J.; Glass, L.; An Der Heiden, U.; Milton, J. Dynamical disease: Identification, temporal aspects and treatment strategies of human illness. *Chaos* **1995**, *5*, 1–7. [CrossRef] [PubMed]
8. Glass, L. Dynamics of Cardiac Arrhythmias. *Phys. Today* **1996**, *49*, 40–45. [CrossRef]
9. Glass, L. Synchronization and rhythmic processes in physiology. *Nature* **2001**, *410*, 277–284. [CrossRef] [PubMed]
10. Strogatz, S.H. *Nonlinear Dynamics and Chaos: With Applications to Physics, Biology, Chemistry, and Engineering*, 2nd ed.; CRC Press: Boca Raton, FL, USA, 2019; 513p.
11. Arnold, V.I. *Geometrical Methods in the Theory of Ordinary Differential Equations*; Springer: New York, NY, USA, 1983; p. 334.
12. Hale, J.K.; Hüseyin, K. *Dynamics and Bifurcations*; Springer: New York, NY, USA, 1991; p. 568.
13. Matsumoto, T.; Komuro, M.; Kokubu, H.; Tokunaga, R. *Bifurcations: Sights, Sounds, and Mathematics*; Springer: Tokyo, Japan, 1993; p. 468.
14. Bers, D.M. *Excitation-Contraction Coupling and Cardiac Contractile Force*, 2nd ed.; Kluwer Academic Publishers: Dordrecht, The Netherlands; Boston, MA, USA, 2001; Volume 237, 427p.
15. Bers, D.M. Cardiac excitation-contraction coupling. *Nature* **2002**, *415*, 198–205. [CrossRef] [PubMed]
16. Bers, D.M. Calcium cycling and signaling in cardiac myocytes. *Annu. Rev. Physiol.* **2008**, *70*, 23–49. [CrossRef] [PubMed]

17. Eisner, D.A.; Caldwell, J.L.; Trafford, A.W.; Hutchings, D.C. The Control of Diastolic Calcium in the Heart: Basic Mechanisms and Functional Implications. *Circ. Res.* **2020**, *126*, 395–412. [CrossRef]
18. Kistamás, K.; Veress, R.; Horváth, B.; Bányász, T.; Nánási, P.P.; Eisner, D.A. Calcium Handling Defects and Cardiac Arrhythmia Syndromes. *Front. Pharmacol.* **2020**, *11*, 72. [CrossRef]
19. Skou, J.C. The influence of some cations on an adenosine triphosphatase from peripheral nerves. *Biochim. Biophys. Acta* **1957**, *23*, 394–401. [CrossRef]
20. Reyes, N.; Gadsby, D.C. Ion permeation through the Na^+,K^+-ATPase. *Nature* **2006**, *443*, 470–474. [CrossRef] [PubMed]
21. Pavlovic, D.; Fuller, W.; Shattock, M.J. Novel regulation of cardiac Na pump via phospholemman. *J. Mol. Cell. Cardiol.* **2013**, *61*, 83–93. [CrossRef] [PubMed]
22. Shattock, M.J.; Ottolia, M.; Bers, D.M.; Blaustein, M.P.; Boguslavskyi, A.; Bossuyt, J.; Bridge, J.H.; Chen-Izu, Y.; Clancy, C.E.; Edwards, A.; et al. Na^+/Ca^{2+} exchange and Na^+/K^+-ATPase in the heart. *J. Physiol.* **2015**, *593*, 1361–1382. [CrossRef] [PubMed]
23. Matchkov, V.V.; Krivoi, I.I. Specialized Functional Diversity and Interactions of the Na,K-ATPase. *Front. Physiol.* **2016**, *7*, 179. [CrossRef]
24. Berry, R.G.; Despa, S.; Fuller, W.; Bers, D.M.; Shattock, M.J. Differential distribution and regulation of mouse cardiac Na^+/K^+-ATPase alpha1 and alpha2 subunits in T-tubule and surface sarcolemmal membranes. *Cardiovasc. Res.* **2007**, *73*, 92–100. [CrossRef]
25. Kimura, J.; Noma, A.; Irisawa, H. Na-Ca exchange current in mammalian heart cells. *Nature* **1986**, *319*, 596–597. [CrossRef]
26. Lytton, J. Na^+/Ca^{2+} exchangers: Three mammalian gene families control Ca^{2+} transport. *Biochem. J.* **2007**, *406*, 365–382. [CrossRef]
27. Fliegel, L.; Dyck, J.R.; Wang, H.; Fong, C.; Haworth, R.S. Cloning and analysis of the human myocardial Na^+/H^+ exchanger. *Mol. Cel. Biochem.* **1993**, *125*, 137–143. [CrossRef]
28. Schwinger, R.H.; Wang, J.; Frank, K.; Müller-Ehmsen, J.; Brixius, K.; McDonough, A.A.; Erdmann, E. Reduced sodium pump alpha1, alpha3, and beta1-isoform protein levels and Na^+,K^+-ATPase activity but unchanged Na^+-Ca^{2+} exchanger protein levels in human heart failure. *Circulation* **1999**, *99*, 2105–2112. [CrossRef]
29. Lingrel, J.; Moseley, A.; Dostanic, I.; Cougnon, M.; He, S.; James, P.; Woo, A.; O'Connor, K.; Neumann, J. Functional roles of the alpha isoforms of the Na,K-ATPase. *Ann. N. Y. Acad. Sci.* **2003**, *986*, 354–359. [CrossRef]
30. Li, Z.; Langhans, S.A. Transcriptional regulators of Na,K-ATPase subunits. *Front. Cell. Dev. Biol.* **2015**, *3*, 66. [CrossRef] [PubMed]
31. Hensley, C.B.; Azuma, K.K.; Tang, M.J.; McDonough, A.A. Thyroid hormone induction of rat myocardial Na(+)-K(+)-ATPase: Alpha 1-, alpha 2-, and beta 1-mRNA and -protein levels at steady state. *Am. J. Physiol.* **1992**, *262*, C484–C492. [CrossRef]
32. Sweadner, K.J.; Herrera, V.L.; Amato, S.; Moellmann, A.; Gibbons, D.K.; Repke, K.R. Immunologic identification of Na+,K(+)-ATPase isoforms in myocardium. Isoform change in deoxycorticosterone acetate-salt hypertension. *Circ. Res.* **1994**, *74*, 669–678. [CrossRef] [PubMed]
33. Shamraj, O.I.; Melvin, D.; Lingrel, J.B. Expression of Na,K-ATPase isoforms in human heart. *Biochem. Biophys. Res. Commun.* **1991**, *179*, 1434–1440. [CrossRef]
34. Zahler, R.; Gilmore-Hebert, M.; Baldwin, J.C.; Franco, K.; Benz, E.J., Jr. Expression of alpha isoforms of the Na,K-ATPase in human heart. *Biochim. Biophys. Acta* **1993**, *1149*, 189–194. [CrossRef]
35. Quednau, B.D.; Nicoll, D.A.; Philipson, K.D. Tissue specificity and alternative splicing of the Na^+/Ca^{2+} exchanger isoforms NCX1, NCX2, and NCX3 in rat. *Am. J. Physiol.* **1997**, *272*, C1250–C1261. [CrossRef] [PubMed]
36. Eisner, D.A.; Caldwell, J.L.; Kistamás, K.; Trafford, A.W. Calcium and Excitation-Contraction Coupling in the Heart. *Circ. Res.* **2017**, *121*, 181–195. [CrossRef] [PubMed]
37. Hurtado, C.; Prociuk, M.; Maddaford, T.G.; Dibrov, E.; Mesaeli, N.; Hryshko, L.V.; Pierce, G.N. Cells expressing unique Na^+/Ca^{2+} exchange (NCX1) splice variants exhibit different susceptibilities to Ca^{2+} overload. *Am. J. Physiol. Heart Circ. Physiol.* **2006**, *290*, H2155–H2162. [CrossRef] [PubMed]
38. Nicoll, D.A.; Longoni, S.; Philipson, K.D. Molecular cloning and functional expression of the cardiac sarcolemmal Na^+-Ca^{2+} exchanger. *Science* **1990**, *250*, 562–565. [CrossRef] [PubMed]
39. Lee, S.L.; Yu, A.S.; Lytton, J. Tissue-specific expression of Na^+-Ca^{2+} exchanger isoforms. *J. Biol. Chem.* **1994**, *269*, 14849–14852. [CrossRef]
40. Hibino, H.; Inanobe, A.; Furutani, K.; Murakami, S.; Findlay, I.; Kurachi, Y. Inwardly rectifying potassium channels: Their structure, function, and physiological roles. *Physiol. Rev.* **2010**, *90*, 291–366. [CrossRef] [PubMed]
41. Reilly, L.; Eckhardt, L.L. Cardiac potassium inward rectifier Kir2: Review of structure, regulation, pharmacology, and arrhythmogenesis. *Heart Rhythm.* **2021**, *18*, 1423–1434. [CrossRef] [PubMed]
42. Jiang, D.; Banh, R.; Gamal El-Din, T.M.; Tonggu, L.; Lenaeus, M.J.; Pomès, R.; Zheng, N.; Catterall, W.A. Open-state structure and pore gating mechanism of the cardiac sodium channel. *Cell* **2021**, *184*, 5151–5162.e5111. [CrossRef] [PubMed]
43. Rook, M.B.; Evers, M.M.; Vos, M.A.; Bierhuizen, M.F. Biology of cardiac sodium channel Nav1.5 expression. *Cardiovasc. Res.* **2012**, *93*, 12–23. [CrossRef]
44. Dixon, J.E.; McKinnon, D. Quantitative analysis of potassium channel mRNA expression in atrial and ventricular muscle of rats. *Circ. Res.* **1994**, *75*, 252–260. [CrossRef]
45. Dixon, J.E.; Shi, W.; Wang, H.S.; McDonald, C.; Yu, H.; Wymore, R.S.; Cohen, I.S.; McKinnon, D. Role of the Kv4.3 K+ channel in ventricular muscle. A molecular correlate for the transient outward current. *Circ. Res.* **1996**, *79*, 659–668. [CrossRef]
46. Gaborit, N.; Le Bouter, S.; Szuts, V.; Varro, A.; Escande, D.; Nattel, S.; Demolombe, S. Regional and tissue specific transcript signatures of ion channel genes in the non-diseased human heart. *J. Physiol.* **2007**, *582*, 675–693. [CrossRef]

47. Näbauer, M.; Beuckelmann, D.J.; Uberfuhr, P.; Steinbeck, G. Regional differences in current density and rate-dependent properties of the transient outward current in subepicardial and subendocardial myocytes of human left ventricle. *Circulation* **1996**, *93*, 168–177. [CrossRef]
48. Kääb, S.; Dixon, J.; Duc, J.; Ashen, D.; Näbauer, M.; Beuckelmann, D.J.; Steinbeck, G.; McKinnon, D.; Tomaselli, G.F. Molecular basis of transient outward potassium current downregulation in human heart failure: A decrease in Kv4.3 mRNA correlates with a reduction in current density. *Circulation* **1998**, *98*, 1383–1393. [CrossRef]
49. Reuter, H. The dependence of slow inward current in Purkinje fibres on the extracellular calcium-concentration. *J. Physiol.* **1967**, *192*, 479–492. [CrossRef]
50. Sanguinetti, M.C.; Jurkiewicz, N.K. Two components of cardiac delayed rectifier K^+ current. Differential sensitivity to block by class III antiarrhythmic agents. *J. Gen Physiol.* **1990**, *96*, 195–215. [CrossRef] [PubMed]
51. Mitra, R.; Morad, M. Two types of calcium channels in guinea pig ventricular myocytes. *Proc. Natl. Acad. Sci. USA* **1986**, *83*, 5340–5344. [CrossRef] [PubMed]
52. Hirano, Y.; Fozzard, H.A.; January, C.T. Characteristics of L- and T-type Ca^{2+} currents in canine cardiac Purkinje cells. *Am. J. Physiol.* **1989**, *256*, H1478–H1492. [CrossRef] [PubMed]
53. Hofmann, F.; Lacinová, L.; Klugbauer, N. Voltage-dependent calcium channels: From structure to function. In *Reviews of Physiology, Biochemistry and Pharmacology*; Springer: Berlin/Heidelberg, Germany, 1999; Volume 139, pp. 33–87.
54. Bohn, G.; Moosmang, S.; Conrad, H.; Ludwig, A.; Hofmann, F.; Klugbauer, N. Expression of T- and L-type calcium channel mRNA in murine sinoatrial node. *FEBS Lett.* **2000**, *481*, 73–76. [CrossRef]
55. Sanguinetti, M.C.; Jiang, C.; Curran, M.E.; Keating, M.T. A mechanistic link between an inherited and an acquired cardiac arrhythmia: HERG encodes the IKr potassium channel. *Cell* **1995**, *81*, 299–307. [CrossRef]
56. Trudeau, M.C.; Warmke, J.W.; Ganetzky, B.; Robertson, G.A. HERG, a human inward rectifier in the voltage-gated potassium channel family. *Science* **1995**, *269*, 92–95. [CrossRef]
57. Sanguinetti, M.C.; Tristani-Firouzi, M. hERG potassium channels and cardiac arrhythmia. *Nature* **2006**, *440*, 463–469. [CrossRef]
58. Stengl, M.; Volders, P.G.; Thomsen, M.B.; Spätjens, R.L.; Sipido, K.R.; Vos, M.A. Accumulation of slowly activating delayed rectifier potassium current (IKs) in canine ventricular myocytes. *J. Physiol.* **2003**, *551*, 777–786. [CrossRef]
59. Nerbonne, J.M.; Kass, R.S. Molecular physiology of cardiac repolarization. *Physiol. Rev.* **2005**, *85*, 1205–1253. [CrossRef]
60. Silva, J.R.; Pan, H.; Wu, D.; Nekouzadeh, A.; Decker, K.F.; Cui, J.; Baker, N.A.; Sept, D.; Rudy, Y. A multiscale model linking ion-channel molecular dynamics and electrostatics to the cardiac action potential. *Proc. Natl. Acad. Sci. USA* **2009**, *106*, 11102–11106. [CrossRef]
61. Huang, C.; Song, Z.; Di, Z.; Qu, Z. Stability of spatially discordant repolarization alternans in cardiac tissue. *Chaos* **2020**, *30*, 123141. [CrossRef] [PubMed]
62. Vandenberg, J.I.; Perozo, E.; Allen, T.W. Towards a Structural View of Drug Binding to hERG K(+) Channels. *Trends Pharm. Sci.* **2017**, *38*, 899–907. [CrossRef] [PubMed]
63. Vandenberg, J.I.; Torres, A.M.; Campbell, T.J.; Kuchel, P.W. The HERG K^+ channel: Progress in understanding the molecular basis of its unusual gating kinetics. *Eur. Biophys. J.* **2004**, *33*, 89–97. [CrossRef] [PubMed]
64. Matsuda, H. Effects of external and internal K^+ ions on magnesium block of inwardly rectifying K^+ channels in guinea-pig heart cells. *J. Physiol.* **1991**, *435*, 83–99. [CrossRef] [PubMed]
65. Ishihara, K. External K(+) dependence of strong inward rectifier K(+) channel conductance is caused not by K(+) but by competitive pore blockade by external Na^+. *J. Gen. Physiol.* **2018**, *150*, 977–989. [CrossRef] [PubMed]
66. Shimizu, W. Update of diagnosis and management of inherited cardiac arrhythmias. *Circ. J.* **2013**, *77*, 2867–2872. [CrossRef]
67. Obeyesekere, M.N.; Antzelevitch, C.; Krahn, A.D. Management of ventricular arrhythmias in suspected channelopathies. *Circ. Arrhythm. Electrophysiol.* **2015**, *8*, 221–231. [CrossRef] [PubMed]
68. Sicouri, S.; Antzelevitch, C. Mechanisms Underlying the Actions of Antidepressant and Antipsychotic Drugs That Cause Sudden Cardiac Arrest. *Arrhythm. Electrophysiol. Rev.* **2018**, *7*, 199–209. [CrossRef]
69. Wilders, R.; Verkerk, A.O. Long QT Syndrome and Sinus Bradycardia-A Mini Review. *Front. Cardiovasc. Med.* **2018**, *5*, 106. [CrossRef]
70. Roden, D.M. Taking the "idio" out of "idiosyncratic": Predicting torsades de pointes. *Pacing Clin. Electrophysiol.* **1998**, *21*, 1029–1034. [CrossRef]
71. Nattel, S.; Maguy, A.; Le Bouter, S.; Yeh, Y.H. Arrhythmogenic ion-channel remodeling in the heart: Heart failure, myocardial infarction, and atrial fibrillation. *Physiol. Rev.* **2007**, *87*, 425–456. [CrossRef] [PubMed]
72. Roden, D.M. Predicting drug-induced QT prolongation and torsades de pointes. *J. Physiol.* **2016**, *594*, 2459–2468. [CrossRef] [PubMed]
73. Wit, A.L. Afterdepolarizations and triggered activity as a mechanism for clinical arrhythmias. *Pacing Clin. Electrophysiol.* **2018**, *41*, 883–896. [CrossRef] [PubMed]
74. Schlotthauer, K.; Bers, D.M. Sarcoplasmic reticulum Ca(2+) release causes myocyte depolarization. Underlying mechanism and threshold for triggered action potentials. *Circ. Res.* **2000**, *87*, 774–780. [CrossRef]
75. Pogwizd, S.M.; Schlotthauer, K.; Li, L.; Yuan, W.; Bers, D.M. Arrhythmogenesis and contractile dysfunction in heart failure: Roles of sodium-calcium exchange, inward rectifier potassium current, and residual beta-adrenergic responsiveness. *Circ. Res.* **2001**, *88*, 1159–1167. [CrossRef]

76. Maruyama, M.; Joung, B.; Tang, L.; Shinohara, T.; On, Y.K.; Han, S.; Choi, E.K.; Kim, D.H.; Shen, M.J.; Weiss, J.N.; et al. Diastolic intracellular calcium-membrane voltage coupling gain and postshock arrhythmias: Role of purkinje fibers and triggered activity. *Circ. Res.* **2010**, *106*, 399–408. [CrossRef]
77. Marban, E.; Robinson, S.W.; Wier, W.G. Mechanisms of arrhythmogenic delayed and early afterdepolarizations in ferret ventricular muscle. *J. Clin. Investig.* **1986**, *78*, 1185–1192. [CrossRef]
78. January, C.T.; Riddle, J.M.; Salata, J.J. A model for early afterdepolarizations: Induction with the Ca^{2+} channel agonist Bay K 8644. *Circ. Res.* **1988**, *62*, 563–571. [CrossRef]
79. January, C.T.; Riddle, J.M. Early afterdepolarizations: Mechanism of induction and block. A role for L-type Ca2+ current. *Circ. Res.* **1989**, *64*, 977–990. [CrossRef]
80. Milberg, P.; Fink, M.; Pott, C.; Frommeyer, G.; Biertz, J.; Osada, N.; Stypmann, J.; Mönnig, G.; Koopmann, M.; Breithardt, G.; et al. Blockade of I(Ca) suppresses early afterdepolarizations and reduces transmural dispersion of repolarization in a whole heart model of chronic heart failure. *Br. J. Pharm.* **2012**, *166*, 557–568. [CrossRef]
81. Zeng, J.; Rudy, Y. Early afterdepolarizations in cardiac myocytes: Mechanism and rate dependence. *Biophys. J.* **1995**, *68*, 949–964. [CrossRef]
82. Qu, Z.; Xie, L.H.; Olcese, R.; Karagueuzian, H.S.; Chen, P.S.; Garfinkel, A.; Weiss, J.N. Early afterdepolarizations in cardiac myocytes: Beyond reduced repolarization reserve. *Cardiovasc. Res.* **2013**, *99*, 6–15. [CrossRef] [PubMed]
83. Gonano, L.A.; Mattiazzi, A. Targeting late ICaL to close the window to ventricular arrhythmias. *J. Gen. Physiol.* **2021**, 153. [CrossRef] [PubMed]
84. Madhvani, R.V.; Angelini, M.; Xie, Y.; Pantazis, A.; Suriany, S.; Borgstrom, N.P.; Garfinkel, A.; Qu, Z.; Weiss, J.N.; Olcese, R. Targeting the late component of the cardiac L-type Ca^{2+} current to suppress early afterdepolarizations. *J. Gen. Physiol.* **2015**, *145*, 395–404. [CrossRef]
85. Kurata, Y.; Tsumoto, K.; Hayashi, K.; Hisatome, I.; Tanida, M.; Kuda, Y.; Shibamoto, T. Dynamical mechanisms of phase-2 early afterdepolarizations in human ventricular myocytes: Insights from bifurcation analyses of two mathematical models. *Am. J. Physiol. Heart Circ. Physiol.* **2017**, *312*, H106–H127. [CrossRef]
86. Nagy, Z.A.; Virág, L.; Tóth, A.; Biliczki, P.; Acsai, K.; Bányász, T.; Nánási, P.; Papp, J.G.; Varró, A. Selective inhibition of sodium-calcium exchanger by SEA-0400 decreases early and delayed after depolarization in canine heart. *Br. J. Pharm.* **2004**, *143*, 827–831. [CrossRef]
87. Milberg, P.; Pott, C.; Frommeyer, G.; Fink, M.; Ruhe, M.; Matsuda, T.; Baba, A.; Klocke, R.; Quang, T.H.; Nikol, S.; et al. Acute inhibition of the Na(+)/Ca(2+) exchanger reduces proarrhythmia in an experimental model of chronic heart failure. *Heart Rhythm.* **2012**, *9*, 570–578. [CrossRef]
88. Pott, C.; Muszynski, A.; Ruhe, M.; Bögeholz, N.; Schulte, J.S.; Milberg, P.; Mönnig, G.; Fabritz, L.; Goldhaber, J.I.; Breithardt, G.; et al. Proarrhythmia in a non-failing murine model of cardiac-specific Na^{+}/Ca^{2+} exchanger overexpression: Whole heart and cellular mechanisms. *Basic Res. Cardiol.* **2012**, *107*, 247. [CrossRef]
89. Xie, Y.; Liao, Z.; Grandi, E.; Shiferaw, Y.; Bers, D.M. Slow [Na]i Changes and Positive Feedback Between Membrane Potential and [Ca]i Underlie Intermittent Early Afterdepolarizations and Arrhythmias. *Circ. Arrhythm. Electrophysiol.* **2015**, *8*, 1472–1480. [CrossRef]
90. Ton, A.T.; Nguyen, W.; Sweat, K.; Miron, Y.; Hernandez, E.; Wong, T.; Geft, V.; Macias, A.; Espinoza, A.; Truong, K.; et al. Arrhythmogenic and antiarrhythmic actions of late sustained sodium current in the adult human heart. *Sci. Rep.* **2021**, *11*, 12014. [CrossRef]
91. Priori, S.G.; Corr, P.B. Mechanisms underlying early and delayed afterdepolarizations induced by catecholamines. *Am. J. Physiol.* **1990**, *258*, H1796–H1805. [CrossRef]
92. Volders, P.G.; Vos, M.A.; Szabo, B.; Sipido, K.R.; de Groot, S.H.; Gorgels, A.P.; Wellens, H.J.; Lazzara, R. Progress in the understanding of cardiac early afterdepolarizations and torsades de pointes: Time to revise current concepts. *Cardiovasc. Res.* **2000**, *46*, 376–392. [CrossRef]
93. Zhao, Z.; Wen, H.; Fefelova, N.; Allen, C.; Baba, A.; Matsuda, T.; Xie, L.H. Revisiting the ionic mechanisms of early afterdepolarizations in cardiomyocytes: Predominant by Ca waves or Ca currents? *Am. J. Physiol. Heart Circ. Physiol.* **2012**, *302*, H1636–H1644. [CrossRef]
94. Choi, B.R.; Burton, F.; Salama, G. Cytosolic Ca^{2+} triggers early afterdepolarizations and Torsade de Pointes in rabbit hearts with type 2 long QT syndrome. *J. Physiol.* **2002**, *543*, 615–631. [CrossRef] [PubMed]
95. Chang, P.C.; Wo, H.T.; Lee, H.L.; Lin, S.F.; Wen, M.S.; Chu, Y.; Yeh, S.J.; Chou, C.C. Role of sarcoplasmic reticulum calcium in development of secondary calcium rise and early afterdepolarizations in long QT syndrome rabbit model. *PLoS ONE* **2015**, *10*, e0123868. [CrossRef] [PubMed]
96. Hodgkin, A.L.; Huxley, A.F. A quantitative description of membrane current and its application to conduction and excitation in nerve. *J. Physiol.* **1952**, *117*, 500–544. [CrossRef]
97. Noble, D. Cardiac action and pacemaker potentials based on the Hodgkin-Huxley equations. *Nature* **1960**, *188*, 495–497. [CrossRef]
98. Luo, C.H.; Rudy, Y. A dynamic model of the cardiac ventricular action potential. II. Afterdepolarizations, triggered activity, and potentiation. *Circ. Res.* **1994**, *74*, 1097–1113. [CrossRef]
99. Luo, C.H.; Rudy, Y. A dynamic model of the cardiac ventricular action potential. I. Simulations of ionic currents and concentration changes. *Circ. Res.* **1994**, *74*, 1071–1096. [CrossRef]

100. Clancy, C.E.; Rudy, Y. Linking a genetic defect to its cellular phenotype in a cardiac arrhythmia. *Nature* **1999**, *400*, 566–569. [CrossRef]
101. Matsuoka, S.; Sarai, N.; Kuratomi, S.; Ono, K.; Noma, A. Role of individual ionic current systems in ventricular cells hypothesized by a model study. *Jpn. J. Physiol.* **2003**, *53*, 105–123. [CrossRef] [PubMed]
102. Shannon, T.R.; Wang, F.; Puglisi, J.; Weber, C.; Bers, D.M. A mathematical treatment of integrated Ca dynamics within the ventricular myocyte. *Biophys. J.* **2004**, *87*, 3351–3371. [CrossRef] [PubMed]
103. Mahajan, A.; Shiferaw, Y.; Sato, D.; Baher, A.; Olcese, R.; Xie, L.H.; Yang, M.J.; Chen, P.S.; Restrepo, J.G.; Karma, A.; et al. A rabbit ventricular action potential model replicating cardiac dynamics at rapid heart rates. *Biophys. J.* **2008**, *94*, 392–410. [CrossRef] [PubMed]
104. Bondarenko, V.E.; Szigeti, G.P.; Bett, G.C.; Kim, S.J.; Rasmusson, R.L. Computer model of action potential of mouse ventricular myocytes. *Am. J. Physiol. Heart Circ. Physiol.* **2004**, *287*, H1378–H1403. [CrossRef] [PubMed]
105. Wang, L.J.; Sobie, E.A. Mathematical model of the neonatal mouse ventricular action potential. *Am. J. Physiol. Heart Circ. Physiol.* **2008**, *294*, H2565–H2575. [CrossRef] [PubMed]
106. Demir, S.S. Computational modeling of cardiac ventricular action potentials in rat and mouse: Review. *Jpn. J. Physiol.* **2004**, *54*, 523–530. [CrossRef]
107. Rozier, K.; Bondarenko, V.E. Mathematical modeling physiological effects of the overexpression of β(2)-adrenoceptors in mouse ventricular myocytes. *Am. J. Physiol. Heart Circ. Physiol.* **2018**, *314*, H643–H658. [CrossRef]
108. Asfaw, T.N.; Tyan, L.; Glukhov, A.V.; Bondarenko, V.E. A compartmentalized mathematical model of mouse atrial myocytes. *Am. J. Physiol. Heart Circ. Physiol.* **2020**, *318*, H485–H507. [CrossRef]
109. Hund, T.J.; Rudy, Y. Rate dependence and regulation of action potential and calcium transient in a canine cardiac ventricular cell model. *Circulation* **2004**, *110*, 3168–3174. [CrossRef]
110. Li, P.; Rudy, Y. A model of canine purkinje cell electrophysiology and Ca(2+) cycling: Rate dependence, triggered activity, and comparison to ventricular myocytes. *Circ. Res.* **2011**, *109*, 71–79. [CrossRef]
111. Decker, K.F.; Heijman, J.; Silva, J.R.; Hund, T.J.; Rudy, Y. Properties and ionic mechanisms of action potential adaptation, restitution, and accommodation in canine epicardium. *Am. J. Physiol. Heart Circ. Physiol.* **2009**, *296*, H1017–H1026. [CrossRef]
112. Kurata, Y.; Hisatome, I.; Matsuda, H.; Shibamoto, T. Dynamical mechanisms of pacemaker generation in IK1-downregulated human ventricular myocytes: Insights from bifurcation analyses of a mathematical model. *Biophys. J.* **2005**, *89*, 2865–2887. [CrossRef]
113. Courtemanche, M.; Ramirez, R.J.; Nattel, S. Ionic mechanisms underlying human atrial action potential properties: Insights from a mathematical model. *Am. J. Physiol.* **1998**, *275*, H301–H321. [CrossRef] [PubMed]
114. ten Tusscher, K.H.; Noble, D.; Noble, P.J.; Panfilov, A.V. A model for human ventricular tissue. *Am. J. Physiol. Heart Circ. Physiol.* **2004**, *286*, H1573–H1589. [CrossRef] [PubMed]
115. ten Tusscher, K.H.; Panfilov, A.V. Alternans and spiral breakup in a human ventricular tissue model. *Am. J. Physiol. Heart Circ. Physiol.* **2006**, *291*, H1088–H1100. [CrossRef] [PubMed]
116. Grandi, E.; Pasqualini, F.S.; Bers, D.M. A novel computational model of the human ventricular action potential and Ca transient. *J. Mol. Cell. Cardiol.* **2010**, *48*, 112–121. [CrossRef] [PubMed]
117. O'Hara, T.; Virág, L.; Varró, A.; Rudy, Y. Simulation of the undiseased human cardiac ventricular action potential: Model formulation and experimental validation. *PLoS Comput. Biol.* **2011**, *7*, e1002061. [CrossRef]
118. Tomek, J.; Bueno-Orovio, A.; Passini, E.; Zhou, X.; Minchole, A.; Britton, O.; Bartolucci, C.; Severi, S.; Shrier, A.; Virag, L.; et al. Development, calibration, and validation of a novel human ventricular myocyte model in health, disease, and drug block. *eLife* **2019**, *8*, e48890. [CrossRef]
119. Noble, D.; Garny, A.; Noble, P.J. How the Hodgkin-Huxley equations inspired the Cardiac Physiome Project. *J. Physiol.* **2012**, *590*, 2613–2628. [CrossRef]
120. Amuzescu, B.; Airini, R.; Epureanu, F.B.; Mann, S.A.; Knott, T.; Radu, B.M. Evolution of mathematical models of cardiomyocyte electrophysiology. *Math. Biosci.* **2021**, *334*, 108567. [CrossRef]
121. Minorsky, N. *Nonlinear Oscillations*; Van Nostrand: Princeton, NJ, USA, 1962; p. 714.
122. Pontryagin, L.S. *Ordinary Differential Equations*; Addison-Wesley: Reading, MA, USA, 1962; p. 298.
123. Kuznetsov, Y. *Elements of Applied Bifurcation Theory*, 3rd ed.; Springer: New York, NY, USA, 2004; p. 631.
124. Yehia, A.R.; Jeandupeux, D.; Alonso, F.; Guevara, M.R. Hysteresis and bistability in the direct transition from 1:1 to 2:1 rhythm in periodically driven single ventricular cells. *Chaos* **1999**, *9*, 916–931. [CrossRef] [PubMed]
125. Michaels, D.C.; Chialvo, D.R.; Matyas, E.P.; Jalife, J. Chaotic activity in a mathematical model of the vagally driven sinoatrial node. *Circ. Res.* **1989**, *65*, 1350–1360. [CrossRef] [PubMed]
126. Okamoto, Y.; Takano, M.; Ohba, T.; Ono, K. Arrhythmogenic coupling between the Na+-Ca^{2+} exchanger and inositol 1,4,5-triphosphate receptor in rat pulmonary vein cardiomyocytes. *J. Mol. Cell. Cardiol.* **2012**, *52*, 988–997. [CrossRef]
127. Okamoto, Y.; Kawamura, K.; Nakamura, Y.; Ono, K. Pathological impact of hyperpolarization-activated chloride current peculiar to rat pulmonary vein cardiomyocytes. *J. Mol. Cell. Cardiol.* **2014**, *66*, 53–62. [CrossRef] [PubMed]
128. Haïssaguerre, M.; Jaïs, P.; Shah, D.C.; Takahashi, A.; Hocini, M.; Quiniou, G.; Garrigue, S.; Le Mouroux, A.; Le Métayer, P.; Clémenty, J. Spontaneous Initiation of Atrial Fibrillation by Ectopic Beats Originating in the Pulmonary Veins. *N. Engl. J. Med.* **1998**, *339*, 659–666. [CrossRef] [PubMed]

129. Umehara, S.; Tan, X.; Okamoto, Y.; Ono, K.; Noma, A.; Amano, A.; Himeno, Y. Mechanisms Underlying Spontaneous Action Potential Generation Induced by Catecholamine in Pulmonary Vein Cardiomyocytes: A Simulation Study. *Int. J. Mol. Sci.* **2019**, *20*, 2913. [CrossRef] [PubMed]
130. Tsumoto, K.; Kurata, Y.; Furutani, K.; Kurachi, Y. Hysteretic Dynamics of Multi-Stable Early Afterdepolarisations with Repolarisation Reserve Attenuation: A Potential Dynamical Mechanism for Cardiac Arrhythmias. *Sci. Rep.* **2017**, *7*, 10771. [CrossRef]
131. Barrio, R.; Martínez, M.; Pueyo, E.; Serrano, S. Dynamical analysis of early afterdepolarization patterns in a biophysically detailed cardiac model. *Chaos* **2021**, *31*, 073137. [CrossRef]
132. Gadsby, D.C.; Cranefield, P.F. Two levels of resting potential in cardiac Purkinje fibers. *J. Gen Physiol.* **1977**, *70*, 725–746. [CrossRef]
133. Cranefield, P.F. Action potentials, afterpotentials, and arrhythmias. *Circ. Res.* **1977**, *41*, 415–423. [CrossRef]
134. Jalife, J.; Antzelevitch, C. Phase resetting and annihilation of pacemaker activity in cardiac tissue. *Science* **1979**, *206*, 695–697. [CrossRef] [PubMed]
135. Walker, M.L.; Wan, X.; Kirsch, G.E.; Rosenbaum, D.S. Hysteresis effect implicates calcium cycling as a mechanism of repolarization alternans. *Circulation* **2003**, *108*, 2704–2709. [CrossRef] [PubMed]
136. Wu, R.; Patwardhan, A. Restitution of action potential duration during sequential changes in diastolic intervals shows multimodal behavior. *Circ. Res.* **2004**, *94*, 634–641. [CrossRef] [PubMed]
137. Berger, R.D. Electrical restitution hysteresis: Good memory or delayed response? *Circ. Res.* **2004**, *94*, 567–569. [CrossRef] [PubMed]
138. Arnol'd, V.I. Loss of stability of self-oscillations close to resonance and versal deformations of equivariant vector fields. *Funct. Anal. Its Appl.* **1977**, *11*, 85–92. [CrossRef]
139. Miake, J.; Marbán, E.; Nuss, H.B. Biological pacemaker created by gene transfer. *Nature* **2002**, *419*, 132–133. [CrossRef]
140. Izhikevich, E.M. Neural excitability, spiking and bursting. *Int. J. Bifurc. Chaos* **2000**, *10*, 1171–1266. [CrossRef]
141. Kitajima, H.; Ioka, E.; Yazawa, T. Generation Mechanism of Alternans in Luo–Rudy Model. *Int. J. Bifurc. Chaos* **2016**, *26*, 1650075. [CrossRef]
142. Guevara, M.R.; Glass, L.; Shrier, A. Phase locking, period-doubling bifurcations, and irregular dynamics in periodically stimulated cardiac cells. *Science* **1981**, *214*, 1350–1353. [CrossRef]
143. Landaw, J.; Qu, Z. Bifurcations Caused by Feedback between Voltage and Intracellular Ion Concentrations in Ventricular Myocytes. *Phys. Rev. Lett.* **2019**, *123*, 218101. [CrossRef]
144. Sun, J.; Amellal, F.; Glass, L.; Billette, J. Alternans and period-doubling bifurcations in atrioventricular nodal conduction. *J. Theor. Biol.* **1995**, *173*, 79–91. [CrossRef] [PubMed]
145. Tsumoto, K.; Ashihara, T.; Naito, N.; Shimamoto, T.; Amano, A.; Kurata, Y.; Kurachi, Y. Specific decreasing of Na(+) channel expression on the lateral membrane of cardiomyocytes causes fatal arrhythmias in Brugada syndrome. *Sci. Rep.* **2020**, *10*, 19964. [CrossRef] [PubMed]
146. Tsumoto, K.; Ashihara, T.; Haraguchi, R.; Nakazawa, K.; Kurachi, Y. Ischemia-related subcellular redistribution of sodium channels enhances the proarrhythmic effect of class I antiarrhythmic drugs: A simulation study. *PLoS ONE* **2014**, *9*, e109271. [CrossRef] [PubMed]
147. Chialvo, D.R.; Gilmour, R.F., Jr.; Jalife, J. Low dimensional chaos in cardiac tissue. *Nature* **1990**, *343*, 653–657. [CrossRef] [PubMed]
148. Frame, L.H.; Simson, M.B. Oscillations of conduction, action potential duration, and refractoriness. A mechanism for spontaneous termination of reentrant tachycardias. *Circulation* **1988**, *78*, 1277–1287. [CrossRef]
149. Qu, Z.; Xie, F.; Garfinkel, A.; Weiss, J.N. Origins of spiral wave meander and breakup in a two-dimensional cardiac tissue model. *Ann. Biomed. Eng.* **2000**, *28*, 755–771. [CrossRef]
150. Qu, Z.; Hu, G.; Garfinkel, A.; Weiss, J.N. Nonlinear and Stochastic Dynamics in the Heart. *Phys. Rep.* **2014**, *543*, 61–162. [CrossRef]
151. AUTO. Available online: http://indy.cs.concordia.ca/auto/ (accessed on 1 November 2021).
152. Doedel, E.; Kernévez, J. *AUTO: Software for Continuation Problems in Ordinary Differential Equations with Applications*; Applied Mathematics Technical Report; California Institute of Technology: Pasadena, CA, USA, 1986.
153. XPP-AUT. Available online: http://www.math.pitt.edu/~{}bard/xpp/xpp.html (accessed on 1 November 2021).
154. Dhooge, A.; Govaerts, W.; Kuznetsov, Y.A. MATCONT: A MATLAB package for numerical bifurcation analysis of ODEs. *ACM Trans. Math. Softw.* **2003**, *29*, 141–164. [CrossRef]
155. Kawakami, H. Bifurcation of periodic responses in forced dynamic nonlinear circuits: Computation of bifurcation values of the system parameters. *IEEE Trans. Circuits Syst.* **1984**, *31*, 248–260. [CrossRef]
156. Tsumoto, K.; Ueta, T.; Yoshinaga, T.; Kawakami, H. Bifurcation analyses of nonlinear dynamical systems: From theory to numerical computations. *Nonlinear Theory Its Appl. IEICE* **2012**, *3*, 458–476. [CrossRef]
157. Parker, T.S.; Chua, L. *Practical Numerical Algorithms for Chaotic Systems*; Springer: New York, NY, USA, 1989; p. 348.
158. Glass, L.; Mackey, M.C. *From Clocks to Chaos: The Rhythms of Life*; Princeton University Press: Princeton, NJ, USA, 1988; p. 248.
159. Shrier, A.; Dubarsky, H.; Rosengarten, M.; Guevara, M.R.; Nattel, S.; Glass, L. Prediction of complex atrioventricular conduction rhythms in humans with use of the atrioventricular nodal recovery curve. *Circulation* **1987**, *76*, 1196–1205. [CrossRef] [PubMed]
160. Keener, J.P. On cardiac arrythmias: AV conduction block. *J. Math. Biol.* **1981**, *12*, 215–225. [CrossRef]
161. Guevara, M.R.; Shrier, A.; Glass, L. Phase resetting of spontaneously beating embryonic ventricular heart cell aggregates. *Am. J. Physiol.* **1986**, *251*, H1298–H1305. [CrossRef] [PubMed]

162. Guevara, M.R.; Shrier, A.; Glass, L. Phase-locked rhythms in periodically stimulated heart cell aggregates. *Am. J. Physiol.* **1988**, *254*, H1–H10. [CrossRef] [PubMed]
163. Keener, J.P.; Glass, L. Global bifurcations of a periodically forced nonlinear oscillator. *J. Math. Biol.* **1984**, *21*, 175–190. [CrossRef]
164. Jalife, J.; Moe, G.K. Phasic effects of vagal stimulation on pacemaker activity of the isolated sinus node of the young cat. *Circ. Res.* **1979**, *45*, 595–608. [CrossRef]
165. Ikeda, N.; Yoshizawa, S.; Sato, T. Difference equation model of ventricular parasystole as an interaction between cardiac pacemakers based on the phase response curve. *J. Theor. Biol.* **1983**, *103*, 439–465. [CrossRef]
166. Glass, L.; Goldberger, A.L.; Bélair, J. Dynamics of pure parasystole. *Am. J. Physiol.* **1986**, *251*, H841–H847. [CrossRef]
167. Landau, M.; Lorente, P.; Michaels, D.; Jalife, J. Bistabilities and annihilation phenomena in electrophysiological cardiac models. *Circ. Res.* **1990**, *66*, 1658–1672. [CrossRef]
168. Guevara, M.R.; Jongsma, H.J. Phase resetting in a model of sinoatrial nodal membrane: Ionic and topological aspects. *Am. J. Physiol.* **1990**, *258*, H734–H747. [CrossRef]
169. Guevara, M.R.; Jongsma, H.J. Three ways of abolishing automaticity in sinoatrial node: Ionic modeling and nonlinear dynamics. *Am. J. Physiol.* **1992**, *262*, H1268–H1286. [CrossRef] [PubMed]
170. Kurata, Y.; Hisatome, I.; Imanishi, S.; Shibamoto, T. Roles of L-type Ca^{2+} and delayed-rectifier K^+ currents in sinoatrial node pacemaking: Insights from stability and bifurcation analyses of a mathematical model. *Am. J. Physiol. Heart Circ. Physiol.* **2003**, *285*, H2804–H2819. [CrossRef] [PubMed]
171. Kurata, Y.; Matsuda, H.; Hisatome, I.; Shibamoto, T. Regional difference in dynamical property of sinoatrial node pacemaking: Role of Na^+ channel current. *Biophys. J.* **2008**, *95*, 951–977. [CrossRef] [PubMed]
172. Kurata, Y.; Matsuda, H.; Hisatome, I.; Shibamoto, T. Roles of hyperpolarization-activated current If in sinoatrial node pacemaking: Insights from bifurcation analysis of mathematical models. *Am. J. Physiol. Heart Circ. Physiol.* **2010**, *298*, H1748–H1760. [CrossRef] [PubMed]
173. Kurata, Y.; Hisatome, I.; Shibamoto, T. Roles of sarcoplasmic reticulum Ca^{2+} cycling and Na^+/Ca^{2+} exchanger in sinoatrial node pacemaking: Insights from bifurcation analysis of mathematical models. *Am. J. Physiol. Heart Circ. Physiol.* **2012**, *302*, H2285–H2300. [CrossRef] [PubMed]
174. Kurata, Y.; Hisatome, I.; Tanida, M.; Shibamoto, T. Effect of hyperpolarization-activated current If on robustness of sinoatrial node pacemaking: Theoretical study on influence of intracellular Na(+) concentration. *Am. J. Physiol. Heart Circ. Physiol.* **2013**, *304*, H1337–H1351. [CrossRef] [PubMed]
175. Pan, Z.; Yamaguchi, R.; Doi, S. Bifurcation analysis and effects of changing ionic conductances on pacemaker rhythm in a sinoatrial node cell model. *Biosystems* **2011**, *106*, 9–18. [CrossRef]
176. Pan, Z.; Doi, S. Global bifurcation structure and variability of pacemaker rhythm in a detailed model of cardiac sinoatrial node cells. *Electron. Commun. Jpn.* **2012**, *95*, 28–39. [CrossRef]
177. Bouchard, S.; Jacquemet, V.; Vinet, A. Automaticity in acute ischemia: Bifurcation analysis of a human ventricular model. *Phys. Rev. E Stat Nonlin Soft Matter Phys.* **2011**, *83*, 011911. [CrossRef]
178. Chay, T.R.; Lee, Y.S. Phase resetting and bifurcation in the ventricular myocardium. *Biophys. J.* **1985**, *47*, 641–651. [CrossRef]
179. Vinet, A.; Chialvo, D.R.; Michaels, D.C.; Jalife, J. Nonlinear dynamics of rate-dependent activation in models of single cardiac cells. *Circ. Res.* **1990**, *67*, 1510–1524. [CrossRef] [PubMed]
180. Kurata, Y.; Matsuda, H.; Hisatome, I.; Shibamoto, T. Effects of pacemaker currents on creation and modulation of human ventricular pacemaker: Theoretical study with application to biological pacemaker engineering. *Am. J. Physiol. Heart Circ. Physiol.* **2007**, *292*, H701–H718. [CrossRef] [PubMed]
181. Gibb, W.J.; Wagner, M.B.; Lesh, M.D. Effects of simulated potassium blockade on the dynamics of triggered cardiac activity. *J. Theor. Biol.* **1994**, *168*, 245–257. [CrossRef] [PubMed]
182. Sato, D.; Xie, L.H.; Nguyen, T.P.; Weiss, J.N.; Qu, Z. Irregularly appearing early afterdepolarizations in cardiac myocytes: Random fluctuations or dynamical chaos? *Biophys. J.* **2010**, *99*, 765–773. [CrossRef] [PubMed]
183. Sato, D.; Xie, L.H.; Sovari, A.A.; Tran, D.X.; Morita, N.; Xie, F.; Karagueuzian, H.; Garfinkel, A.; Weiss, J.N.; Qu, Z. Synchronization of chaotic early afterdepolarizations in the genesis of cardiac arrhythmias. *Proc. Natl. Acad. Sci. USA* **2009**, *106*, 2983–2988. [CrossRef] [PubMed]
184. Luo, C.H.; Rudy, Y. A model of the ventricular cardiac action potential. Depolarization, repolarization, and their interaction. *Circ. Res.* **1991**, *68*, 1501–1526. [CrossRef] [PubMed]
185. Landaw, J.; Qu, Z. Memory-induced nonlinear dynamics of excitation in cardiac diseases. *Phys. Rev. E* **2018**, *97*, 042414. [CrossRef] [PubMed]
186. Huang, X.; Song, Z.; Qu, Z. Determinants of early afterdepolarization properties in ventricular myocyte models. *PLoS Comput. Biol.* **2018**, *14*, e1006382. [CrossRef]
187. Qu, Z. Chaos in the genesis and maintenance of cardiac arrhythmias. *Prog. Biophys. Mol. Biol.* **2011**, *105*, 247–257. [CrossRef]
188. Kimrey, J.; Vo, T.; Bertram, R. Canard analysis reveals why a large Ca^{2+} window current promotes early afterdepolarizations in cardiac myocytes. *PLoS Comput. Biol.* **2020**, *16*, e1008341. [CrossRef]
189. Amuzescu, B.; Georgescu, A.; Nistor, G.; Popescu, M.; Svab, I.; Flonta, M.L.; Corlan, A.D. Stability and sustained oscillations in a ventricular cardiomyocyte model. *Interdiscip Sci.* **2012**, *4*, 1–18. [CrossRef] [PubMed]

190. Tran, D.X.; Sato, D.; Yochelis, A.; Weiss, J.N.; Garfinkel, A.; Qu, Z. Bifurcation and chaos in a model of cardiac early afterdepolarizations. *Phys. Rev. Lett.* **2009**, *102*, 258103. [CrossRef] [PubMed]
191. Asakura, K.; Cha, C.Y.; Yamaoka, H.; Horikawa, Y.; Memida, H.; Powell, T.; Amano, A.; Noma, A. EAD and DAD mechanisms analyzed by developing a new human ventricular cell model. *Prog. Biophys. Mol. Biol.* **2014**, *116*, 11–24. [CrossRef] [PubMed]
192. Himeno, Y.; Asakura, K.; Cha, C.Y.; Memida, H.; Powell, T.; Amano, A.; Noma, A. A human ventricular myocyte model with a refined representation of excitation-contraction coupling. *Biophys. J.* **2015**, *109*, 415–427. [CrossRef]
193. Iyer, V.; Mazhari, R.; Winslow, R.L. A computational model of the human left-ventricular epicardial myocyte. *Biophys. J.* **2004**, *87*, 1507–1525. [CrossRef]
194. Heijman, J.; Volders, P.G.; Westra, R.L.; Rudy, Y. Local control of β-adrenergic stimulation: Effects on ventricular myocyte electrophysiology and Ca(2+)-transient. *J. Mol. Cell. Cardiol.* **2011**, *50*, 863–871. [CrossRef]
195. Morotti, S.; Edwards, A.G.; McCulloch, A.D.; Bers, D.M.; Grandi, E. A novel computational model of mouse myocyte electrophysiology to assess the synergy between Na+ loading and CaMKII. *J. Physiol.* **2014**, *592*, 1181–1197. [CrossRef]
196. Hegyi, B.; Morotti, S.; Liu, C.; Ginsburg, K.S.; Bossuyt, J.; Belardinelli, L.; Izu, L.T.; Chen-Izu, Y.; Bányász, T.; Grandi, E.; et al. Enhanced Depolarization Drive in Failing Rabbit Ventricular Myocytes: Calcium-Dependent and β-Adrenergic Effects on Late Sodium, L-Type Calcium, and Sodium-Calcium Exchange Currents. *Circ. Arrhythm. Electrophysiol.* **2019**, *12*, e007061. [CrossRef]
197. Kurata, Y.; Tanida, M.; Kuda, Y.; Shibamoto, T. Dynamical mechanisms underlying abnormal electrical activities of cardiac myocytes: Insights from bifurcation analyses of mathematical models for human ventricular myocytes. *J. Kanazawa Med. Univ.* **2014**, *39*, 59–66.
198. Kurata, Y.; Tsumoto, K.; Hayashi, K.; Hisatome, I.; Kuda, Y.; Tanida, M. Multiple Dynamical Mechanisms of Phase-2 Early Afterdepolarizations in a Human Ventricular Myocyte Model: Involvement of Spontaneous SR Ca(2+) Release. *Front. Physiol.* **2019**, *10*, 1545. [CrossRef]
199. Kitajima, H.; Yazawa, T. Bifurcation Analysis on a Generation of Early Afterdepolarization in a Mathematical Cardiac Model. *Int. J. Bifurc. Chaos* **2021**, *31*, 2150179. [CrossRef]
200. Ten Tusscher, K.H.; Bernus, O.; Hren, R.; Panfilov, A.V. Comparison of electrophysiological models for human ventricular cells and tissues. *Prog. Biophys. Mol. Biol.* **2006**, *90*, 326–345. [CrossRef] [PubMed]
201. Roden, D.M.; Hoffman, B.F. Action potential prolongation and induction of abnormal automaticity by low quinidine concentrations in canine Purkinje fibers. Relationship to potassium and cycle length. *Circ. Res.* **1985**, *56*, 857–867. [CrossRef] [PubMed]
202. Jost, N.; Virág, L.; Bitay, M.; Takács, J.; Lengyel, C.; Biliczki, P.; Nagy, Z.; Bogáts, G.; Lathrop, D.A.; Papp, J.G.; et al. Restricting excessive cardiac action potential and QT prolongation: A vital role for IKs in human ventricular muscle. *Circulation* **2005**, *112*, 1392–1399. [CrossRef] [PubMed]
203. Takanaka, C.; Singh, B.N. Barium-induced nondriven action potentials as a model of triggered potentials from early afterdepolarizations: Significance of slow channel activity and differing effects of quinidine and amiodarone. *J. Am. Coll Cardiol.* **1990**, *15*, 213–221. [CrossRef]
204. Burashnikov, A.; Antzelevitch, C. Acceleration-induced action potential prolongation and early afterdepolarizations. *J. Cardiovasc. Electrophysiol.* **1998**, *9*, 934–948. [CrossRef]
205. Viswanathan, P.C.; Rudy, Y. Cellular arrhythmogenic effects of congenital and acquired long-QT syndrome in the heterogeneous myocardium. *Circulation* **2000**, *101*, 1192–1198. [CrossRef]
206. Silva, J.; Rudy, Y. Subunit interaction determines IKs participation in cardiac repolarization and repolarization reserve. *Circulation* **2005**, *112*, 1384–1391. [CrossRef]
207. Britton, O.J.; Bueno-Orovio, A.; Virág, L.; Varró, A.; Rodriguez, B. The Electrogenic Na(+)/K(+) Pump Is a Key Determinant of Repolarization Abnormality Susceptibility in Human Ventricular Cardiomyocytes: A Population-Based Simulation Study. *Front. Physiol.* **2017**, *8*, 278. [CrossRef]
208. Kügler, P.; Bulelzai, M.A.; Erhardt, A.H. Period doubling cascades of limit cycles in cardiac action potential models as precursors to chaotic early Afterdepolarizations. *BMC Syst. Biol.* **2017**, *11*, 42. [CrossRef]
209. Furutani, K.; Tsumoto, K.; Chen, I.S.; Handa, K.; Yamakawa, Y.; Sack, J.T.; Kurachi, Y. Facilitation of I Kr current by some hERG channel blockers suppresses early afterdepolarizations. *J. Gen. Physiol.* **2019**, *151*, 214–230. [CrossRef] [PubMed]
210. Fitzhugh, R. Thresholds and plateaus in the Hodgkin-Huxley nerve equations. *J. Gen. Physiol.* **1960**, *43*, 867–896. [CrossRef] [PubMed]
211. Erhardt, A.H. Early Afterdepolarisations Induced by an Enhancement in the Calcium Current. *Processes* **2019**, *7*, 20. [CrossRef]
212. Sato, D.; Clancy, C.E.; Bers, D.M. Dynamics of sodium current mediated early afterdepolarizations. *Heliyon* **2017**, *3*, e00388. [CrossRef] [PubMed]
213. Kügler, P.; Erhardt, A.H.; Bulelzai, M.A.K. Early afterdepolarizations in cardiac action potentials as mixed mode oscillations due to a folded node singularity. *PLoS ONE* **2018**, *13*, e0209498. [CrossRef]
214. Erhardt, A.H.; Solem, S. On complex dynamics in a Purkinje and a ventricular cardiac cell model. *Commun. Nonlinear Sci. Numer. Simul.* **2021**, *93*, 105511. [CrossRef]
215. Vo, T.; Bertram, R. Why pacing frequency affects the production of early afterdepolarizations in cardiomyocytes: An explanation revealed by slow-fast analysis of a minimal model. *Phys Rev. E* **2019**, *99*, 052205. [CrossRef]

216. Doi, S.; Nabetani, S.; Kumagai, S. Complex nonlinear dynamics of the Hodgkin-Huxley equations induced by time scale changes. *Biol. Cybern* **2001**, *85*, 51–64. [CrossRef]
217. Song, Z.; Ko, C.Y.; Nivala, M.; Weiss, J.N.; Qu, Z. Calcium-voltage coupling in the genesis of early and delayed afterdepolarizations in cardiac myocytes. *Biophys. J.* **2015**, *108*, 1908–1921. [CrossRef]
218. Borowski, P.; Kuske, R.; Li, Y.X.; Cabrera, J.L. Characterizing mixed mode oscillations shaped by noise and bifurcation structure. *Chaos* **2010**, *20*, 043117. [CrossRef]
219. Yaru, L.; Shenquan, L. Characterizing mixed-mode oscillations shaped by canard and bifurcation structure in a three-dimensional cardiac cell model. *Nonlinear Dyn.* **2021**, *103*, 2881–2902. [CrossRef]

biomolecules

Review

Automatic Activity Arising in Cardiac Muscle Sleeves of the Pulmonary Vein

Pierre Bredeloux [1,*], Come Pasqualin [1], Romain Bordy [1], Veronique Maupoil [1] and Ian Findlay [2]

1 EA4245, Transplantation, Immunologie et Inflammation, Groupe Physiologie des Cellules Cardiaques et Vasculaires, Université de Tours, 37200 Tours, France; come.pasqualin@univ-tours.fr (C.P.); romain.bordy@univ-tours.fr (R.B.); veronique.maupoil@univ-tours.fr (V.M.)

2 Laboratoire de Pharmacologie, Faculté de Pharmacie, Université de Tours, 37200 Tours, France; ian.findlay@univ-tours.fr

* Correspondence: pierre.bredeloux@univ-tours.fr; Tel.: +33-247-367-205

Abstract: Ectopic activity in the pulmonary vein cardiac muscle sleeves can both induce and maintain human atrial fibrillation. A central issue in any study of the pulmonary veins is their difference from the left atrial cardiac muscle. Here, we attempt to summarize the physiological phenomena underlying the occurrence of ectopic electrical activity in animal pulmonary veins. We emphasize that the activation of multiple signaling pathways influencing not only myocyte electrophysiology but also the means of excitation–contraction coupling may be required for the initiation of triggered or automatic activity. We also gather information regarding not only the large-scale structure of cardiac muscle sleeves but also recent studies suggesting that cellular heterogeneity may contribute to the generation of arrythmogenic phenomena and to the distinction between pulmonary vein and left atrial heart muscle.

Keywords: pulmonary veins; cardiomyocytes; automatic activity; catecholamine

Citation: Bredeloux, P.; Pasqualin, C.; Bordy, R.; Maupoil, V.; Findlay, I. Automatic Activity Arising in Cardiac Muscle Sleeves of the Pulmonary Vein. *Biomolecules* **2022**, *12*, 23. https://doi.org/10.3390/biom12010023

Academic Editors: Yosuke Okamoto and Kyoichi Ono

Received: 6 December 2021
Accepted: 22 December 2021
Published: 24 December 2021

1. Introduction

Although it has been known since the late 19th century [1], interest in the study of cardiac muscle into mammalian thoracic vein exploded with the discovery within human pulmonary veins (PV) of sites of focal ectopic electrical discharges triggering and maintaining atrial fibrillation (AF) [2].

In this context, it should be emphasized that the PV retain a venous structure, including endothelium and vascular smooth muscle layers. Exterior to these is the layer of striated cardiac muscle [3]. To our knowledge, no physical interaction (tissue abutment and/or gap junction connections) between cardiac and vascular smooth muscle in the PV has been reported. The possibility of paracrine interactions between epithelium, smooth and cardiac muscle, however, cannot be excluded.

Recently there have been special issues of both Circulation Research [4] and Cardiovascular Research [5] dedicated to the subject of AF. The interested reader should consult these for an up-to-date synthesis of the mechanisms, therapies, laboratory studies and animal models appropriate for translation to clinical situations.

In parallel with laboratory studies intended for translation to the clinic, basic physiological research has been conducted to characterize differences in the electrophysiology, pharmacology and excitation–contraction coupling of atrial muscle and that of the PV sleeves in an attempt to understand the phenomena underlying the occurrence of PV ectopic electrical activity promoting AF. It is these studies that we will consider here. We also emphasize that this represents a synthetic rather than an exhaustive review.

2. Pulmonary Veins: An Anatomical Substrate Favorable to the Initiation and Conduction of Ectopic Electrical Activities

2.1. Embryological Development of Pulmonary Veins Cardiac Muscle Sleeves

The clearest evidence for differences between cardiac muscle in the left atrium (LA) and the PV comes from studies of embryological development in the mouse [6,7]. The tracing of transcription factors Pitx2 and Tbx5 permitting the identification of entirely different origins of the PV myocardium from that of the LA. In brief, PV myocardium arises from mesenchyme tissue at the base of the PV after both the development of the LA and the establishment of the venous connection between the heart and the lung. Differentiation of this mesenchyme to the myocyte phenotype is followed over a very short period (<24 h) by proliferation and migration along the vein to intra-lobular vessels in the lung.

In the human heart, the development of PV myocardium follows similar principles [8,9]. Only after the connection of the early embryonic single PV to the LA does a myocardial sleeve develop by proliferation and migration of NKX2-5 positive and TH18 negative cells along the vein. In the 110 day human embryo, the single initial PV has divided to show four separate PV ostia draining into the LA body [8]. All four veins show cardiac muscle sleeves which at this stage extend as far as the hilum but not within the developing lung. In the adult human PV, cardiac muscle sleeves only extend 1–2 cm from the PV ostia. This is a common feature for adult PV in other large mammals, including the pig, sheep and dog. Adult rodents (mice, rats, guinea pig and rabbit) on the other hand retain PV cardiac muscle sleeves along their entire lengths including into veins in the interior of the lungs.

Taken together, these data indicate that the PV myocytes have a different embryological origin from those of the LA. It is therefore quite possible that these two populations of cells also exhibit different physiological functioning that would result in a greater propensity of PV cardiomyocytes to triggering electrical ectopic activities that promote AF.

2.2. The Organization of Cardiac Muscle in the Pulmonary Veins and Left Atria

The structure of human PV myocardium has been widely studied (for review see Sánchez-Quintana et al. [10]). In brief, human PV myocardial sleeves appeared to be organized as a complex mesh of muscular fibers. The LA roof between the right and left-sided PV also shows a non-uniform architecture of circular and longitudinally orientated fibers. In addition, fibrosis may create and enhance discontinuities between the myofibrils of the PV and LA roof. Similar complexity of fiber orientation has been found in the PV of the dog [11], mouse [12] and rat [3,13]. In the rat, PV consecutive cardiomyocyte layers are almost orthogonally orientated, whereas a unidirectional orientation of consecutive cardiomyocyte layers are seen in the LA [13].

At the cellular level, some studies have shown that there was no difference in size, shape or capacitance between PV and LA cardiomyocytes in dog, rabbit and rat [14–16]. On the other hand, other studies indicated that rat PV cardiomyocytes were larger and had a higher capacitance than those of the LA [13,17,18]. These differences might be explained by the considerable heterogeneity found within the PV cardiomyocytes population [13,18]. In particular that the incidence and organization of t-tubules in PV myocytes was highly variable from cell to cell, ranging from their virtual absence to well-organized tubular systems. These different types of myocytes were not randomly distributed in the muscle sleeves, but rather in clusters of similar cells. Clusters of PV myocytes with one type of tubular network could be surrounded by PV myocytes with another type of tubule network. The myocytes on the borders of these clusters were directly connected. All of this was in contrast to the regular transversally organized t-tubules network observed in ventricular myocytes irrespective of the species and the relative homogeneity in atrial myocytes [19]. In large mammals, the atrial cell t-tubule network is described as well-developed and extensive [20,21] whereas in small mammals the atrial cell tubular network is generally sparse and not transversally organized [22]. This cellular heterogeneity, associated with the diverse orientation of PV myocardial fibers may favor anisotropic conduction of electrical activity and facilitate reentry and the formation of rotors of ectopic foci in the PV.

2.3. Innervation of the Pulmonary Veins

The human PV myocardium is supplied by nerves originating from the epicardial neural plexus, formed by branches from the left and right vagus nerves, recurrent laryngeal nerves and both sympathetic trunks at the cervico-thoracic and thoracic levels [23–26]. Three epicardial neural plexi supply the four PV [26]. These ganglia contain both adrenergic and cholinergic elements [27]. Ganglionic plexi of the dog also contain both sympathetic and parasympathetic elements [28]. They also found a greater nerve density around the PV ostia where the majority of foci of ectopic electrical activity were located in humans [2], as well as a decline of nerve density from the external to the internal layer of the PV wall. The latter is consistent with human observation: nerve filaments and ganglia are located in the adventitia [29]. In the mouse, catecholamine fibers were concentrated in the PV rather than in the LA appendage [30]. In the rat, the density of catecholamine fibers in the PV increased with post-natal age [31]. These data suggest that autonomic innervation might be an important actor in the regulation of PV excitability and in the genesis of abnormal automatism triggered in the PV, as depicted in the following sections.

3. Pulmonary Vein Electrophysiology

3.1. Resting Membrane Potential in the Pulmonary Vein Myocytes

The PV cardiac muscle is reported to have a less negative resting membrane potential (RMP) than that of the LA. This was described first in the dog where it was associated with a reduced action potential (AP) amplitude and lesser maximum rate of phase 0 depolarization [11,15,32]. Relative to the LA, a lesser PV RMP has been confirmed in the rat [17,18,33,34], the mouse [35] and the guinea pig [36,37]. A consistent observation across diverse species is the decreased expression of the background inwardly rectified channel Kir2.1 in the cardiac muscle of the PV [14,16,18,37–39]. Other contributions to the basal conductance in at least the rat PV cardiac muscle, include an enhanced background Na^+ permeability [17] and a hyperpolarization activated Cl^- current [40,41]. Neither of these were recorded in the LA. In the dog and the guinea pig, agonist independent GIRK channels could provide an additional basal K^+ conductance [15,37]. These differences in the expression of ion channels involved in the regulation of cardiomyocyte membrane potential probably contribute to the greater arrhythmogenic potential of PV by facilitating the initiation of abnormal AP. Interestingly, in both the rat and in the guinea pig there may be a gradient of basal RMP along the length of the PV [34,42], suggesting further heterogeneity in the distribution of ion channels.

3.2. Conduction of Electrical Activity in the Pulmonary Veins

Differences in AP recorded from PV and LA myocardia vary across a range of species and usually involve differences in action potential duration (APD) [15,33,35,42,43]. In ex vivo preparations which included the right atrium, sinus rhythm evoked AP were recorded in the right atrium and along the PV [42,44].

While only APD differentiated the PV and LA AP in the rat under basal conditions, their reactions to selective adrenergic agonists were markedly different. The β_1-adrenergic receptor agonist isoprenaline reduced APD in both tissues. In the PV, it also hyperpolarized plasma membrane and increased AP amplitude. The α_1-adrenergic receptor agonist cirazoline only increased APD in the LA while it depolarized the PV myocytes membrane leading to loss of the AP and thus the tissue excitability [33]. These particular observations led us to investigate the conduction of AP in intact rat LA and PV preparations with either a linear multi-electrode array or simultaneous intracellular microelectrode recordings in the PV and LA [44]. Electrical pacing applied to the apex of the LA elicited electrical activity that propagated across the LA and along the PV. Superfusion of cirazoline led to the loss of electrical activity detectible by the electrode array in the PV. The intracellular microelectrode revealed that this resulted from depolarization of the cell membrane and the reduction of AP amplitude to a small electrotonic wave. Conduction and full overshooting AP were still visible in the LA.

In the intact rat LA and PV preparation, the gradient of depolarization along the length of the PV could sometimes lead to the loss of conduction in distal regions of the vein, which was restored by hyperpolarization evoked by acetylcholine [34]. Egorov et al. [39] found that longitudinal stretch of the PV provoked depolarization, and reduction of AP amplitude to small electrotonic waves. This was reversed by the Cl^- current blockers DIDS and DCPIB provoking hyperpolarization. In isolated PV myocytes, they identified I_{swell} which was blocked by DIDS and DCPIB. They also found a potential pathological role for I_{swell} in the PV of hypertensive rats [39]. In normotensive rat PV, I_{swell} was associated with Caveolin3 in the myocyte membrane. Down regulation of Caveolin3 in hypertensive rats was associated with a greatly enhanced sensitivity of PV conduction to stretch.

The evidence is therefore clear for stretch inducing depolarization in the PV. The mechanism under-laying depolarization resulting from α-adrenergic stimulation remains to be determined. In a different preparation of the rat superior vena cava, depolarization by phenylephrine was associated with reduction of the background Kir 2.1 current [45].

These recent data from the literature suggest that electrical conduction within PV is highly dependent on the activity of the autonomic nervous system and the level of stretch applied to the preparation. Furthermore, it suggests for the first time that it is possible to induce electrical disconnection of the PV by one or more pharmacological approaches, thus offering innovative research perspectives for the identification of molecular targets allowing the future development of more specific pharmacological treatments of AF by preventing the conduction of PV ectopic electrical activity to the LA.

4. Excitation–Contraction Coupling

In cardiomyocytes, excitation–contraction coupling is initiated by AP triggering the opening of L-type calcium channels (LCC) and calcium entry, which induces in return the release of calcium from the sarcoplasmic reticulum through ryanodine receptors (RyR). This process is known as calcium induced calcium release (CICR). Then, the Ca^{2+} released from the reticulum sarcoplasmic engage the myofilaments triggering contraction [46].

There are differences in CICR between ventricular, atrial and PV cardiomyocytes. First, L-type calcium current density is smaller in PV than in LA cardiomyocytes in rat and dog [15,18]. A T-type calcium current may also be present in PV cardiomyocytes of rabbit and rat, which may contribute to CICR [47,48].

Second, in rat the heterogeneity of cellular t-tubule organization, as described in Section 2.2, has direct consequences on the distribution of these voltage gated calcium channels which are largely expressed on t-tubules. Conversely, the spatial distribution of type 2 ryanodine receptors (RyR2) located on the sarcoplasmic reticulum (SR) membrane does not differ between ventricular, LA and PV cardiomyocytes. Thus, the number of calcium release units constituted by LCC in close vicinity to RyR2 clusters will present a great variability in PV cardiomyocytes contrary to ventricular and LA ones.

The amplitude and the spatiotemporal shape of calcium transients in PV cardiomyocytes do show considerable heterogeneity compared to those of the ventricle and atrium [13]. In dog, such differences have not been observed between PV and LA [49].

Thereby, the cellular and tissue heterogeneities may lead to non-uniformity of excitation–contraction coupling within the PV myocardial sleeves leading more easily to the generation of arrhythmogenic calcium waves, which are able to propagate within the PV. Moreover, it may also have consequences in terms of PV cardiomyocytes APD heterogeneity. Thus, localized discordance of APD and consequent refractory period could facilitate the existence of ectopic foci, mostly due to re-entry phenomena [13].

β-adrenergic stimulation substantially modifies excitation–contraction coupling and presents subtle differences between rat atria and ventricular cardiomyocytes [50]. In the PV, the consequences of cardiomyocyte heterogeneity in this context remain to be explored.

5. Pulmonary Veins: A Source of Spontaneous, Triggered and Catecholaminergic Automatic Activity

Spontaneous calcium events are often involved in the firing of abnormal electrical activity. Indeed, beside the singular electrophysiology of PV myocytes, some studies described intense spontaneous calcium events in rabbit and rat [13,51] whereas no difference was found in dogs [49]. In rat, calcium sparks occur more frequently in PV than in LA myocytes, which in turn leads to a higher frequency of spontaneous calcium waves. Their amplitude, duration and width were also greater in PV than in LA cardiomyocytes. Moreover, these waves were associated with depolarizing currents larger than those in LA cardiomyocytes [13]. Altogether, these results strongly suggest the involvement of calcium handling in the generation of several arrythmogenic mechanisms.

5.1. Spontaneous Electrical Activity

Under basal conditions, without electrical stimulation, no [33] or rare spontaneous activities [34] were observed in rat isolated PV. Some sporadic and transient depolarizations [18], but no pacemaker activity [16,41], were observed in rat isolated PV cardiomyocytes. The same has been observed in dog PV tissue preparations and isolated cardiomyocytes [11,15,52]. This absence of spontaneous activity under physiological conditions might be explained by the absence of HCN4, which underlies I_f in the pacemaker cells of the sinoatrial node, in either rat [53] or dog PV [15].

In guinea pig [36,37,41,42] and rabbit PV [14,41,54], spontaneous AP were observed but their incidence varied in different studies. Their frequency was lower than sinus rhythm and thus they were masked in vivo. The spontaneous AP were influenced by both cholinergic and adrenergic stimulation, as well as by stretch. Acetylcholine stopped spontaneous activity, whereas isoproterenol [14,42] and low levels of stretch (<0.3 g) [55,56] increased AP frequency. PV spontaneous AP were also blocked by the Ca^{2+} channel blocker nifedipine in the rabbit [14], and by the inhibitor of the Na^+-Ca^{2+} exchanger (NCX) SEA0400, and ryanodine in the guinea pig [36]. Ca^{2+} release from the SR activating the forward mode of NCX could then be involved in the evocation of spontaneous AP upon the background of the reduced density of IK_1 in the PV [14,37]. Takagi et al. [41] showed a hyperpolarization-activated inward current in guinea PV myocytes which was suppressed by Cs^+ which block I_f current. This suggests the presence of this current in guinea pig PV, although to our knowledge, neither the presence of HCN4 protein and/or mRNA nor the application of the more specific I_f blocker ivabradine have been reported for the guinea pig PV. On the contrary, in rabbit PV there was negligible hyperpolarization-activated inward current [41] and little HCN4 mRNA [53].

In the mouse, half of the PV preparations showed spontaneous electrical activity [35]. This appeared as either constant AP firing or repetitive bursts of AP. Both were abolished by either acetylcholine or adenosine. No spontaneous activity was observed in mouse LA.

5.2. Triggered Activity

Because most ex vivo PV preparations from different animal species were quiescent, laboratory investigations centered upon the means to evoke arrythmogenic activity. Most of these studies involved the use of high-frequency electrical stimulation of the cardiac muscle sleeves to either induce ectopic triggered activity or to induce the classic arrythmogenic features of cardiac muscle that are early-after-depolarizations (EAD) or delayed-after-depolarizations (DAD). The hope in these investigations was to demonstrate the character or simply the incidence of these phenomena, which could be specific to PV rather than a general condition of the LA myocardium.

In human and dog PV, a protocol of high-frequency pacing, a pause and then a single stimulus would result in tachycardia. In the human PV, this post-pause activation could arise from a sinus beat [57]. In the dog PV the post-pause single stimulus was not effective under control conditions, but required the superfusion of norepinephrine (NE) or isoprotenerol and acetylcholine to provoke tachycardia [57,58]. Ryanodine abolished both EAD formation during the sympathetic stimulation and triggered bursts during the combination of sympathetic and parasympathetic stimulation [58].

In what could be clearly a species difference, a post-pause stimulus train applied to the rabbit PV actually required ryanodine to evoke sustained triggered activity. This effect was enhanced by isoproterenol and reduced by the calcium depletion of the sarcoplasmic reticulum with cyclopiazonic acid [59].

In the guinea pig, ryanodine blocked pacing triggered extra systoles in the PV [60]. In the rat, atrial remodeling and especially increases in tissue fibrosis are pre-requisites for pacing to induce either tachycardia or fibrillation [61–64].

In a series of experiments conducted upon the canine PV, Patterson et al. [52] combined direct stimulation of cardiac muscle to evoke AP in the vein with high-frequency short duration stimulation to the ganglionic plexus innervating the vein. The coincidence of ganglionic stimulation with an evoked AP resulted in the reduction of AP duration and the triggering of tachycardia. As for pacing induced tachycardia [58], the effects of ganglionic stimulation were blocked by atropine, atenolol and ryanodine [52]. Human ganglionic plexus ablation has gone on to prove an adjunct to PV ablation in the clinic [65–67]. It also suggests an involvement of both sympathetic and parasympathetic nervous systems in abnormal electrical activity.

5.3. Catecholaminergic Automatic Activity

Ex vivo rat PV responded to NE with bursts of spontaneous contractions. This catecholaminergic automatic activity (CAA) which was independent of electrical stimulation, required the simultaneous activation of α_1 and β_1 adrenergic receptors and was not observed in the LA [68]. Intracellular microelectrode recording showed that the response of the PV quiescent membrane potential to NE was biphasic. First, a hyperpolarization resulting from activation of β_1 adrenergic receptors was observed. It was followed by depolarization resulting from the activation of α_1 adrenergic receptors that preceded the onset of CAA. Automatic activity presented as bursts of slow rising phase and low amplitude AP. Moreover, automatic AP showed variable form and frequency during these bursts. Intervals were characterized by hyperpolarization at the end of the burst and slow depolarization to the onset of the next. The effects of NE upon the quiescent membrane potential in the LA were slight and without evocation of CAA [33]. Doisne et al. [33] also showed that upon a continuous background of α-adrenergic receptors stimulation, different doses of β-adrenergic receptors agonist evoked different forms of automatic activity (Figure 1). These ranged from activities with varying burst and inter-burst intervals to activity reduced to isolated low-frequency AP arising from a relatively depolarized membrane potential. The highest dose of isoprenaline actually abolished automatic activity. Figure 2 illustrates that different forms of automatic activity in the rat PV can result from slight changes in the concentration of NE. Rather than blocking CAA, tetrodotoxin evoked changes by reducing burst duration and increasing interval length [17].

In an intact PV-LA preparation of the rat, Egorov et al. [34] also showed that NE evoked a biphasic response in the PV membrane potential, which could result in automatic activity. This automatic activity presented as continuous firing in the majority of cases. They revealed complex interactions between automatic AP arising from separate sites in the vein and with electrical pacing of the LA [34]. The reactions of the rat PV to adrenaline were also dose dependent [69]. Low adrenaline concentrations induced depolarization and loss of electrical conduction, whereas high concentrations led to hyperpolarization, restoration of intra-PV conduction and the induction of automatic AP activity. Moreover, moderate stretch applied to the PV facilitated the development of arrythmogenic activity induced by a high concentration of adrenaline [69]. The predisposition of rat PV to show CAA evoked by NA has been found to develop with age and the post-natal development of the sympathetic innervation [31].

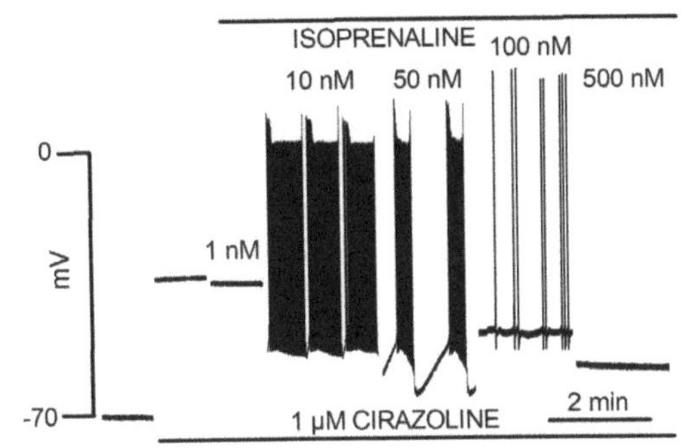

Figure 1. Automatic activity in cardiac muscle of the PV produced by the combination of α- and β-adrenergic receptor stimulation. Traces represent isolated segments of an otherwise continuous recording of membrane potential in cardiac muscle of one PV during the superfusion of 1 μM Cirazoline and different concentrations of isoprenaline. Modified from Doisne et al. [33].

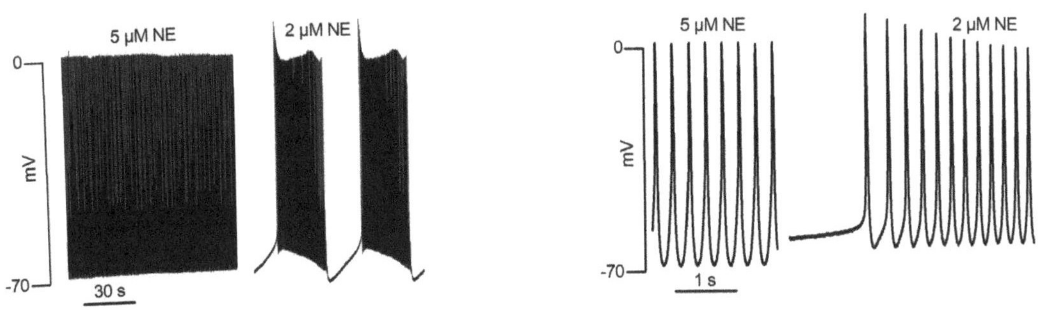

(**a**) (**b**)

Figure 2. An intracellular microelectrode recording from an isolated PV of the rat. These traces form parts of an otherwise continuous recording during one penetration. (**a**) Automatic activity which had been evoked by the superfusion of 5 μM NE presented as continuous firing (3.8 Hz) of action potentials arising from a diastolic membrane potential of −69 mV, overshooting to +4 mV with a maximum rate of phase 0 depolarization of 5.4 V/s. Ten minutes after reducing the concentration of NE to 2 μM, continuous firing was replaced by automatic activity representing as bursts separated by silent intervals. Details of individual automatic action potentials recorded under these conditions are shown upon an expanded scale in (**b**).

These data may contribute, at least in part, to the explanation of the increased incidence of AF observed with age and in patients with certain chronic cardiovascular pathologies, such as hypertension and heart failure.

In isolated rat PV cardiomyocytes, spontaneous and sustained AP induced by NE were recorded with the perforated patch clamp technique [18]. In these cells, NE induced automaticity was blocked by SEA0400, a selective NCX inhibitor, as well as by 2-APB, a functional antagonist of IP_3R [18], suggesting a functional coupling between the NCX and IP_3R in the t-tubule micro-domain. These data were then incorporated into a mathematical model of the PV cardiomyocyte [70]. However, in another study, NE failed to induce automatic activity in rat isolated PV cardiomyocytes [16], perhaps because of differences between the patch clamp methods.

In quiescent mouse PV [30,35], NE also induced automatic activity that appeared in the different forms (continuous or bursts of AP) already observed in spontaneously active preparations under basal conditions (Section 5.1). NE was without effect on the LA. However, Potekhina et al. [30] showed that automatic activity could also be evoked by isoprenaline or phenylephrine alone, which seems to be clearly a species-specific feature. Both substances caused hyperpolarization of the PV membrane potential. Acetylcholine and adenosine inhibited NE-induced activity in the mouse PV [35].

NE was also shown to induce a CAA in the distal end of quiescent guinea pig PV preparations [42]. There was a gradual transition in the form of the spontaneous AP, from sharp early repolarizations at the distal end to atrium like AP in the proximal PV. This NE-induced automatic activity in guinea pig PV depended, as in the rat [33,68], on both α_1 and β_1-adrenergic receptors activation [71].

A question that follows from the induction of automatic activity in muscle sleeves of the PV is that in order for such activity to evoke and/or sustain AF, it has to be conducted to the atrium.

CAA in the rat PV [33,34,69] is characterized by slow AP [33] that are not blocked by doses of tetrodotoxin sufficient to abolish AP in the LA [17]. The combination of these elements suggests that they represent slow Ca^{2+} current rather than fast Na^{+} current induced AP which characterizes electrical activity of non-pacemaker cardiac muscle. Therefore, it is likely that conduction of these slow AP is limited by the space constant. The results of Egorov et al. [34,69] showed some examples of low amplitude automatic AP evoked by either NE or adrenaline in the distal vein resulting in clear sharp over-shooting AP in the PV ostium. No details have been provided regarding the rates of AP rising phase. Although in rats, the mapping of foci in the vein and their success or failure to conduct and convert to atrial AP remains to be determined; mapping in the mouse showed that foci located in the PV ostium did show transmission to the LA [30].

5.4. Other Thoracic Veins

Although this review focuses on the physiology of the PV myocardial sleeves, similar results have also been observed in the rat superior vena cava (SVC) and azygos vein [31,45,72], in which myocardial sleeves also extend and constitute a source of arrhythmia in humans [2]. The effects of age on the induction of CAA in the SVC and PV was also explored [31]. CAA occurred mainly in the SVC of young animals and its incidence decreased with age to become rare in adult animals. On the other hand, CAA was rare in the PV of young animals and its incidence increased with age to become dominant in the adult.

6. Conclusions

The mechanisms involved in the onset of arrhythmia within PV have been widely studied. In particular those that lead to the development of abnormal action potentials but which are not specific to the PV. Indeed, PV cardiac muscle electrophysiological and calcium handling properties as well as innervation make it particularly suitable for triggered activity.

However, in addition to these well-known arrythmogenic mechanisms, the PV structure and singular cellular physiological properties open the door for other ways to trigger arrhythmia. CAA illustrates one arrythmogenic mechanism specific to the PV. The fine mechanics of CAA induction remains to be determined but requires further investigation of the cell physiology of PV cardiomyocytes and especially their regulation by adrenergic pathways. To which can be added, the singularity of fiber orientation and the heterogeneity of the cell population which is specific to PV myocardium contributing to its arrythmogenic potential.

Finally, recent data from basic research suggest that it is possible to induce electrical disconnection of the PV from the LA by one or more pharmacological approaches. This offers, for the first time, innovative research perspectives for the identification of molecular targets allowing the future development of more specific pharmacological treatments of AF by preventing the conduction of PV ectopic electrical activity to the LA.

Author Contributions: All authors researched and wrote sections of the manuscript. P.B. and I.F. assembled the text. All authors have read and agreed to the published version of the manuscript.

Funding: This research received no external funding.

Acknowledgments: We apologize in advance to colleagues whose work was not directly cited in this review.

Conflicts of Interest: The authors declare no conflict of interest.

References

1. Brunton, T.L.; Fayrer, J. Note on Independent Pulsation of the Pulmonary Veins and Vena Cava. *Proc. R. Soc. Lond.* **1877**, *25*, 174–176. [CrossRef]
2. Haïssaguerre, M.; Jaïs, P.; Shah, D.C.; Takahashi, A.; Hocini, M.; Quiniou, G.; Garrigue, S.; le Mouroux, A.; le Métayer, P.; Clémenty, J. Spontaneous Initiation of Atrial Fibrillation by Ectopic Beats Originating in the Pulmonary Veins. *N. Engl. J. Med.* **1998**, *339*, 659–666. [CrossRef] [PubMed]
3. Hashizume, H.; Tango, M.; Ushiki, T. Three-Dimensional Cytoarchitecture of Rat Pulmonary Venous Walls: A Light and Scanning Electron Microscopic Study. *Anat. Embryol.* **1998**, *198*, 473–480. [CrossRef] [PubMed]
4. Ellinor, P.T. Compendium on Atrial Fibrillation. *Circ. Res.* **2020**, *127*, 1–3. [CrossRef]
5. Guzik, T.J. (Ed.) Spotlight Issue on atrial fibrillation from bench to community: Issues in translating new basic ideas to clinical practice. *Cardiovasc. Res.* **2021**, *117*, e88–e90.
6. Mommersteeg, M.T.M.; Hoogaars, W.M.H.; Prall, O.W.J.; de Gier-de Vries, C.; Wiese, C.; Clout, D.E.W.; Papaioannou, V.E.; Brown, N.A.; Harvey, R.P.; Moorman, A.F.M.; et al. Molecular Pathway for the Localized Formation of the Sinoatrial Node. *Circ. Res.* **2007**, *100*, 354–362. [CrossRef]
7. Mommersteeg, M.T.M.; Domínguez, J.N.; Wiese, C.; Norden, J.; de Gier-de Vries, C.; Burch, J.B.E.; Kispert, A.; Brown, N.A.; Moorman, A.F.M.; Christoffels, V.M. The Sinus Venosus Progenitors Separate and Diversify from the First and Second Heart Fields Early in Development. *Cardiovasc. Res.* **2010**, *87*, 92–101. [CrossRef]
8. Douglas, Y.L.; Jongbloed, M.R.M.; Gittenberger-de Groot, A.C.; Evers, D.; Dion, R.A.E.; Voigt, P.; Bartelings, M.M.; Schalij, M.J.; Ebels, T.; DeRuiter, M.C. Histology of Vascular Myocardial Wall of Left Atrial Body after Pulmonary Venous Incorporation. *Am. J. Cardiol.* **2006**, *97*, 662–670. [CrossRef]
9. Sizarov, A.; Anderson, R.H.; Christoffels, V.M.; Moorman, A.F.M. Three-Dimensional and Molecular Analysis of the Venous Pole of the Developing Human Heart. *Circulation* **2010**, *122*, 798–807. [CrossRef]
10. Sánchez-Quintana, D.; López-Mínguez, J.R.; Pizarro, G.; Murillo, M.; Cabrera, J.A. Triggers and Anatomical Substrates in the Genesis and Perpetuation of Atrial Fibrillation. *Curr. Cardiol. Rev.* **2012**, *8*, 310–326. [CrossRef]
11. Hocini, M.; Ho, S.Y.; Kawara, T.; Linnenbank, A.C.; Potse, M.; Shah, D.; Jaïs, P.; Janse, M.J.; Haïssaguerre, M.; de Bakker, J.M.T. Electrical Conduction in Canine Pulmonary Veins: Electrophysiological and Anatomic Correlation. *Circulation* **2002**, *105*, 2442–2448. [CrossRef]
12. Kracklauer, M.P.; Feng, H.-Z.; Jiang, W.; Lin, J.L.-C.; Lin, J.J.-C.; Jin, J.-P. Discontinuous Thoracic Venous Cardiomyocytes and Heart Exhibit Synchronized Developmental Switch of Troponin Isoforms. *FEBS J.* **2013**, *280*, 880–891. [CrossRef]
13. Pasqualin, C.; Yu, A.; Malécot, C.O.; Gannier, F.; Cognard, C.; Godin-Ribuot, D.; Morand, J.; Bredeloux, P.; Maupoil, V. Structural Heterogeneity of the Rat Pulmonary Vein Myocardium: Consequences on Intracellular Calcium Dynamics and Arrhythmogenic Potential. *Sci. Rep.* **2018**, *8*, 3244. [CrossRef]
14. Chen, Y.-J.; Chen, S.-A.; Chen, Y.-C.; Yeh, H.-I.; Chang, M.-S.; Lin, C.-I. Electrophysiology of Single Cardiomyocytes Isolated from Rabbit Pulmonary Veins: Implication in Initiation of Focal Atrial Fibrillation. *Basic Res. Cardiol.* **2002**, *97*, 26–34. [CrossRef]
15. Ehrlich, J.R.; Cha, T.-J.; Zhang, L.; Chartier, D.; Melnyk, P.; Hohnloser, S.H.; Nattel, S. Cellular Electrophysiology of Canine Pulmonary Vein Cardiomyocytes: Action Potential and Ionic Current Properties. *J. Physiol.* **2003**, *551*, 801–813. [CrossRef]
16. Bond, R.C.; Choisy, S.C.; Bryant, S.M.; Hancox, J.C.; James, A.F. Ion Currents, Action Potentials, and Noradrenergic Responses in Rat Pulmonary Vein and Left Atrial Cardiomyocytes. *Physiol. Rep.* **2020**, *8*, e14432. [CrossRef]
17. Malécot, C.O.; Bredeloux, P.; Findlay, I.; Maupoil, V. A TTX-Sensitive Resting Na^+ Permeability Contributes to the Catecholaminergic Automatic Activity in Rat Pulmonary Vein. *J. Cardiovasc. Electrophysiol.* **2015**, *26*, 311–319. [CrossRef]
18. Okamoto, Y.; Takano, M.; Ohba, T.; Ono, K. Arrhythmogenic Coupling between the Na^+-Ca^{2+} Exchanger and Inositol 1,4,5-Triphosphate Receptor in Rat Pulmonary Vein Cardiomyocytes. *J. Mol. Cell. Cardiol.* **2012**, *52*, 988–997. [CrossRef]
19. Blatter, L.A.; Kanaporis, G.; Martinez-Hernandez, E.; Oropeza-Almazan, Y.; Banach, K. Excitation-Contraction Coupling and Calcium Release in Atrial Muscle. *Pflüg. Arch.-Eur. J. Physiol.* **2021**, *473*, 317–329. [CrossRef]
20. Dibb, K.M.; Clarke, J.D.; Eisner, D.A.; Richards, M.A.; Trafford, A.W. A Functional Role for Transverse (t-) Tubules in the Atria. *J. Mol. Cell. Cardiol.* **2013**, *58*, 84–91. [CrossRef]

21. Richards, M.A.; Clarke, J.D.; Saravanan, P.; Voigt, N.; Dobrev, D.; Eisner, D.A.; Trafford, A.W.; Dibb, K.M. Transverse Tubules Are a Common Feature in Large Mammalian Atrial Myocytes Including Human. *Am. J. Physiol. Heart Circ. Physiol.* **2011**, *301*, H1996–H2005. [CrossRef]
22. Brandenburg, S.; Pawlowitz, J.; Fakuade, F.E.; Kownatzki-Danger, D.; Kohl, T.; Mitronova, G.Y.; Scardigli, M.; Neef, J.; Schmidt, C.; Wiedmann, F.; et al. Axial Tubule Junctions Activate Atrial Ca^{2+} Release Across Species. *Front. Physiol.* **2018**, *9*, 1227. [CrossRef]
23. Orts Llorca, F.; Domenech Mateu, J.M.; Puerta Fonolla, J. Innervation of the Sinu-Atrial Node and Neighbouring Regions in Two Human Embryos. *J. Anat.* **1979**, *128*, 365–375.
24. Gardner, E.; O'Rahilly, R. The Nerve Supply and Conducting System of the Human Heart at the End of the Embryonic Period Proper. *J. Anat.* **1976**, *121*, 571–587.
25. Tan, A.Y.; Li, H.; Wachsmann-Hogiu, S.; Chen, L.S.; Chen, P.-S.; Fishbein, M.C. Autonomic Innervation and Segmental Muscular Disconnections at the Human Pulmonary Vein-Atrial Junction: Implications for Catheter Ablation of Atrial-Pulmonary Vein Junction. *J. Am. Coll. Cardiol.* **2006**, *48*, 132–143. [CrossRef]
26. Vaitkevicius, R.; Saburkina, I.; Rysevaite, K.; Vaitkeviciene, I.; Pauziene, N.; Zaliunas, R.; Schauerte, P.; Jalife, J.; Pauza, D.H. Nerve Supply of the Human Pulmonary Veins: An Anatomical Study. *Heart Rhythm* **2009**, *6*, 221–228. [CrossRef]
27. Tan, A.Y.; Chen, P.-S.; Chen, L.S.; Fishbein, M.C. Autonomic Nerves in Pulmonary Veins. *Heart Rhythm* **2007**, *4*, S57–S60. [CrossRef]
28. Gao, C.; Wang, F.; Jiang, R.; Zhang, J.; Mou, H.; Yin, Y. A Region-Specific Quantitative Profile of Autonomic Innervation of the Canine Left Atrium and Pulmonary Veins. *Auton. Neurosci. Basic Clin.* **2011**, *162*, 42–47. [CrossRef]
29. Chevalier, P.; Tabib, A.; Meyronnet, D.; Chalabreysse, L.; Restier, L.; Ludman, V.; Aliès, A.; Adeleine, P.; Thivolet, F.; Burri, H.; et al. Quantitative Study of Nerves of the Human Left Atrium. *Heart Rhythm* **2005**, *2*, 518–522. [CrossRef]
30. Potekhina, V.M.; Averina, O.A.; Razumov, A.A.; Kuzmin, V.S.; Rozenshtraukh, L.V. The Local Repolarization Heterogeneity in the Murine Pulmonary Veins Myocardium Contributes to the Spatial Distribution of the Adrenergically Induced Ectopic Foci. *J. Physiol. Sci. JPS* **2019**, *69*, 1041–1055. [CrossRef]
31. Kuzmin, V.S.; Ivanova, A.D.; Potekhina, V.M.; Samoilova, D.V.; Ushenin, K.S.; Shvetsova, A.A.; Petrov, A.M. The Susceptibility of the Rat Pulmonary and Caval Vein Myocardium to the Catecholamine-Induced Ectopy Changes Oppositely in Postnatal Development. *J. Physiol.* **2021**, *599*, 2803–2821. [CrossRef] [PubMed]
32. Po, S.S.; Li, Y.; Tang, D.; Liu, H.; Geng, N.; Jackman, W.M.; Scherlag, B.; Lazzara, R.; Patterson, E. Rapid and Stable Re-Entry within the Pulmonary Vein as a Mechanism Initiating Paroxysmal Atrial Fibrillation. *J. Am. Coll. Cardiol.* **2005**, *45*, 1871–1877. [CrossRef] [PubMed]
33. Doisne, N.; Maupoil, V.; Cosnay, P.; Findlay, I. Catecholaminergic Automatic Activity in the Rat Pulmonary Vein: Electrophysiological Differences between Cardiac Muscle in the Left Atrium and Pulmonary Vein. *Am. J. Physiol. Heart Circ. Physiol.* **2009**, *297*, H102–H108. [CrossRef] [PubMed]
34. Egorov, Y.V.; Kuz'min, V.S.; Glukhov, A.V.; Rosenshtraukh, L.V. Electrophysiological Characteristics, Rhythm, Disturbances and Conduction Discontinuities Under Autonomic Stimulation in the Rat Pulmonary Vein Myocardium. *J. Cardiovasc. Electrophysiol.* **2015**, *26*, 1130–1139. [CrossRef] [PubMed]
35. Tsuneoka, Y.; Kobayashi, Y.; Honda, Y.; Namekata, I.; Tanaka, H. Electrical Activity of the Mouse Pulmonary Vein Myocardium. *J. Pharmacol. Sci.* **2012**, *119*, 287–292. [CrossRef]
36. Namekata, I.; Tsuneoka, Y.; Takahara, A.; Shimada, H.; Sugimoto, T.; Takeda, K.; Nagaharu, M.; Shigenobu, K.; Kawanishi, T.; Tanaka, H. Involvement of the Na(+)/Ca(2+) Exchanger in the Automaticity of Guinea-Pig Pulmonary Vein Myocardium as Revealed by SEA0400. *J. Pharmacol. Sci.* **2009**, *110*, 111–116. [CrossRef]
37. Tsuneoka, Y.; Irie, M.; Tanaka, Y.; Sugimoto, T.; Kobayashi, Y.; Kusakabe, T.; Kato, K.; Hamaguchi, S.; Namekata, I.; Tanaka, H. Permissive Role of Reduced Inwardly-Rectifying Potassium Current Density in the Automaticity of the Guinea Pig Pulmonary Vein Myocardium. *J. Pharmacol. Sci.* **2017**, *133*, 195–202. [CrossRef]
38. Melnyk, P.; Ehrlich, J.R.; Pourrier, M.; Villeneuve, L.; Cha, T.-J.; Nattel, S. Comparison of Ion Channel Distribution and Expression in Cardiomyocytes of Canine Pulmonary Veins versus Left Atrium. *Cardiovasc. Res.* **2005**, *65*, 104–116. [CrossRef]
39. Egorov, Y.V.; Lang, D.; Tyan, L.; Turner, D.; Lim, E.; Piro, Z.D.; Hernandez, J.J.; Lodin, R.; Wang, R.; Schmuck, E.G.; et al. Caveolae-Mediated Activation of Mechanosensitive Chloride Channels in Pulmonary Veins Triggers Atrial Arrhythmogenesis. *J. Am. Heart Assoc.* **2019**, *8*, e012748. [CrossRef]
40. Okamoto, Y.; Kawamura, K.; Nakamura, Y.; Ono, K. Pathological Impact of Hyperpolarization-Activated Chloride Current Peculiar to Rat Pulmonary Vein Cardiomyocytes. *J. Mol. Cell. Cardiol.* **2014**, *66*, 53–62. [CrossRef]
41. Takagi, D.; Okamoto, Y.; Ohba, T.; Yamamoto, H.; Ono, K. Comparative Study of Hyperpolarization-Activated Currents in Pulmonary Vein Cardiomyocytes Isolated from Rat, Guinea Pig, and Rabbit. *J. Physiol. Sci. JPS* **2020**, *70*, 6. [CrossRef]
42. Cheung, D.W. Electrical Activity of the Pulmonary Vein and Its Interaction with the Right Atrium in the Guinea-Pig. *J. Physiol.* **1981**, *314*, 445–456. [CrossRef]
43. Miyauchi, Y.; Hayashi, H.; Miyauchi, M.; Okuyama, Y.; Mandel, W.J.; Chen, P.-S.; Karagueuzian, H.S. Heterogeneous Pulmonary Vein Myocardial Cell Repolarization Implications for Reentry and Triggered Activity. *Heart Rhythm* **2005**, *2*, 1339–1345. [CrossRef]
44. Bredeloux, P.; Findlay, I.; Pasqualin, C.; Hocini, M.; Bernus, O.; Maupoil, V. Selective Inhibition of Electrical Conduction within the Pulmonary Veins by A1-Adrenergic Receptors Activation in the Rat. *Sci. Rep.* **2020**, *10*, 5390. [CrossRef]

45. Ivanova, A.D.; Filatova, T.S.; Abramochkin, D.V.; Atkinson, A.; Dobrzynski, H.; Kokaeva, Z.G.; Merzlyak, E.M.; Pustovit, K.B.; Kuzmin, V.S. Attenuation of Inward Rectifier Potassium Current Contributes to the A1-Adrenergic Receptor-Induced Proarrhythmicity in the Caval Vein Myocardium. *Acta Physiol. Oxf. Engl.* **2021**, *231*, e13597. [CrossRef]
46. Fabiato, A.; Fabiato, F. Calcium and Cardiac Excitation-Contraction Coupling. *Annu. Rev. Physiol.* **1979**, *41*, 473–484. [CrossRef]
47. Chen, Y.-C.; Chen, S.-A.; Chen, Y.-J.; Tai, C.-T.; Chan, P.; Lin, C.-I. T-Type Calcium Current in Electrical Activity of Cardiomyocytes Isolated from Rabbit Pulmonary Vein. *J. Cardiovasc. Electrophysiol.* **2004**, *15*, 567–571. [CrossRef]
48. Malécot, C.O. Low Voltage-Activated Channels in Rat Pulmonary Vein Cardiomyocytes: Coexistence of a Non-Selective Cationic Channel and of T-Type Ca Channels. *Pflugers Arch.* **2020**, *472*, 1019–1029. [CrossRef]
49. Coutu, P.; Chartier, D.; Nattel, S. Comparison of Ca^{2+}-Handling Properties of Canine Pulmonary Vein and Left Atrial Cardiomyocytes. *Am. J. Physiol. Heart Circ. Physiol.* **2006**, *291*, H2290–H2300. [CrossRef]
50. Bootman, M.D.; Smyrnias, I.; Thul, R.; Coombes, S.; Roderick, H.L. Atrial Cardiomyocyte Calcium Signalling. *Biochim. Biophys. Acta BBA-Mol. Cell Res.* **2011**, *1813*, 922–934. [CrossRef]
51. Logantha, S.J.R.J.; Cruickshank, S.F.; Rowan, E.G.; Drummond, R.M. Spontaneous and Electrically Evoked Ca2+ Transients in Cardiomyocytes of the Rat Pulmonary Vein. *Cell Calcium* **2010**, *48*, 150–160. [CrossRef]
52. Patterson, E.; Po, S.S.; Scherlag, B.J.; Lazzara, R. Triggered Firing in Pulmonary Veins Initiated by in Vitro Autonomic Nerve Stimulation. *Heart Rhythm* **2005**, *2*, 624–631. [CrossRef]
53. Yamamoto, M.; Dobrzynski, H.; Tellez, J.; Niwa, R.; Billeter, R.; Honjo, H.; Kodama, I.; Boyett, M.R. Extended Atrial Conduction System Characterised by the Expression of the HCN4 Channel and Connexin45. *Cardiovasc. Res.* **2006**, *72*, 271–281. [CrossRef]
54. Jones, S.A.; Yamamoto, M.; Tellez, J.O.; Billeter, R.; Boyett, M.R.; Honjo, H.; Lancaster, M.K. Distinguishing Properties of Cells From the Myocardial Sleeves of the Pulmonary Veins. *Circ. Arrhythm. Electrophysiol.* **2008**, *1*, 39–48. [CrossRef]
55. Chang, S.-L.; Chen, Y.-C.; Chen, Y.-J.; Wangcharoen, W.; Lee, S.-H.; Lin, C.-I.; Chen, S.-A. Mechanoelectrical Feedback Regulates the Arrhythmogenic Activity of Pulmonary Veins. *Heart Br. Card. Soc.* **2007**, *93*, 82–88. [CrossRef]
56. Hamaguchi, S.; Hikita, K.; Tanaka, Y.; Tsuneoka, Y.; Namekata, I.; Tanaka, H. Enhancement of Automaticity by Mechanical Stretch of the Isolated Guinea Pig Pulmonary Vein Myocardium. *Biol. Pharm. Bull.* **2016**, *39*, 1216–1219. [CrossRef]
57. Patterson, E.; Jackman, W.M.; Beckman, K.J.; Lazzara, R.; Lockwood, D.; Scherlag, B.J.; Wu, R.; Po, S. Spontaneous Pulmonary Vein Firing in Man: Relationship to Tachycardia-Pause Early Afterdepolarizations and Triggered Arrhythmia in Canine Pulmonary Veins in Vitro. *J. Cardiovasc. Electrophysiol.* **2007**, *18*, 1067–1075. [CrossRef]
58. Patterson, E.; Lazzara, R.; Szabo, B.; Liu, H.; Tang, D.; Li, Y.-H.; Scherlag, B.J.; Po, S.S. Sodium-Calcium Exchange Initiated by the Ca^{2+} Transient: An Arrhythmia Trigger within Pulmonary Veins. *J. Am. Coll. Cardiol.* **2006**, *47*, 1196–1206. [CrossRef]
59. Honjo, H.; Boyett, M.R.; Niwa, R.; Inada, S.; Yamamoto, M.; Mitsui, K.; Horiuchi, T.; Shibata, N.; Kamiya, K.; Kodama, I. Pacing-Induced Spontaneous Activity in Myocardial Sleeves of Pulmonary Veins after Treatment with Ryanodine. *Circulation* **2003**, *107*, 1937–1943. [CrossRef]
60. Takahara, A.; Sugimoto, T.; Kitamura, T.; Takeda, K.; Tsuneoka, Y.; Namekata, I.; Tanaka, H. Electrophysiological and Pharmacological Characteristics of Triggered Activity Elicited in Guinea-Pig Pulmonary Vein Myocardium. *J. Pharmacol. Sci.* **2011**, *115*, 176–181. [CrossRef]
61. Iwasaki, Y.-K.; Kato, T.; Xiong, F.; Shi, Y.-F.; Naud, P.; Maguy, A.; Mizuno, K.; Tardif, J.-C.; Comtois, P.; Nattel, S. Atrial Fibrillation Promotion with Long-Term Repetitive Obstructive Sleep Apnea in a Rat Model. *J. Am. Coll. Cardiol.* **2014**, *64*, 2013–2023. [CrossRef] [PubMed]
62. Iwasaki, Y.-K.; Yamashita, T.; Sekiguchi, A.; Hayami, N.; Shimizu, W. Importance of Pulmonary Vein Preferential Fibrosis for Atrial Fibrillation Promotion in Hypertensive Rat Hearts. *Can. J. Cardiol.* **2016**, *32*, 767–776. [CrossRef] [PubMed]
63. Dai, H.; Wang, X.; Yin, S.; Zhang, Y.; Han, Y.; Yang, N.; Xu, J.; Sun, L.; Yuan, Y.; Sheng, L.; et al. Atrial Fibrillation Promotion in a Rat Model of Rheumatoid Arthritis. *J. Am. Heart Assoc.* **2017**, *6*, e007320. [CrossRef] [PubMed]
64. Ye, T.; Liu, X.; Qu, C.; Zhang, C.; Fo, Y.; Guo, Y.; Chen, X.; Shi, S.; Yang, B. Chronic Inhibition of the Sigma-1 Receptor Exacerbates Atrial Fibrillation Susceptibility in Rats by Promoting Atrial Remodeling. *Life Sci.* **2019**, *235*, 116837. [CrossRef]
65. Stavrakis, S.; Po, S. Ganglionated Plexi Ablation: Physiology and Clinical Applications. *Arrhythmia Electrophysiol. Rev.* **2017**, *6*, 186–190. [CrossRef]
66. Kampaktsis, P.N.; Oikonomou, E.K.; Choi, D.Y.; Cheung, J.W. Efficacy of Ganglionated Plexi Ablation in Addition to Pulmonary Vein Isolation for Paroxysmal versus Persistent Atrial Fibrillation: A Meta-Analysis of Randomized Controlled Clinical Trials. *J. Interv. Card. Electrophysiol. Int. J. Arrhythm. Pacing* **2017**, *50*, 253–260. [CrossRef]
67. Hanna, P.; Dacey, M.J.; Brennan, J.; Moss, A.; Robbins, S.; Achanta, S.; Biscola, N.P.; Swid, M.A.; Rajendran, P.S.; Mori, S.; et al. Innervation and Neuronal Control of the Mammalian Sinoatrial Node a Comprehensive Atlas. *Circ. Res.* **2021**, *128*, 1279–1296. [CrossRef]
68. Maupoil, V.; Bronquard, C.; Freslon, J.-L.; Cosnay, P.; Findlay, I. Ectopic Activity in the Rat Pulmonary Vein Can Arise from Simultaneous Activation of Alpha1- and Beta1-Adrenoceptors. *Br. J. Pharmacol.* **2007**, *150*, 899–905. [CrossRef]
69. Egorov, Y.V.; Rosenshtraukh, L.V.; Glukhov, A.V. Arrhythmogenic Interaction Between Sympathetic Tone and Mechanical Stretch in Rat Pulmonary Vein Myocardium. *Front. Physiol.* **2020**, *11*, 237. [CrossRef]
70. Umehara, S.; Tan, X.; Okamoto, Y.; Ono, K.; Noma, A.; Amano, A.; Himeno, Y. Mechanisms Underlying Spontaneous Action Potential Generation Induced by Catecholamine in Pulmonary Vein Cardiomyocytes: A Simulation Study. *Int. J. Mol. Sci.* **2019**, *20*, E2913. [CrossRef]

71. Irie, M.; Tsuneoka, Y.; Shimobayashi, M.; Hasegawa, N.; Tanaka, Y.; Mochizuki, S.; Ichige, S.; Hamaguchi, S.; Namekata, I.; Tanaka, H. Involvement of Alpha- and Beta-Adrenoceptors in the Automaticity of the Isolated Guinea Pig Pulmonary Vein Myocardium. *J. Pharmacol. Sci.* **2017**, *133*, 247–253. [CrossRef]
72. Ivanova, A.D.; Kuzmin, V.S. Electrophysiological Characteristics of the Rat Azygos Vein under Electrical Pacing and Adrenergic Stimulation. *J. Physiol. Sci. JPS* **2018**, *68*, 617–628. [CrossRef]

biomolecules

MDPI

Review

Molecular, Subcellular, and Arrhythmogenic Mechanisms in Genetic RyR2 Disease

Ewan Douglas Fowler [1] and Spyros Zissimopoulos [2,*]

1 School of Biosciences, College of Biomedical and Life Sciences, Cardiff University, Cardiff CF10 3AX, UK; fowlered@cardiff.ac.uk
2 Institute of Life Science, Swansea University Medical School, Swansea University, Swansea SA2 8PP, UK
* Correspondence: spyros.zissimopoulos@swansea.ac.uk

Abstract: The ryanodine receptor (RyR2) has a critical role in controlling Ca^{2+} release from the sarcoplasmic reticulum (SR) throughout the cardiac cycle. RyR2 protein has multiple functional domains with specific roles, and four of these RyR2 protomers are required to form the quaternary structure that comprises the functional channel. Numerous mutations in the gene encoding RyR2 protein have been identified and many are linked to a wide spectrum of arrhythmic heart disease. Gain of function mutations (GoF) result in a hyperactive channel that causes excessive spontaneous SR Ca^{2+} release. This is the predominant cause of the inherited syndrome catecholaminergic polymorphic ventricular tachycardia (CPVT). Recently, rare hypoactive loss of function (LoF) mutations have been identified that produce atypical effects on cardiac Ca^{2+} handling that has been termed calcium release deficiency syndrome (CRDS). Aberrant Ca^{2+} release resulting from both GoF and LoF mutations can result in arrhythmias through the Na^+/Ca^{2+} exchange mechanism. This mini-review discusses recent findings regarding the role of RyR2 domains and endogenous regulators that influence RyR2 gating normally and with GoF/LoF mutations. The arrhythmogenic consequences of GoF/LoF mutations will then be discussed at the macromolecular and cellular level.

Keywords: ryanodine receptor; catecholaminergic polymorphic ventricular tachycardia; calcium release deficiency syndrome; delayed afterdepolarizations; early afterdepolarizations; calcium sparks; long QT syndrome; arrhythmias

Citation: Fowler, E.D.; Zissimopoulos, S. Molecular, Subcellular, and Arrhythmogenic Mechanisms in Genetic RyR2 Disease. *Biomolecules* **2022**, *12*, 1030. https://doi.org/10.3390/biom12081030

Academic Editors: Yosuke Okamoto and Kyoichi Ono

Received: 4 July 2022
Accepted: 24 July 2022
Published: 26 July 2022

1. Introduction

Cardiac contraction is initiated by a wave of electrical activity (action potential) triggering a rapid rise in intracellular Ca^{2+} concentration, which activates sarcomere contraction, followed by repolarization and Ca^{2+} removal that allows the heart to relax and refill [1]. Gene mutations in any of the proteins involved in excitation–contraction coupling (ECC) can lead to arrhythmogenic cardiac disease, most commonly, long QT and Brugada syndromes (repolarization abnormalities), catecholaminergic polymorphic ventricular tachycardia (CPVT) (dysregulated intracellular Ca^{2+}), and dilated and hypertrophic cardiomyopathies (sarcomere dysfunction) [2]. Cardiomyocyte Ca^{2+} homeostasis is indeed an arrhythmogenic substrate [3]. For contraction of the adult human heart, plasma membrane depolarization induces Ca^{2+} influx via voltage-gated Ca^{2+} channels (also known as L-type Ca^{2+} channel or dihydropyridine receptor) located on invaginations of the sarcolemma called transverse (T)-tubules. This by itself is insufficient to trigger sarcomere contraction, but the rapid rise in Ca^{2+} activates ryanodine receptors (RyR2) located on the immediately opposing junctional sarcoplasmic reticulum (SR) through which a much larger Ca^{2+} release occurs in the form of Ca^{2+} sparks [4]. The near-synchronous activation of ~10^4 Ca^{2+} sparks at the start of the action potential produces the cytosolic Ca^{2+} transient that activates the contractile apparatus within cardiac myocytes. For relaxation, cytosolic Ca^{2+} concentration returns to baseline with Ca^{2+} extrusion outside of the cell via the Na^+/Ca^{2+} exchanger

(NCX) to remove an amount equal to the extracellular Ca^{2+} entry source, and Ca^{2+} uptake into the SR via the sarco/endoplasmic reticulum Ca^{2+} ATPase (SERCA) to remove an amount equal to the SR Ca^{2+} release source. Imbalances in cardiomyocyte Ca^{2+} cycling, namely the Ca^{2+} fluxes across the plasma and SR membrane, can alter the action potential and lead to arrhythmia.

The role of RyR2 in cardiac pathophysiology came to prominence more than 20 years ago. The first report implicating acquired RyR2 dysfunction in heart failure was published in 2000 [5], whereas a year later came the first studies associating mutations in the *RYR2* gene with catecholaminergic polymorphic ventricular tachycardia (CPVT) and arrhythmogenic right ventricular dysplasia type 2 (ARVD2) [6–8]. Recent advances in our knowledge of the molecular and arrhythmogenic mechanisms in genetic RyR2 disease are explored in detail in this mini-review.

2. RyR2 Structure—Function Relationships

The RyR is the largest known ion channel, consisting of four identical subunits of ~5000 amino acids each, which combine to form a single Ca^{2+} permeable pore. Three RyR isoforms are expressed in mammals, RyR1 (the predominant isotype in skeletal muscle), RyR2 (the predominant isotype in cardiac muscle), and RyR3. Considering its size, complexity, and responsibility for achieving both short-(beat-to-beat) and long-term maintenance of heart contraction, it is not surprising that many factors fine-tune RyR2 function. In the following, we aim to describe determinants of RyR2 regulation as interacting layers consisting of: the pore-forming domains, domains that directly interact with the pore, domains that indirectly regulate the pore through intermediaries, or interacting regulatory proteins.

Recent high-resolution cryo-electron microscopy (EM) studies reveal that the long RyR polypeptide chain is organized in distinct structural domains that are predominantly α-helical and face the cytoplasm [9,10] (Figure 1). They include the N-terminus domain (NTD) composed of sub-domains A/B/C, three SPRY domains including RyR repeat (RY) 1&2, junctional solenoid (JSol), bridging solenoid (BSol) consisting of sub-domains B&C flanking RY3&4, and core solenoid (CSol). According to a different nomenclature by Peng et al. [10], JSol is known as the "handle domain", BSol is known as the "helical domain" but it was previously referred to as the "central domain" [11], and CSol is also known as the "central domain". To avoid confusion with "central domain" terminology, we adopt the nomenclature by des Georges et al. [9]. The C-terminal ~500 residues contain the transmembrane domain (TMD) composed of six segments that anchor the protein on the SR membrane, followed by the short cytosolic C-terminus domain (CTD). The pore is formed by the S6 lining helix supported by S5 and the S5-S6 loop containing the selectivity filter.

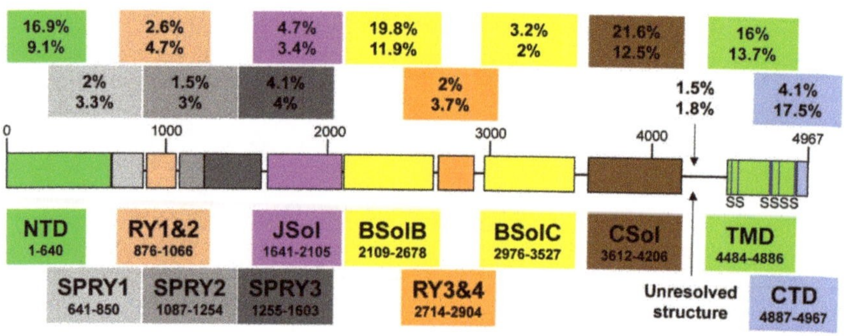

Figure 1. A schematic illustration of human RyR2 domain organization and boundaries. Color-coded distribution of mutations associated with disease (top value) and their frequency within individual structural domains (bottom value). The six transmembrane segments (S) are represented by a vertical line. See Figure 2 below for a corresponding cartoon representation showing the arrangement of domains in 3D (using same color coding).

RyR2 Channel Self-Assembly

Although RyR2 is a macromolecular complex with cytosolic, SR integral, and luminal accessory proteins (see Section 3.1), the 4-subunit RyR2 self-assembly is necessary and sufficient to form a fully functional Ca^{2+} release channel. To understand how RyR2 is auto-regulated, we consider a reductionist, "bottom-up" approach starting from the pore. Expression of the last two TM segments together with the CTD forms a tetrameric cation permeable pore that is constitutively open [12]. The CTD self-associates to support RyR tetramerization required for the formation of functional channels [13,14]. On top of the pore sits the CSol acting as the channels' gatekeeper. The CSol makes extensive contacts with the pore-forming region including the S2S3 loop, the extended cytosolic part of S6, and the CTD [9,10]. However, the expression of large constructs encompassing the CSol, TMD, and CTD form ion channels that remain largely open and leaky [15–17]. Clearly, the remaining domains are needed to support the closure of the pore and fully regulate the channel. They are not making direct contact with the pore-forming region, but their effects are mediated via the CSol. Indeed, the NTD, Jsol, Bsol, and CSol are involved in multiple intra- and inter-subunit interactions with each other [9,10,18–20]. The NTD also interacts with itself to promote channel closure as well as to support the formation of functional tetrameric RyR2 channels [21–23]. RyR2 activity is fine-tuned by the peripheral SPRY and RY domains, and their interactions with JSol and BSol [9,10].

3. Physiological RyR2 Regulators

The principal physiological RyR2 regulator is Ca^{2+} acting from both the cytosolic and SR luminal side. RyR2 is mostly inactive at low diastolic (~100 nM) cytosolic Ca^{2+} concentration ([Ca^{2+}]). Membrane depolarization causes a rapid rise in [Ca^{2+}] from nanomolar to micromolar levels in the narrow dyadic cleft formed between T-tubule and SR junctional membrane, which increases RyR2 activity in a sigmoidal function of cytosolic Ca^{2+} concentration. RyR2 can be fully activated at ~100 μM Ca^{2+} (reviewed in [24,25]). Recent cryo-EM and mutagenesis studies suggest that the cytosolic Ca^{2+}-binding site is located at the CSol–CTD interface with residues E3848, E3922, and T4931 (coordinates for human RyR2) critical for Ca^{2+} interaction [9,26–28]. Interestingly, although the Ca^{2+}-bound RyR structure reveals a contraction of the CSol that facilitates the dilation of the S6 bundle, the pore remains closed [9,26]. This could be due to the in vitro experimental conditions of cryo-EM. RyR2 is also regulated by SR luminal Ca^{2+} (reviewed in [24,25]). Although luminal Ca^{2+} can indirectly activate RyR2 by acting on the cytosolic Ca^{2+}-binding site following passage through the pore, a separate site accessible only to luminal Ca^{2+} is likely to exist. Mutagenesis analysis has indicated that residue E4872 within S6 is necessary for luminal Ca^{2+} activation without affecting cytosolic Ca^{2+} modulation [29].

3.1. Regulation of RyR2 by Accessory Proteins

RyR2 is also regulated by accessory proteins and post-translational modifications (phosphorylation and oxidation) (reviewed in [25,30]). Here, we focus on three RyR2-associated proteins, FK506-binding protein (FKBP), calmodulin (CaM), and calsequestrin (CSQ2) because of their implication in genetic cardiac disease.

3.1.1. FKBP12/12.6

In native cardiac tissue, both of the two FKBP homologues of 12 and 12.6 kDa are found associated with RyR2 depending on the animal species [31]. RyR2 affinity is much greater for FKBP12.6 than FKBP12; however, native RyR2 occupancy by FKBP12.6 is minimal due to the latter being expressed at very low levels [31–33]. Functional studies generally suggest that FKBP12.6 promotes RyR2 channel closure and prevents SR Ca^{2+} leak (reviewed in [34]). Cryo-EM RyR structures in complex with FKBP12/12.6 indicate that the binding site is composed of multiple determinants contributed by the SPRY1, SPRY3, NTD, and JSol domains [26,35,36]. FKBP12.6 stabilizes RyR2 in a closed state by inducing the relaxation of the CSol and strengthening the NTD-NTD inter-subunit interaction [26].

3.1.2. Calmodulin (CaM)

CaM is a high-affinity, Ca^{2+}-binding, dumbbell-shaped protein containing four EF-hand motifs, two of each on both N and C lobes. Both the Ca^{2+} free (apo-CaM) and Ca^{2+} bound forms of CaM interact with RyR2 to inhibit channel activity, although inhibition is more pronounced at high Ca^{2+} [37,38]. The cryo-EM RyR2 structures in complex with apo-CaM and Ca^{2+}-CaM indicate that they bind to distinct sites [39]. The N-lobe of apo-CaM binds BSolB, whereas its C-lobe binds at the interface between JSol and CSol. Ca^{2+}-CaM shifts position and interacts only with the CSol. Ca^{2+}-CaM stabilizes RyR2 in a closed state by inducing conformational changes in the CSol to release the pulling force for the dilation of the S6 bundle. Notably, CaM mutations have been linked with arrhythmia and sudden cardiac death, with most cases diagnosed as long QT syndrome and CPVT (reviewed in [40,41]). Mutations in all three *CALM* genes, which produce identical CaM peptide sequences, have been reported. Most of the disease-associated CaM mutations studied to date seem to reduce CaM's inhibitory action on RyR2.

3.1.3. Calsequestrin-2 (CSQ2)

CSQ2 is a low-affinity, high-capacity Ca^{2+}-binding protein of ~45 kDa located within the SR lumen, where it buffers the free Ca^{2+} concentration to ~1 mM. It is associated with RyR2 via the SR integral proteins triadin and junctin, as well as directly interacting with the S1S2 luminal loop of RyR2 [42,43]. It is generally believed that CSQ2 exerts an inhibitory effect on RyR2 Ca^{2+} release (reviewed in [44]). CSQ2 knockout mouse models present with stress-induced arrhythmia indicating that CSQ2 plays an important role in the cardiac function [45,46]. Indeed, *CASQ2* is the second most mutated gene in CPVT accounting for ~5% of cases [41]. Initially described as the recessive form of CPVT (CPVT2), autosomal dominant mutations have also been reported [47–49]. CPVT-linked CSQ2 mutations may result in impaired Ca^{2+} buffering, multimer formation and/or RyR2 regulation [41,44,50].

To summarize, we propose a 4-layer model for cytosolic regulation of the RyR2 pore (Figure 2). The first layer directly blanketing the pore consists of the CSol. The second layer consisting of NTD, Jsol, and BSol transmits stimulatory/inhibitory signals to the pore via the CSol. The third layer consists of the SPRY and RY domains acting on the second layer. The final fourth layer consists of cytoplasmic ligands, namely accessory proteins and protein kinases. While small ligands such as Ca^{2+} may have access to the pore-forming region, the large size of proteins precludes them from doing so. The effects of cytoplasmic proteins feed into layers 1/2/3 to alter their conformation and RyR2 inter-domain interactions. For example, FKBP12.6 binding (to SPRY1, SPRY3, NTD, and JSol domains) directly affects layers 2&3, which in turn transmit an allosteric signal to the pore via layer 1. The effect of phosphorylation (mediated by protein kinases) also feeds into layers 1/2/3. S2808 and S2814, whose phosphorylation is implicated in RyR2 physiology and disease are located within RY3&4. The luminal regulation of the RyR2 pore is simpler. The effects of luminal ligands can be direct to the pore (e.g., luminal Ca^{2+}) or via the small RyR2 luminal loops.

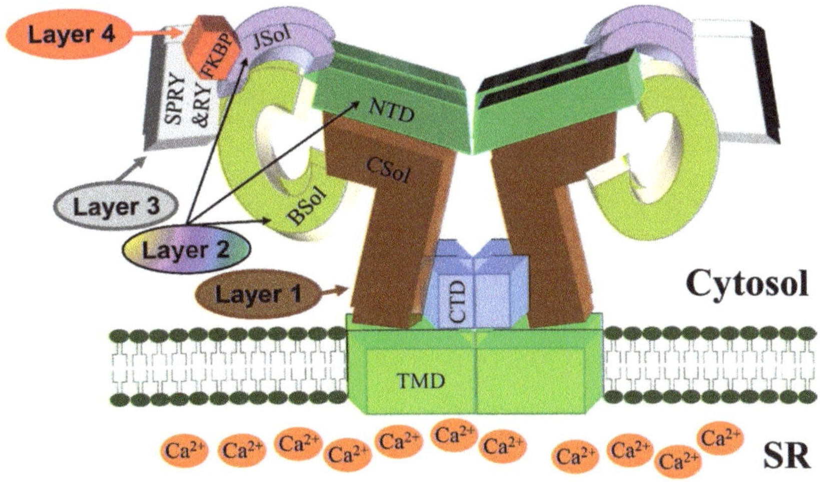

Figure 2. A schematic representation of the RyR2 3D architecture depicting 4 regulatory layers from the cytoplasmic side. The pore-forming region, comprised of the TMD and CTD, is modulated by the CSol (Layer 1), NTD, JSol and BSol (Layer 2), SPRY1/2/3 and RY1/2/3/4 (Layer 3), and accessory proteins (Layer 4), e.g., FKBP. The colors of regions in the RyR2 3D structure correspond to the colors in the RyR2 sequence in Figure 1.

4. Cardiac Disease Associated with RyR2 Mutations

RyR2 mutations transmitted in an autosomal dominant manner account for ~70% of CPVT cases (CPVT1) [41,50]. CPVT is characterized by episodes of potentially lethal bidirectional or polymorphic ventricular tachycardia triggered by emotional or physical stress. CPVT is manifested in the absence of structural abnormalities, while electrocardiogram (ECG) recordings are normal at rest. It is estimated to affect 1:10,000 mostly young, seemingly healthy individuals. According to European Society of Cardiology guidelines [51], the first-line therapy for CPVT is beta-blockers to suppress the catecholamine stimulation of the β-adrenergic pathway. It was very recently reported that non-selective β-blockers (nadolol, propranolol) were more efficient at reducing the risk of life-threatening arrhythmic events compared with selective β-blockers [52]. If beta-blocker therapy is ineffective, additional treatment with flecainide is recommended. Flecainide is a Na^+ channel inhibitor that may also inhibit RyR2 Ca^{2+} release [53] (see also [54,55]). Left cardiac sympathetic denervation is considered for patients who continue to experience arrhythmia despite optimal medication [51]. The implantable cardioverter defibrillator can be considered the last line of defense for patients who have survived a cardiac arrest or those who do not respond to optimal medication. However, its use is debated due to questionable efficacy and inappropriate shocks that may have a pro-arrhythmic effect [50].

Mutations in the *RYR2* gene are also causative of ARVD2 [8,56]. The clinical presentation of ARVD2 is very similar to CPVT including stress-induced ventricular tachycardia, but it is also associated with progressive degeneration and fibro-fatty replacement of the right ventricle. In addition, RyR2 mutations have also been implicated in non-stress-induced cardiac disease including long QT syndrome [57,58], torsade de pointes [59,60], left ventricular noncompaction [61,62], dilated cardiomyopathy [63,64], and hypertrophic cardiomyopathy [65]. Very recently, a new disease termed RyR2 Ca^{2+} release deficiency syndrome (CRDS) was described, underlined by RyR2 loss of function (LoF) mutations [66,67]. CRDS is characterized by ventricular arrhythmias and sudden cardiac death but a negative exercise stress testing for CPVT. A novel classification for genetic RyR2-related cardiac disease was recently proposed as a typical and atypical CPVT [41]. Typical CPVT as described above is due to CSQ2 and gain of function (GoF) RyR2 mutations. Atypical CPVT covers non-stress as well as stress-induced disease other than bidirectional or polymorphic ventricular tachycardia, due to RyR2 LoF or mutations in other genes such as *CALM*.

5. Molecular Mechanisms of Genetic RyR2 Disease

To date, more than 350 mutations have been reported including splicing, deletions, insertions, and nonsense mutations, but the vast majority are missense (for an up-to-date list, see the Human Gene Mutation Database (HGMD®), http://www.hgmd.cf.ac.uk/ac/gene.php?gene=RYR2) (accessed on 25 July 2022). Pending further characterization, most of these mutations are currently classified as variants of unknown significance [68]. Mutations are found throughout the RyR2 peptide sequence but tend to concentrate on five structural domains, NTD (16.9%), BSolB (19.8%), CSol (21.6%), TMD (16%), and CTD (4.1%) (Figure 1 and Table 1). This is more evident once the amino acid length of each domain is taken into consideration. If we consider how many amino acid variants have been reported within each individual domain, NTD (9.1%), BSolB (11.9%), CSol (12.5%), TMD (13.7%), and CTD (17.5%), the small CTD is, in fact, the most frequently mutated structural element. In terms of disease diagnosis, there is no apparent correlation between the location of mutations on the RyR2 structure and the clinical phenotype presenting as typical or atypical CPVT. On the other hand, the few deletions, insertions, and nonsense mutations reported to date are often associated with atypical CPVT or cardiomyopathy irrespective of their location on the RyR2 structure (e.g., [61,63,67,69]).

Table 1. RyR2 domain distribution of disease-linked mutations.

Domain	Coordinates (Amino Acids)	Length (Amino Acids)	Number of Mutations	Mutation Frequency Overall (%)	Mutation Frequency within Domain (%)
NTD	1–640	640	58	16.9	9.1
SPRY1	641–850	209	7	2.0	3.3
RY1&2	876–1066	190	9	2.6	4.7
SPRY2	1087–1254	167	5	1.5	3.0
SPRY3	1255–1603	348	14	4.1	4.0
JSol	1641–2105	464	16	4.7	3.4
BSolB	2109–2678	569	68	19.8	11.9
RY3&4	2714–2904	190	7	2.0	3.7
BSolC	2976–3527	551	11	3.2	2.0
CSol	3612–4206	594	74	21.6	12.5
Unresolved	4207–4483	276	5	1.5	1.8
TMD	4484–4886	402	55	16.0	13.7
CTD	4887–4967	80	14	4.1	17.5

Most pathological RyR2 missense mutations are GoF and causative of typical CPVT. They may show increased sensitivity to either the cytosolic [70,71] or luminal Ca^{2+} activation [72]. It has also been suggested that the GoF phenotype is due to reduced FKBP12.6 affinity (unmasked following RyR2 phosphorylation) [73] (see also [74,75]). Similarly, reduced CaM affinity has also been proposed to contribute to the GoF phenotype [76]. Moreover, altered RyR2 inter-domain interactions have been shown to underlie the GoF phenotype [19,20,22,77].

Based on their observations, three labs have put forward their own hypotheses to explain how RyR2 GoF mutations result in aberrant cardiomyocyte Ca^{2+} cycling. The first hypothesis by Marks and colleagues is that mutations impair the binding of FKBP12.6 which helps maintain the channel closed in the diastolic phase [73]. The second hypothesis by Yano and colleagues is that mutations impair the inter-domain interaction between the NTD and BSolB to destabilize the closed conformation of the channel [20]. The third hypothesis by Chen and colleagues is that mutations enhance the sensitivity to SR luminal Ca^{2+} thereby increasing the propensity for arrhythmogenic spontaneous Ca^{2+} waves [72]. However, these three hypotheses can be refined to include additional observations. The first hypothesis can be expanded to include other accessory proteins such as CaM [76]. The second hypothesis can be generalized to include altered RyR2 intra- and inter-subunit interactions, such as the JSol-BSolB interaction [19], NTD self-association [22], and NTD-CSol interaction [77]. The third hypothesis can be expanded to include mutations that enhance the sensitivity to cytosolic Ca^{2+} [70,71]. In addition, these three hypotheses are not mutually exclusive, with individual mutations potentially operating via different mechanisms. Moreover, the same mutation may affect more than one mechanism. For example, the R2474S is found to disrupt the NTD-BSolB interaction [20], which in turn impairs the RyR2 interaction with CaM [76].

Unlike GoF mutations, LoF mutations are associated with CRDS and other forms of atypical CPVT. Initially only a handful, >20 have been reported in the last couple of years as the focus has shifted to patients with atypical CPVT. Most of them are missense mutations, but for deletion and a large duplication/insertion, found within the CSol and TMD (Table 2). A common defect of LoF mutations is the perturbation of cytosolic and/or luminal Ca^{2+} activation [66,78]. Alternatively, they may affect the CSol-TMD interaction or the Ca^{2+}-conducting properties of the pore. Other LoF mutations (e.g., those occurring within the CTD) may impair RyR2 tetramerization that is necessary for the formation of a functional channel, as reported for L433P (which has both GoF and LoF characteristics) [22].

Table 2. RyR2 loss-of-function mutations.

Mutation	Domain	Disease	References
Duplication [1]	-	CRDS	[67]
L433P [2]	NTD	ARVD2	[22]
G570D	NTD	CRDS	[79]
Q2275H	BSolB	CRDS	[80]
Q3774L	CSol	CRDS	[66]
Q3925E	CSol	CRDS	[79]
I3995V	CSol	CRDS	[66]
M4109R	CSol	CRDS	[79]
D4112N	CSol	CRDS	[66]
A4142T	CSol	CRDS	[81]
E4146D	CSol	long QT syndrome	[82]
E4146K	CSol	CRDS	[83]
R4147K	CSol	CRDS	[79]
S4168P	CSol	long QT syndrome	[82]
T4196I	CSol	CRDS	[66]
A4203V	CSol	CRDS	[79]
A4204V	CSol	CRDS	[79]
E4451del	Unresolved	CRDS	[80]
F4499C	TMD	CRDS	[80]

Table 2. *Cont.*

Mutation	Domain	Disease	References
K4594Q	TMD	long QT syndrome	[82]
V4606E	TMD	CRDS	[80]
R4608Q	TMD	CRDS	[80]
R4608W	TMD	CRDS	[80]
D4646A	TMD	CRDS	[66]
I4855M	TMD	left ventricular noncompaction	[62]
A4860G [3]	TMD	CRDS	[66,78]
Q4879H	TMD	CRDS	[66]
G4935R	TMD	CRDS	[83]
S4938F	TMD	torsade de pointes	[59]

Notes: [1] Homozygous duplication of the promoter region and exons 1–4. [2] L433P has both GoF and LoF characteristics. [3] Idiopathic ventricular fibrillation that was reclassified as CRDS.

Thus, mutations can be either GoF or LoF and result in typical or atypical CPVT, respectively, irrespective of their location on the RyR2 structure. We propose the following four non-mutually exclusive molecular mechanisms for channel deregulation in genetic RyR2 disease. The first mechanism involves altered cytosolic and/or luminal Ca^{2+} activation, which may account for both GoF (e.g., R4497C [70] and N4104K [72] increase cytosolic and luminal Ca^{2+} sensitivity, respectively) and LoF mutations (e.g., Q3774L [66] and A4860G [78] decrease cytosolic and luminal Ca^{2+} sensitivity, respectively). The second mechanism involves altered RyR2 intra- and inter-subunit interactions. For example, if the domain X interacts with domain Y to stabilize the closed channel, a mutation that disrupts the inter-domain X-Y interaction will be GoF (e.g., R2474S [20], R176Q [18]), whereas a mutation that enhances it will be LoF. On the other hand, if domain X interacts with domain Y to stabilize the open channel, a mutation that disrupts the inter-domain X-Y interaction will be LoF, whereas a mutation that enhances it will be GoF (e.g., R420Q [77]). The third mechanism involves altered interactions with accessory proteins (e.g., S2246L reducing FKBP12.6 binding [73]) as well as phosphorylation by protein kinases, which in turn will affect the second or third mechanisms. Finally, the fourth mechanism involves altered Ca^{2+} conductance and/or permeability, which may account for LoF mutations within the pore-forming S6 helix and the selectivity filter.

6. RyR2 Subcellular Geometry Influences Cardiac Ca^{2+} Release

6.1. *RyR2 Clusters and Functional Ca^{2+} Release Units (CRU)*

Early investigations using electron microscopy established that square-shape RyR2 tetramers (~30 × 30 nm) organize as discrete clusters in a quatrefoil crystalline arrangement on junctional SR membrane [84]. Purified RyR2-FKBP12.6 complexes can self-assemble *in vitro* into oblique multi channel structures that appear to involve SPRY1 and FKBP12.6 interactions between neighboring clamp regions [85]. Recent advances in super-resolution optical microscopy have further enabled the detection and characterization of RYR2 organization at near single molecule level in both 2D [86,87] and 3D [88], which was previously not possible due to diffraction limitations of conventional confocal microscopes. These recent investigations in rodent, sheep, and human myocytes have shown that RyR2 clusters typically consist of ~15–20 channels that can be sparsely packed [86,88–90]. This is much smaller than previous estimates of ~100 RyR2 per cluster, calculated from thin sectioning that relied on assumptions of complete filling and circular geometry [91]. An emergent property of these clusters is the apparent allosteric regulation between channels, through some form of coupled gating or cross-talk, that enhances the fidelity of Ca^{2+} spark activation when $[Ca^{2+}]$ is increased, e.g., evoked by an action potential, but suppresses RyR2 opening at diastolic $[Ca^{2+}]$ [92]. Ca^{2+} spark amplitude was found to be fairly insensitive to cluster sizes greater than ~20 RyR2 (quantified indirectly using fluorescently-labeled FKBP12.6) [93]. Although larger clusters should increase the peak Ca^{2+} flux rate in proportion to the number of open channels times the trans-SR $[Ca^{2+}]$ gradient, this will more rapidly deplete junctional SR $[Ca^{2+}]$, thus reducing the driving force for Ca^{2+} release and cause a gradual loss of regenerative capacity and finally Ca^{2+} spark termination [94].

Ca^{2+} concentration decreases rapidly as it diffuses further from its source, therefore the probability of one cluster causing Ca^{2+}-induced Ca^{2+} release (CICR) in distant clusters falls sharply with distance, and this limits uncontrolled activation. However, if two or more clusters are sufficiently close then the increase in cytosolic $[Ca^{2+}]$ can be enough to activate its neighbor, such that closely spaced RyR2 clusters (~100–150 nm edge-to-edge) are predicted to act together as functional Ca^{2+} release units (CRU) in the formation of Ca^{2+} sparks [86,92]. Ca^{2+} spark frequency showed a strong positive relationship with cluster size, indicating that larger clusters are more likely to initiate Ca^{2+} sparks [93] and support the notion that multiple clusters are involved in producing Ca^{2+} sparks.

6.2. Pathological Fragmentation of CRU

The break-up and fragmentation of CRU into smaller, dislocated clusters could contribute to the slower Ca^{2+} release kinetics and increased Ca^{2+} leak in inherited or acquired heart disease. Using stimulated emission-depletion (STED) microscopy, Macquaide et al. found that in a sheep model of chronic atrial fibrillation (AF) the mean distance between cluster centroids was shorter, without a significant change in the number of RyR2 per cluster, compared to healthy controls [89]. Consequently, more clusters fell within the CRU boundary (defined here as $\leq$150 nm) which increased the mean CRU area, but at a lower overall RyR2 density, suggesting greater fragmentation of CRU in this sheep model of AF. Reduced RyR2 density was also found in CRU from rats with right ventricular failure, using expansion microscopy combined with Airyscan super-resolution imaging [95]. Sheard et al. used geometrically realistic experimental maps of CRU to model the effect of cluster morphology on Ca^{2+} release and found the disrupted CRU often failed to fully activate within the typical time-course of a Ca^{2+} spark (~20 ms) [95]. The mechanism for CRU breakup remains to be established, but could include loss of accessory proteins tethering RyR2 clusters together, or simply CRU failing to keep up with cellular remodeling (e.g., hypertrophy in the case of heart failure and possibly RyR2 mutations [65]). Yin et al. used RyR2 immunolabeling and STED microscopy to investigate cluster morphology in mice with the CPVT-causing RyR2-R420Q$^{+/-}$ missense mutation. RyR2 clusters were smaller on average in knock-in than WT controls, despite total RyR2 protein being similar between groups [77]. Junctophilin-2 may play a role in determining RyR2 clustering because overexpression increased the size of clusters [96]. Yin et al. found a reduced association between junctophilin-2 and RyR2 in RyR2-R420Q$^{+/-}$ knock-in mice measured by co-immunoprecipitation, suggesting this could contribute to the redistribution of cluster size that results from this mutation [77]. In contrast to work in the experimental sheep heart failure [89], Munro et al. found RyR2 cluster and CRU properties to be unaltered in atrial myocytes from human patients with persistent AF versus patients without AF [90]. Here, the non-AF samples came from patients requiring coronary artery bypass graft surgery and may have experienced some degree of prior cardiac remodeling, which could contribute to the lack of CRU disruption observed. Alternatively, the sheep AF model may not fully recapitulate human atrial disease.

7. Arrhythmogenic Consequences of RyR2 Dysfunction

7.1. Ca^{2+} Sparks

Ca^{2+} sparks are brief events lasting ~30 ms that increase cytosolic Ca^{2+} by ~100 nM and have a spatial spread of ~2 μm full width at half maximum [4]. Unlike action potential evoked Ca^{2+} release, spontaneous Ca^{2+} sparks do not require Ca^{2+} entry through L-type

Ca^{2+} channels to occur, but instead occur stochastically, usually at a low frequency between beats, due to the small but finite open probability of RyR2. RyR2 opening causes a localized increase in Ca^{2+} in the immediate cluster vicinity [97] that initiates strong positive CICR feedback, causing further channels in the CRU to open. The prevailing control theories of how Ca^{2+} sparks terminate all include some form of local depletion of junctional SR Ca^{2+}, as observed experimentally [98], which diminishes CICR from Ca^{2+} released from the SR into the dyad [94,99].

The remainder of this mini-review focuses on how changes in RyR2 behavior, associated with specific mutations, RyR2 modulators, and beta-adrenergic signaling, produce diverse phenotypes and potentially fatal cardiac arrhythmias with involvement from Ca^{2+} sparks and Ca^{2+} waves. The ability of influx and efflux pathways in the heart to "autoregulate" Ca^{2+} levels is essential for understanding how these effects arise (see [100] for a detailed description). In brief, a change in the balance of influx and efflux pathways that favors net Ca^{2+} entry increases Ca^{2+} loading, whereas an increase in net Ca^{2+} efflux will cause unloading. This occurs because the steep load dependence of SR Ca^{2+} release [101] means that a transient reduction in SR load will cause: (1) a smaller Ca^{2+} transient amplitude, (2) less Ca^{2+}-dependent inactivation of I_{Ca} (allowing more Ca^{2+} entry), (3) decreased activation of NCX (less Ca^{2+} removal), (4) reduced diastolic Ca^{2+} release, and (5) SR Ca^{2+} load will tend towards its previous state (if SR load is transiently increased then points 1–4 can be flipped).

As described in Section 5, the majority of pathological RyR2 mutations result in GoF phenotype, that manifests in cardiac myocytes as increased trigger ability of Ca^{2+} sparks to cytosolic or luminal Ca^{2+}, from a low resting level (~1/s/100 μm in normal mouse ventricular myocytes) to much greater levels with pathological consequences. GoF RyR2 mutations and RyR2 sensitizing agents, such as caffeine, promote Ca^{2+} sparks, Ca^{2+} waves, and non-spark mediated leak, and, consequently, SR Ca^{2+} load is depressed during basal conditions (e.g., RyR2-R4496C [102], RyR2-R420Q [77]). Lower resting SR load probably explains why CPVT patients with GoF mutations typically have normal resting ECG [103] because the reduced SR load will somewhat limit spontaneous Ca^{2+} release. Arrhythmias typically manifest during periods of increased sympathetic activity, which increases Ca^{2+} entry and re-uptake pathways and shifts the SR towards greater operating Ca^{2+} levels that greatly promotes spontaneous Ca^{2+} release [102]. Ca^{2+} release refractoriness, measured using the ratio of the second:first Ca^{2+} spark occurring from the same CRU using a chemical modification of RyR2 with low concentration ryanodine [104], was reduced in a transgenic mouse model of the GoF CPVT-causing RyR2-R420Q$^{+/-}$ mutation under both basal and isoproterenol stimulation, indicating increased CICR sensitivity [105]. Reduced refractoriness was also observed in the same RyR2 CPVT model using two-photon photolysis of a caged Ca^{2+} chelator [106] and in other models of GoF CPVT due to missense mutations (CSQ2-R33Q [107]) or knockout of CSQ2 [108].

7.2. Spontaneous Ca^{2+} Waves

An increase in Ca^{2+} spark frequency increases the likelihood that a spontaneous regenerative process of Ca^{2+}-induced Ca^{2+} release, called a Ca^{2+} wave, will occur. Ca^{2+} waves propagate by the saltatory triggering of neighboring CRU and firing of Ca^{2+} sparks. For Ca^{2+} waves to propagate it is necessary that cytosolic [Ca^{2+}] at nearby clusters increases sufficiently to activate nearby CRU spaced ~0.5 μm apart in the same Z-line [92], or across the much larger inter-Z line spacing (~1.8–2.0 μm). An increase in cytosolic or SR Ca^{2+} increases the interactions between CRU and the occurrence of propagating Ca^{2+} waves [109]. The likelihood of Ca^{2+} wave generation is predicted to be highly dependent on Z-disk spacing, by influencing the initial Ca^{2+} flux needed to raise Ca^{2+} above the threshold for CRU activation at nearby sites [110]. Shorter inter-sarcomere spacing has been reported in isolated ventricular myocytes from left ventricular heart failure in the spontaneously hypertensive rat [111] and right ventricular failure due to pulmonary arterial hypertension [112]. The shorter sarcomere spacing in right ventricular failure was partially reversed by chronic in vivo treatment with the β_1-adrenergic receptor blocker, metoprolol [113], and this was proposed to contribute to the observed reduction in Ca^{2+} wave frequency and triggered activity, despite similar Ca^{2+} spark frequency between treated and untreated cells.

'Rogue' RyR2 located on the network SR outside of SR junctions could, in principle, facilitate Ca^{2+} wave propagation by reducing the longitudinal inter-CRU distance. Rogue RyR2 occur infrequently in normal ventricular myocytes [86] and Ca^{2+} sparks occur almost exclusively from RyR2 clusters in Z-lines [114], suggesting they may have a minor role in Ca^{2+} wave propagation in normal cardiomyocytes. There was no difference in the proportion of non-Z-line RyR2 in human persistent AF [90]; however, the proportion of RyR2 clusters between Z-lines was greater in atrial myocytes from sheep with chronic AF [89], so it is feasible that rogue RyR2 could contribute to increased spontaneous Ca^{2+} waves in a pathological context. Rogue RyR2 could be involved in substantial non-Ca^{2+} spark mediated leak, involving brief openings of small size clusters that would minimally raise cytosolic Ca^{2+} and would be below the optical limit of detection using fluorescent Ca^{2+} sensitive dyes [115]. Increased Ca^{2+} spark-independent leak in failing rabbit ventricular myocytes contributed to reduced SR Ca^{2+} load [101] and would be expected to impair contractility due to the steep dependence of evoked Ca^{2+} release on SR Ca^{2+} load during normal excitation–contraction coupling.

7.3. Delayed Afterdepolarizations (DADs)

The Na^+/Ca^{2+} exchanger is one of the main Ca^{2+} removal pathways that restore cytosolic Ca^{2+} during diastolic Ca^{2+} waves back towards resting levels. Near resting membrane potentials, NCX works in the forward mode to remove 1 Ca^{2+} ion from the cytosol in return for 3 Na^+ ions entering. This net electrogenic action produces a transient inward current that depolarizes the resting membrane potential, causing events called delayed afterdepolarizations (DADs). If DADs reach the threshold for Na^+ channel activation, then triggered action potentials can occur and activate surrounding cells resulting in ectopic electrical activity in the heart that can degenerate into ventricular tachycardias that can lead to sudden cardiac death. DADs are believed to be the underlying event responsible for arrhythmia in CPVT [116]. Fortunately, there is a large mismatch between the current generated by DADs in a single cell (the source) that is shared between all the surrounding electrically-connected cells (the sink), and therefore occasional Ca^{2+} waves in cells in the heart are unlikely to propagate the electrical activity further. Using optogenetic activation of channelrhodopsin-2 expressing cardiac-specific transgenic mice, it was estimated that ~2000 myocytes must depolarize near-simultaneously to initiate premature ventricular contractions [117]. The stochastic nature of DADs makes the probability of this happening by chance highly unlikely. However, this probability increases if Ca^{2+} regulation is deranged by factors causing RyR2 GoF phenotype and if there is coupling to entrain oscillations between cells. Indeed, Borile et al. found that during beta-adrenergic stimulation in ventricular heart slices from mice harboring the CPVT-causing RyR2-R2474S heterozygous missense mutation, cells surrounding those that exhibited DADs also had a greater probability of Ca^{2+} waves occurring almost synchronously [118].

7.4. Early Afterdepolarizations (EADs)

Early afterdepolarizations (EADs) occur during repolarization due to a mismatch between outward K^+ currents and inward currents (such as I_{Na}, I_{Ca}, I_{NCX}) and are believed to be a major source of ventricular tachyarrhythmias [119]. Similar to DADs, spontaneous Ca^{2+} release during repolarization increases inward I_{NCX} current and has been recognized as being involved in EADs for many years [120] (and see [121] for a recent review). I_{NCX} is a non-regenerative current (current decreases as membrane potential increases) and was believed to contribute mainly by delaying repolarization and allowing time for reactivation of other inward currents (such as I_{Na} or I_{Ca}) to drive the upstroke of the EAD [120,122,123]. Stochastic SR Ca^{2+} release, in the form of 'late Ca^{2+} sparks' [124], occurs during the normal Ca^{2+} transient decay triggered by I_{Ca} during the AP plateau and increased cytosolic Ca^{2+}. The frequency of late Ca^{2+} sparks was greater in failing rabbit ventricular myocytes [125,126] and could initiate propagating wave-like 'Ca^{2+} ripples'. Under conditions of increased Ca^{2+} load, late Ca^{2+} sparks-mediated I_{NCX} could take a dominant role in driving the depolarization [125]. This conclusion is supported by computer modeling by Zhong et al. demonstrating that increased late Ca^{2+} release due to hyperactive RyR2 (increased Ca^{2+} sensitivity and reduced refractoriness) contributes to EAD genesis by increasing I_{NCX} [127]. Late Ca^{2+} sparks can fire more than once during a Ca^{2+} transient, after an apparent refractory period that is likely limited by the rate of junctional Ca^{2+} refilling [104,124], providing a mechanism to sustain or increase I_{NCX} at positive membrane potentials. The involvement of SR Ca^{2+} release in EADs is further apparent in the case of conditional SERCA2a knockout mouse ventricular myocytes [128]. The SR Ca^{2+} store is essentially absent in knockout mice, demonstrated by the absence of Ca^{2+} sparks, meaning contraction is solely maintained by an increase in trans-sarcolemmal Ca^{2+} entry through I_{Ca} and a slight decrease in K^+ current density. Despite such major electrophysiological adaptations, myocytes from SERCA2a knockout mice do not exhibit greater EAD incidence than control mice. It is worth noting that the major ion channels implicated in causing EADs are highly coupled to each other (Figure 3), and that knocking-out or inhibiting one part of the pathway will affect the others and tend to exaggerate its (patho)physiological role. Alternatively, combining experimental findings with mathematical modeling is a powerful way to dissect the bidirectional coupling between $Ca^{2+} \leftrightarrow$ membrane potential that is difficult to interpret from experiments alone [123,125].

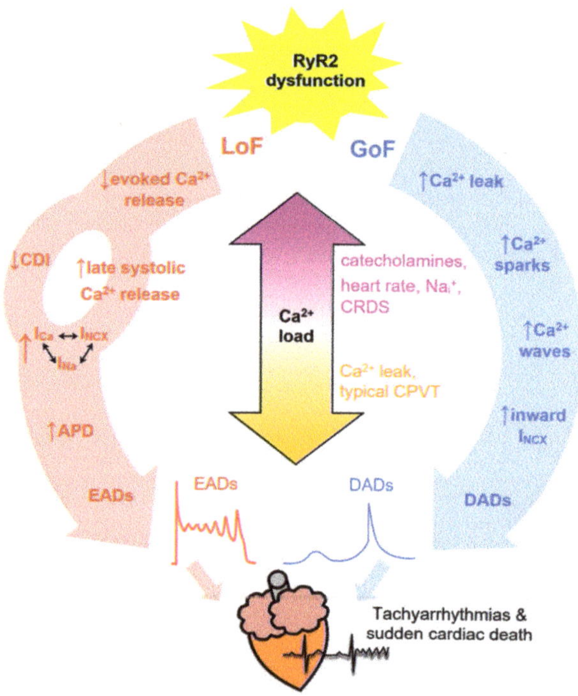

Figure 3. Simplified schematic to illustrate the proposed pathways linking RyR2 loss-of-function (LoF, red) and gain-of-function (GoF, blue) abnormalities to pro-arrhythmic behavior. In GoF, excess Ca^{2+} leak leads to diastolic Ca^{2+} waves and DADs which tend to decrease the Ca^{2+} load. In LoF, evoked Ca^{2+} release is impaired resulting in Ca^{2+} loading and less Ca^{2+}-dependent inactivation (CDI) of I_{Ca}, ultimately resulting in prolonged systolic Ca^{2+} release and EADs. Increasing Ca^{2+} load worsens Ca^{2+} regulation and hastens arrhythmia development.

Factors that increase RyR2 Ca^{2+} sensitivity, or decrease refractoriness, are likely to increase the frequency of late Ca^{2+} sparks. Low dose caffeine caused a biphasic Ca^{2+} transient decay in rat ventricular myocytes which, although not directly measured using confocal microscopy, would be consistent with RyR2 sensitization promoting cytosolic Ca^{2+} triggered late Ca^{2+} sparks [129]. Post-translational modification of RyR2 by PKA and/or CaMKII phosphorylation may increase the RyR2 Ca^{2+} sensitivity and has been attributed as a possible cause of the hyperactive RyR2 observed in heart failure, although the physiological consequences and precise locations of phosphorylation are still the matter of ongoing investigation (for a recent review see [130]). Guo et al. investigated the effect of overexpression of CaMKIIδc on RyR2 Ca^{2+} release properties in transgenic mice on a phospholamban knockout background to avoid the confounding influence of altered SR Ca^{2+} load (because CaMKII can also phosphorylate phospholamban and stimulate SERCA2a Ca^{2+} uptake) [131]. They found that diastolic Ca^{2+} sparks had a faster rate of rising and shorter refractory period between consecutive Ca^{2+} sparks, possibly suggesting more RyR2 channels opening during each event due to increased Ca^{2+} sensitivity. The Ca^{2+} transient decay was slower and more late Ca^{2+} sparks occurred in CaMKIIδc overexpressing mice than phospholamban knockout controls, which could be caused by RyR2 Ca^{2+} hypersensitivity, but could also be due to CaMKIIδc facilitation of I_{Ca} [132] that would also prolong the Ca^{2+} transient and promote late Ca^{2+} sparks [124] Liu et al. found abnormal Ca^{2+} release in ventricular myocytes from mice with the CPVT-causing CSQ2-R33Q mutation that results in a GoF RyR2 phenotype, suggesting systolic Ca^{2+} handling could also be affected by GoF mutations [133].

The rare LoF RyR2 mutations described in Table 2 are associated with atypical electrical behavior including EADs. Zhao et al. investigated the properties of excitation-contraction coupling in transgenic mice with the LoF RyR2-A4860G$^{+/-}$ missense mutation [134]. Under basal conditions, the amplitude and rate of rising of the Ca^{2+} transient were depressed and SR Ca^{2+} load was greater in RyR2-A4860G$^{+/-}$ ventricular myocytes compared to WT cells. Continued Ca^{2+} entry during the action potential via I_{Ca} with less efficient evoked SR Ca^{2+} release was proposed to increase net Ca^{2+} entry, overloading the SR, and resulting in frequent prolonged systolic Ca^{2+} release and EADs under basal conditions, and at greater frequency during stimulation with isoproterenol. They proposed that the extended Ca^{2+} release during EADs help unload the SR in this LoF model, although this will be balanced somewhat by the increased Ca^{2+} entry through I_{Ca} at depolarized membrane potentials. Alternatively, it was recently proposed that the reduced Ca^{2+} transient rate of rise in RyR2-A4860G$^{+/-}$ could cause less Ca^{2+}-dependent inactivation and therefore prolong I_{Ca} during the action potential plateau, which would increase action potential duration both directly and indirectly (through increased I_{NCX}), although this remains to be demonstrated experimentally [135]. Sun et al. found that transgenic mice carrying the LoF RyR2-D4646A$^{+/-}$ mutation also showed increased EADs and SR Ca^{2+} load in isolated myocytes and re-entrant arrhythmias following burst pacing in isolated hearts, although this occurred without a change in steady state Ca^{2+} transient amplitude or rate of rise, and a markedly increased I_{Ca} density [66]. These initial findings suggest the principal arrhythmogenic substrate may be different in LoF compared to GoF RyR2 disorders (although there is likely some cross-over in mechanisms) (Figure 3).

8. Conclusions

The list of mutations in genes encoding RyR2 and its regulatory proteins associated with pro-arrhythmic phenotype continues to grow. These are revealing new insights into the structure/function relationship of RyR2 that is communicated both within and between subdomain interactions of this critical ion channel. It is now generally accepted that there is no common molecular defect, no single unifying hypothesis to explain how RyR2 GoF mutations result in typical CPVT. Moreover, molecular characterization of site-specific mutations may help explain the apparent diverging phenotypes in GoF and LoF RyR2 mutations. However, as LoF mutations have only fairly recently been the subject of attention, further work is needed to establish the extent to which their arrhythmogenic substrates are unique, or whether there is also some overlap with GoF mutations, as this may have important implications for the optimal therapeutic approach. Improved capabilities to investigate subcellular changes in RyR2 behavior and Ca^{2+} release properties in the native environment using advanced imaging technologies offers an exciting future direction that promises to better our understanding of how cardiac excitation-contraction coupling is regulated in health and disease.

Funding: E.D.F. was supported by funding from the British Heart Foundation, grant number FS/IBSRF/21/25071. S.Z. was supported by funding from the British Heart Foundation, grant number PG/21/10657.

Conflicts of Interest: The authors declare no conflict of interest.

References

1. Bers, D. Cardiac excitation-contraction coupling. *Nature* **2002**, *415*, 198–205. [CrossRef] [PubMed]
2. Tester, D.J.; Ackerman, M.J. Cardiomyopathic and channelopathic causes of sudden unexplained death in infants and children. *Annu. Rev. Med.* **2009**, *60*, 69–84. [CrossRef] [PubMed]
3. Eisner, D.A.; Caldwell, J.L.; Kistamas, K.; Trafford, A.W. Calcium and Excitation-Contraction Coupling in the Heart. *Circ. Res.* **2017**, *121*, 181–195. [CrossRef] [PubMed]
4. Cheng, H.; Lederer, W.J.; Cannell, M.B. Calcium sparks: Elementary events underlying excitation-contraction coupling in heart muscle. *Science* **1993**, *262*, 740–744. [CrossRef]
5. Marx, S.; Reiken, S.; Hisamatsu, Y.; Jayaraman, T.; Burkhoff, D.; Rosemblit, N.; Marks, A. PKA phosphorylation dissociates FKBP12.6 from the calcium release channel (ryanodine receptor): Defective regulation in failing hearts. *Cell* **2000**, *101*, 365–376. [CrossRef]
6. Laitinen, P.; Brown, K.; Piippo, K.; Swan, H.; Devaney, J.; BrahmBhatt, B.; Donarum, E.; Marino, M.; Tiso, N.; Viitasalo, M.; et al. Mutations of the cardiac ryanodine receptor (RyR2) gene in familial polymorphic ventricular tachycardia. *Circulation* **2001**, *103*, 485–490. [CrossRef]
7. Priori, S.; Napolitano, C.; Tiso, N.; Memmi, M.; Vignati, G.; Bloise, R.; Sorrentino, V.; Danieli, G. Mutations in the cardiac ryanodine receptor gene (hRyR2) underlie catecholaminergic polymorphic ventricular tachycardia. *Circulation* **2001**, *103*, 196–200. [CrossRef]
8. Tiso, N.; Stephan, D.; Nava, A.; Bagattin, A.; Devaney, J.; Stanchi, F.; Larderet, G.; Brahmbhatt, B.; Brown, K.; Bauce, B.; et al. Identification of mutations in the cardiac ryanodine receptor gene in families affected with arrhythmogenic right ventricular cardiomyopathy type 2 (ARVD2). *Hum. Mol. Genet.* **2001**, *10*, 189–194. [CrossRef]
9. Des Georges, A.; Clarke, O.B.; Zalk, R.; Yuan, Q.; Condon, K.J.; Grassucci, R.A.; Hendrickson, W.A.; Marks, A.R.; Frank, J. Structural Basis for Gating and Activation of RyR1. *Cell* **2016**, *167*, 145–157.e117. [CrossRef]
10. Peng, W.; Shen, H.; Wu, J.; Guo, W.; Pan, X.; Wang, R.; Chen, S.R.; Yan, N. Structural basis for the gating mechanism of the type 2 ryanodine receptor RyR2. *Science* **2016**, *354*, aah5324. [CrossRef]
11. Yano, M.; Yamamoto, T.; Ikeda, Y.; Matsuzaki, M. Mechanisms of Disease: Ryanodine receptor defects in heart failure and fatal arrhythmia. *Nat. Clin. Pract. Cardiovasc. Med.* **2006**, *3*, 43–52. [CrossRef]
12. Euden, J.; Mason, S.A.; Williams, A.J. Functional characterization of the cardiac ryanodine receptor pore-forming region. *PLoS ONE* **2013**, *8*, e66542. [CrossRef]
13. Gao, L.; Tripathy, A.; Lu, X.; Meissner, G. Evidence for a role of C-terminal amino acid residues in skeletal muscle Ca^{2+} release channel (ryanodine receptor) function. *FEBS Lett.* **1997**, *412*, 223–226. [CrossRef]
14. Stewart, R.; Zissimopoulos, S.; Lai, F. Oligomerization of the cardiac ryanodine receptor C-terminal tail. *Biochem. J.* **2003**, *376*, 795–799. [CrossRef]
15. Bhat, M.; Zhao, J.; Takeshima, H.; Ma, J. Functional calcium release channel formed by the carboxyl-terminal portion of ryanodine receptor. *Biophys. J.* **1997**, *73*, 1329–1336. [CrossRef]
16. George, C.; Jundi, H.; Thomas, N.; Scoote, M.; Walters, N.; Williams, A.; Lai, F. Ryanodine receptor regulation by intramolecular interaction between cytoplasmic and transmembrane domains. *Mol. Biol. Cell* **2004**, *15*, 2627–2638. [CrossRef]
17. Xu, X.; Bhat, M.; Nishi, M.; Takeshima, H.; Ma, J. Molecular cloning of cDNA encoding a *Drosophila* ryanodine receptor and functional studies of the carboxyl-terminal calcium release channel. *Biophys. J.* **2000**, *78*, 1270–1281. [CrossRef]
18. Seidel, M.; de Meritens, C.R.; Johnson, L.; Parthimos, D.; Bannister, M.; Thomas, N.L.; Ozekhome-Mike, E.; Lai, F.A.; Zissimopoulos, S. Identification of an amino-terminus determinant critical for ryanodine receptor/Ca^{2+} release channel function. *Cardiovasc. Res.* **2021**, *117*, 780–791. [CrossRef]
19. Suetomi, T.; Yano, M.; Uchinoumi, H.; Fukuda, M.; Hino, A.; Ono, M.; Xu, X.; Tateishi, H.; Okuda, S.; Doi, M.; et al. Mutation-linked defective interdomain interactions within ryanodine receptor cause aberrant Ca^{2+} release leading to catecholaminergic polymorphic ventricular tachycardia. *Circulation* **2011**, *124*, 682–694. [CrossRef]

20. Uchinoumi, H.; Yano, M.; Suetomi, T.; Ono, M.; Xu, X.; Tateishi, H.; Oda, T.; Okuda, S.; Doi, M.; Kobayashi, S.; et al. Catecholaminergic polymorphic ventricular tachycardia is caused by mutation-linked defective conformational regulation of the ryanodine receptor. *Circ. Res.* **2010**, *106*, 1413–1424. [CrossRef]
21. Kimlicka, L.; Lau, K.; Tung, C.C.; Van Petegem, F. Disease mutations in the ryanodine receptor N-terminal region couple to a mobile intersubunit interface. *Nat. Commun.* **2013**, *4*, 1506. [CrossRef]
22. Seidel, M.; Thomas, N.L.; Williams, A.J.; Lai, F.A.; Zissimopoulos, S. Dantrolene rescues aberrant N-terminus inter-subunit interactions in mutant pro-arrhythmic cardiac ryanodine receptors. *Cardiovasc. Res.* **2015**, *105*, 118–128. [CrossRef]
23. Zissimopoulos, S.; Viero, C.; Seidel, M.; Cumbes, B.; White, J.; Cheung, I.; Stewart, R.; Jeyakumar, L.H.; Fleischer, S.; Mukherjee, S.; et al. N-terminus oligomerization regulates the function of cardiac ryanodine receptors. *J. Cell Sci.* **2013**, *126*, 5042–5051. [CrossRef]
24. Laver, D.R. Regulation of the RyR channel gating by Ca^{2+} and Mg^{2+}. *Biophys. Rev.* **2018**, *10*, 1087–1095. [CrossRef]
25. Meissner, G. The structural basis of ryanodine receptor ion channel function. *J. Gen. Physiol.* **2017**, *149*, 1065–1089. [CrossRef]
26. Chi, X.; Gong, D.; Ren, K.; Zhou, G.; Huang, G.; Lei, J.; Zhou, Q.; Yan, N. Molecular basis for allosteric regulation of the type 2 ryanodine receptor channel gating by key modulators. *Proc. Natl. Acad. Sci. USA* **2019**, *116*, 25575–25582. [CrossRef]
27. Murayama, T.; Ogawa, H.; Kurebayashi, N.; Ohno, S.; Horie, M.; Sakurai, T. A tryptophan residue in the caffeine-binding site of the ryanodine receptor regulates Ca^{2+} sensitivity. *Commun. Biol.* **2018**, *1*, 98. [CrossRef]
28. Xu, L.; Chirasani, V.R.; Carter, J.S.; Pasek, D.A.; Dokholyan, N.V.; Yamaguchi, N.; Meissner, G. Ca^{2+}-mediated activation of the skeletal-muscle ryanodine receptor ion channel. *J. Biol. Chem.* **2018**, *293*, 19501–19509. [CrossRef]
29. Chen, W.; Wang, R.; Chen, B.; Zhong, X.; Kong, H.; Bai, Y.; Zhou, Q.; Xie, C.; Zhang, J.; Guo, A.; et al. The ryanodine receptor store-sensing gate controls Ca^{2+} waves and Ca^{2+}-triggered arrhythmias. *Nat. Med.* **2014**, *20*, 184–192. [CrossRef]
30. Zissimopoulos, S.; Lai, F. Ryanodine receptor structure, function and pathophysiology. In *Calcium: A Matter of Life or Death*; Elsevier: Amsterdam, The Netherlands, 2007; Volume 41, pp. 287–342.
31. Zissimopoulos, S.; Seifan, S.; Maxwell, C.; Williams, A.J.; Lai, F.A. Disparities in the association of the ryanodine receptor and the FK506-binding proteins in mammalian heart. *J. Cell Sci.* **2012**, *125*, 1759–1769. [CrossRef]
32. Guo, T.; Cornea, R.L.; Huke, S.; Camors, E.; Yang, Y.; Picht, E.; Fruen, B.R.; Bers, D.M. Kinetics of FKBP12.6 binding to ryanodine receptors in permeabilized cardiac myocytes and effects on Ca sparks. *Circ. Res.* **2010**, *106*, 1743–1752. [CrossRef] [PubMed]
33. Jeyakumar, L.; Ballester, L.; Cheng, D.; McIntyre, J.; Chang, P.; Olivey, H.; Rollins-Smith, L.; Barnett, J.; Murray, K.; Xin, H.-B.; et al. FKBP binding characteristics of cardiac microsomes from diverse vertebrates. *Biochem. Biophys. Res. Commun.* **2001**, *281*, 979–986. [CrossRef] [PubMed]
34. MacMillan, D. FK506 binding proteins: Cellular regulators of intracellular Ca^{2+} signalling. *Eur. J. Pharmacol.* **2013**, *700*, 181–193. [CrossRef] [PubMed]
35. Yan, Z.; Bai, X.C.; Yan, C.; Wu, J.; Li, Z.; Xie, T.; Peng, W.; Yin, C.C.; Li, X.; Scheres, S.H.; et al. Structure of the rabbit ryanodine receptor RyR1 at near-atomic resolution. *Nature* **2015**, *517*, 50–55. [CrossRef]
36. Zalk, R.; Clarke, O.B.; des Georges, A.; Grassucci, R.A.; Reiken, S.; Mancia, F.; Hendrickson, W.A.; Frank, J.; Marks, A.R. Structure of a mammalian ryanodine receptor. *Nature* **2015**, *517*, 44–49. [CrossRef]
37. Balshaw, D.; Xu, L.; Yamaguchi, N.; Pasek, D.; Meissner, G. Calmodulin binding and inhibition of cardiac muscle calcium release channel (ryanodine receptor). *J. Biol. Chem.* **2001**, *276*, 20144–20153. [CrossRef]
38. Fruen, B.; Bardy, J.; Byrem, T.; Strasburg, G.; Louis, C. Differential Ca^{2+} sensitivity of skeletal and cardiac muscle ryanodine receptors in the presence of calmodulin. *Am. J. Physiol.* **2000**, *279*, C724–C733. [CrossRef]
39. Gong, D.; Chi, X.; Wei, J.; Zhou, G.; Huang, G.; Zhang, L.; Wang, R.; Lei, J.; Chen, S.R.W.; Yan, N. Modulation of cardiac ryanodine receptor 2 by calmodulin. *Nature* **2019**, *572*, 347–351. [CrossRef]
40. Chazin, W.J.; Johnson, C.N. Calmodulin Mutations Associated with Heart Arrhythmia: A Status Report. *Int. J. Mol. Sci.* **2020**, *21*, 1418. [CrossRef]
41. Priori, S.G.; Mazzanti, A.; Santiago, D.J.; Kukavica, D.; Trancuccio, A.; Kovacic, J.C. Precision Medicine in Catecholaminergic Polymorphic Ventricular Tachycardia: JACC Focus Seminar 5/5. *J. Am. Coll. Cardiol.* **2021**, *77*, 2592–2612. [CrossRef]
42. Handhle, A.; Ormonde, C.E.; Thomas, N.L.; Bralesford, C.; Williams, A.J.; Lai, F.A.; Zissimopoulos, S. Calsequestrin interacts directly with the cardiac ryanodine receptor luminal domain. *J. Cell Sci.* **2016**, *129*, 3983–3988. [CrossRef]
43. Zhang, L.; Kelley, J.; Schmeisser, G.; Kobayashi, Y.; Jones, L. Complex formation between Junctin, Triadin, Calsequestrin, and the ryanodine receptor: Proteins of the cardiac junctional sarcoplasmic reticulum membrane. *J. Biol. Chem.* **1997**, *272*, 23389–23397. [CrossRef]
44. Sibbles, E.T.; Waddell, H.M.M.; Mereacre, V.; Jones, P.P.; Munro, M.L. The function and regulation of calsequestrin-2: Implications in calcium-mediated arrhythmias. *Biophys. Rev.* **2022**, *14*, 329–352. [CrossRef]
45. Knollmann, B.C.; Chopra, N.; Hlaing, T.; Akin, B.; Yang, T.; Ettensohn, K.; Knollmann, B.E.; Horton, K.D.; Weissman, N.J.; Holinstat, I.; et al. Casq2 deletion causes sarcoplasmic reticulum volume increase, premature Ca^{2+} release, and catecholaminergic polymorphic ventricular tachycardia. *J. Clin. Investig.* **2006**, *116*, 2510–2520. [CrossRef]
46. Song, L.; Alcalai, R.; Arad, M.; Wolf, C.M.; Toka, O.; Conner, D.A.; Berul, C.I.; Eldar, M.; Seidman, C.E.; Seidman, J.G. Calsequestrin 2 (CASQ2) mutations increase expression of calreticulin and ryanodine receptors, causing catecholaminergic polymorphic ventricular tachycardia. *J. Clin. Investig.* **2007**, *117*, 1814–1823. [CrossRef]

47. Gray, B.; Bagnall, R.D.; Lam, L.; Ingles, J.; Turner, C.; Haan, E.; Davis, A.; Yang, P.C.; Clancy, C.E.; Sy, R.W.; et al. A novel heterozygous mutation in cardiac calsequestrin causes autosomal dominant catecholaminergic polymorphic ventricular tachycardia. *Heart Rhythm.* **2016**, *13*, 1652–1660. [CrossRef]
48. Ng, K.; Titus, E.W.; Lieve, K.V.; Roston, T.M.; Mazzanti, A.; Deiter, F.H.; Denjoy, I.; Ingles, J.; Till, J.; Robyns, T.; et al. An International Multicenter Evaluation of Inheritance Patterns, Arrhythmic Risks, and Underlying Mechanisms of CASQ2-Catecholaminergic Polymorphic Ventricular Tachycardia. *Circulation* **2020**, *142*, 932–947. [CrossRef]
49. Titus, E.W.; Deiter, F.H.; Shi, C.; Wojciak, J.; Scheinman, M.; Jura, N.; Deo, R.C. The structure of a Calsequestrin filament reveals mechanisms of familial arrhythmia. *Nat. Struct. Mol. Biol.* **2020**, *27*, 1142–1151. [CrossRef]
50. Wleklinski, M.J.; Kannankeril, P.J.; Knollmann, B.C. Molecular and tissue mechanisms of catecholaminergic polymorphic ventricular tachycardia. *J. Physiol.* **2020**, *598*, 2817–2834. [CrossRef]
51. Priori, S.G.; Blomstrom-Lundqvist, C.; Mazzanti, A.; Blom, N.; Borggrefe, M.; Camm, J.; Elliott, P.M.; Fitzsimons, D.; Hatala, R.; Hindricks, G.; et al. 2015 ESC Guidelines for the management of patients with ventricular arrhythmias and the prevention of sudden cardiac death: The Task Force for the Management of Patients with Ventricular Arrhythmias and the Prevention of Sudden Cardiac Death of the European Society of Cardiology (ESC). Endorsed by: Association for European Paediatric and Congenital Cardiology (AEPC). *Eur. Heart J.* **2015**, *36*, 2793–2867. [CrossRef]
52. Mazzanti, A.; Kukavica, D.; Trancuccio, A.; Memmi, M.; Bloise, R.; Gambelli, P.; Marino, M.; Ortiz-Genga, M.; Morini, M.; Monteforte, N.; et al. Outcomes of Patients with Catecholaminergic Polymorphic Ventricular Tachycardia Treated with beta-Blockers. *JAMA Cardiol.* **2022**, *7*, 504–512. [CrossRef]
53. Kryshtal, D.O.; Blackwell, D.J.; Egly, C.L.; Smith, A.N.; Batiste, S.M.; Johnston, J.N.; Laver, D.R.; Knollmann, B.C. RYR2 Channel Inhibition Is the Principal Mechanism of Flecainide Action in CPVT. *Circ. Res.* **2021**, *128*, 321–331. [CrossRef]
54. Bannister, M.L.; MacLeod, K.T.; George, C.H. Moving in the right direction: Elucidating the mechanisms of interaction between flecainide and the cardiac ryanodine receptor. *Br. J. Pharmacol.* **2022**, *179*, 2558–2563. [CrossRef]
55. Benitah, J.P.; Gomez, A.M. Is the Debate on the Flecainide Action on the RYR2 in CPVT Closed? *Circ. Res.* **2021**, *128*, 332–334. [CrossRef]
56. Roux-Buisson, N.; Gandjbakhch, E.; Donal, E.; Probst, V.; Deharo, J.C.; Chevalier, P.; Klug, D.; Mansencal, N.; Delacretaz, E.; Cosnay, P.; et al. Prevalence and significance of rare RYR2 variants in arrhythmogenic right ventricular cardiomyopathy/dysplasia: Results of a systematic screening. *Heart Rhythm* **2014**, *11*, 1999–2009. [CrossRef]
57. Shigemizu, D.; Aiba, T.; Nakagawa, H.; Ozaki, K.; Miya, F.; Satake, W.; Toda, T.; Miyamoto, Y.; Fujimoto, A.; Suzuki, Y.; et al. Exome Analyses of Long QT Syndrome Reveal Candidate Pathogenic Mutations in Calmodulin-Interacting Genes. *PLoS ONE* **2015**, *10*, e0130329. [CrossRef]
58. Tester, D.J.; Kopplin, L.J.; Will, M.L.; Ackerman, M.J. Spectrum and prevalence of cardiac ryanodine receptor (RyR2) mutations in a cohort of unrelated patients referred explicitly for long QT syndrome genetic testing. *Heart Rhythm* **2005**, *2*, 1099–1105. [CrossRef]
59. Fujii, Y.; Itoh, H.; Ohno, S.; Murayama, T.; Kurebayashi, N.; Aoki, H.; Blancard, M.; Nakagawa, Y.; Yamamoto, S.; Matsui, Y.; et al. A type 2 ryanodine receptor variant associated with reduced Ca^{2+} release and short-coupled torsades de pointes ventricular arrhythmia. *Heart Rhythm* **2017**, *14*, 98–107. [CrossRef]
60. Touat-Hamici, Z.; Blancard, M.; Ma, R.; Lin, L.; Iddir, Y.; Denjoy, I.; Leenhardt, A.; Yuchi, Z.; Guicheney, P. A SPRY1 domain cardiac ryanodine receptor variant associated with short-coupled torsade de pointes. *Sci. Rep.* **2021**, *11*, 5243. [CrossRef]
61. Ohno, S.; Omura, M.; Kawamura, M.; Kimura, H.; Itoh, H.; Makiyama, T.; Ushinohama, H.; Makita, N.; Horie, M. Exon 3 deletion of RYR2 encoding cardiac ryanodine receptor is associated with left ventricular non-compaction. *Europace* **2014**, *16*, 1646–1654. [CrossRef]
62. Roston, T.M.; Guo, W.; Krahn, A.D.; Wang, R.; Van Petegem, F.; Sanatani, S.; Chen, S.R.; Lehman, A. A novel RYR2 loss-of-function mutation (I4855M) is associated with left ventricular non-compaction and atypical catecholaminergic polymorphic ventricular tachycardia. *J. Electrocardiol.* **2017**, *50*, 227–233. [CrossRef] [PubMed]
63. Costa, S.; Medeiros-Domingo, A.; Gasperetti, A.; Breitenstein, A.; Steffel, J.; Guidetti, F.; Flammer, A.; Odening, K.; Ruschitzka, F.; Duru, F.; et al. Familial dilated cardiomyopathy associated with a novel heterozygous RYR2 early truncating variant. *Cardiol. J.* **2021**, *28*, 173–175. [CrossRef] [PubMed]
64. Haas, J.; Frese, K.S.; Peil, B.; Kloos, W.; Keller, A.; Nietsch, R.; Feng, Z.; Muller, S.; Kayvanpour, E.; Vogel, B.; et al. Atlas of the clinical genetics of human dilated cardiomyopathy. *Eur. Heart J.* **2015**, *36*, 1123–1135. [CrossRef] [PubMed]
65. Alvarado, F.J.; Bos, J.M.; Yuchi, Z.; Valdivia, C.R.; Hernandez, J.J.; Zhao, Y.T.; Henderlong, D.S.; Chen, Y.; Booher, T.R.; Marcou, C.A.; et al. Cardiac hypertrophy and arrhythmia in mice induced by a mutation in ryanodine receptor 2. *JCI Insight* **2019**, *5*, e126544. [CrossRef] [PubMed]
66. Sun, B.; Yao, J.; Ni, M.; Wei, J.; Zhong, X.; Guo, W.; Zhang, L.; Wang, R.; Belke, D.; Chen, Y.X.; et al. Cardiac ryanodine receptor calcium release deficiency syndrome. *Sci. Transl. Med.* **2021**, *13*, eaba7287. [CrossRef]
67. Tester, D.J.; Kim, C.S.J.; Hamrick, S.K.; Ye, D.; O'Hare, B.J.; Bombei, H.M.; Fitzgerald, K.K.; Haglund-Turnquist, C.M.; Atkins, D.L.; Nunez, L.A.O.; et al. Molecular characterization of the calcium release channel deficiency syndrome. *JCI Insight* **2020**, *5*, e135952. [CrossRef]

68. Olubando, D.; Hopton, C.; Eden, J.; Caswell, R.; Lowri Thomas, N.; Roberts, S.A.; Morris-Rosendahl, D.; Venetucci, L.; Newman, W.G. Classification and correlation of RYR2 missense variants in individuals with catecholaminergic polymorphic ventricular tachycardia reveals phenotypic relationships. *J. Hum. Genet.* **2020**, *65*, 531–539. [CrossRef]
69. Duvekot, J.C.; Baas, A.F.; Volker-Touw, C.M.L.; Bikker, H.; Schroer, C.; Breur, J. Early Lethal Noncompaction Cardiomyopathy in Siblings with Compound Heterozygous RYR2 Variant. *Can. J. Cardiol.* **2021**, *37*, 1864–1866. [CrossRef]
70. Fernandez-Velasco, M.; Rueda, A.; Rizzi, N.; Benitah, J.P.; Colombi, B.; Napolitano, C.; Priori, S.G.; Richard, S.; Gomez, A.M. Increased Ca^{2+} sensitivity of the ryanodine receptor mutant RyR2R4496C underlies catecholaminergic polymorphic ventricular tachycardia. *Circ. Res.* **2009**, *104*, 201–209. [CrossRef]
71. Xiao, Z.; Guo, W.; Sun, B.; Hunt, D.J.; Wei, J.; Liu, Y.; Wang, Y.; Wang, R.; Jones, P.P.; Back, T.G.; et al. Enhanced Cytosolic Ca^{2+} Activation Underlies a Common Defect of Central Domain Cardiac Ryanodine Receptor Mutations Linked to Arrhythmias. *J. Biol. Chem.* **2016**, *291*, 24528–24537. [CrossRef]
72. Jiang, D.; Xiao, B.; Yang, D.; Wang, R.; Choi, P.; Zhang, L.; Cheng, H.; Chen, S. RyR2 mutations linked to ventricular tachycardia and sudden death reduce the threshold for store-overload-induced Ca^{2+} release (SOICR). *Proc. Natl. Acad. Sci. USA* **2004**, *101*, 13062–13067. [CrossRef] [PubMed]
73. Wehrens, X.; Lenhart, S.; Huang, F.; Vest, J.; Reiken, S.; Mohler, P.; Sun, J.; Guatimosim, S.; Song, L.-S.; Rosemblit, N.; et al. FKBP12.6 deficiency and defective calcium release channel (ryanodine receptor) function linked to exercise-induced sudden cardiac death. *Cell* **2003**, *113*, 829–840. [CrossRef]
74. Jiang, D.; Wang, R.; Xiao, B.; Kong, H.; Hunt, D.; Choi, P.; Zhang, L.; Chen, S. Enhanced store overload-induced Ca^{2+} release and channel sensitivity to luminal Ca^{2+} activation are common defects of RyR2 mutations linked to ventricular tachycardia and sudden death. *Circ. Res.* **2005**, *97*, 1173–1181. [CrossRef] [PubMed]
75. Zissimopoulos, S.; Thomas, N.L.; Jamaluddin, W.W.; Lai, F.A. FKBP12.6 binding of ryanodine receptors carrying mutations associated with arrhythmogenic cardiac disease. *Biochem. J.* **2009**, *419*, 273–278. [CrossRef]
76. Nakamura, Y.; Yamamoto, T.; Kobayashi, S.; Tamitani, M.; Hamada, Y.; Fukui, G.; Xu, X.; Nishimura, S.; Kato, T.; Uchinoumi, H.; et al. Ryanodine receptor-bound calmodulin is essential to protect against catecholaminergic polymorphic ventricular tachycardia. *JCI Insight* **2019**, *4*, e126112. [CrossRef]
77. Yin, L.; Zahradnikova, A., Jr.; Rizzetto, R.; Boncompagni, S.; Rabesahala de Meritens, C.; Zhang, Y.; Joanne, P.; Marques-Sule, E.; Aguilar-Sanchez, Y.; Fernandez-Tenorio, M.; et al. Impaired Binding to Junctophilin-2 and Nanostructural Alteration in CPVT Mutation. *Circ. Res.* **2021**, *129*, e35–e52. [CrossRef]
78. Jiang, D.; Chen, W.; Wang, R.; Zhang, L.; Chen, S. Loss of luminal Ca^{2+} activation in the cardiac ryanodine receptor is associated with ventricular fibrillation and sudden death. *Proc. Natl. Acad. Sci. USA* **2007**, *104*, 18309–18314. [CrossRef]
79. Li, Y.; Wei, J.; Guo, W.; Sun, B.; Estillore, J.P.; Wang, R.; Yoruk, A.; Roston, T.M.; Sanatani, S.; Wilde, A.A.M.; et al. Human RyR2 (Ryanodine Receptor 2) Loss-of-Function Mutations: Clinical Phenotypes and In Vitro Characterization. *Circ. Arrhythm. Electrophysiol.* **2021**, *14*, e010013. [CrossRef]
80. Roston, T.M.; Wei, J.; Guo, W.; Li, Y.; Zhong, X.; Wang, R.; Estillore, J.P.; Peltenburg, P.J.; Noguer, F.R.I.; Till, J.; et al. Clinical and Functional Characterization of Ryanodine Receptor 2 Variants Implicated in Calcium-Release Deficiency Syndrome. *JAMA Cardiol.* **2022**, *7*, 84–92. [CrossRef]
81. Ormerod, J.O.M.; Ormondroyd, E.; Li, Y.; Taylor, J.; Wei, J.; Guo, W.; Wang, R.; Sarton, C.N.S.; McGuire, K.; Dreau, H.M.P.; et al. Provocation Testing and Therapeutic Response in a Newly Described Channelopathy: RyR2 Calcium Release Deficiency Syndrome. *Circ. Genom. Precis. Med.* **2022**, *15*, e003589. [CrossRef]
82. Hirose, S.; Murayama, T.; Tetsuo, N.; Hoshiai, M.; Kise, H.; Yoshinaga, M.; Aoki, H.; Fukuyama, M.; Wuriyanghai, Y.; Wada, Y.; et al. Loss-of-function mutations in cardiac ryanodine receptor channel cause various types of arrhythmias including long QT syndrome. *Europace* **2022**, *24*, 497–510. [CrossRef]
83. Zhong, X.; Guo, W.; Wei, J.; Tang, Y.; Liu, Y.; Zhang, J.Z.; Tan, V.H.; Zhang, L.; Wang, R.; Jones, P.P.; et al. Identification of loss-of-function RyR2 mutations associated with idiopathic ventricular fibrillation and sudden death. *Biosci. Rep.* **2021**, *41*, BSR20210209. [CrossRef]
84. Franzini-Armstrong, C.; Protasi, F. Ryanodine receptors of striated muscles: A complex channel capable of multiple interactions. *Physiol. Rev.* **1997**, *77*, 699–729. [CrossRef]
85. Cabra, V.; Murayama, T.; Samsó, M. Ultrastructural Analysis of Self-Associated RyR2s. *Biophys. J.* **2016**, *110*, 2651–2662. [CrossRef]
86. Baddeley, D.; Jayasinghe, I.D.; Lam, L.; Rossberger, S.; Cannell, M.B.; Soeller, C. Optical single-channel resolution imaging of the ryanodine receptor distribution in rat cardiac myocytes. *Proc. Natl. Acad. Sci. USA* **2009**, *106*, 22275–22280. [CrossRef]
87. Hou, Y.; Jayasinghe, I.; Crossman, D.J.; Baddeley, D.; Soeller, C. Nanoscale analysis of ryanodine receptor clusters in dyadic couplings of rat cardiac myocytes. *J. Mol. Cell. Cardiol.* **2015**, *80*, 45–55. [CrossRef]
88. Shen, X.; den Brink, J.; Hou, Y.; Colli, D.; Le, C.; Kolstad, T.R.; MacQuaide, N.; Carlson, C.R.; Kekenes-Huskey, P.M.; Edwards, A.G.; et al. 3D dSTORM imaging reveals novel detail of ryanodine receptor localization in rat cardiac myocytes. *J. Physiol.* **2019**, *597*, 399–418. [CrossRef]
89. Macquaide, N.; Tuan, H.-T.M.; Hotta, J.-I.; Sempels, W.; Lenaerts, I.; Holemans, P.; Hofkens, J.; Jafri, M.S.; Willems, R.; Sipido, K.R. Ryanodine receptor cluster fragmentation and redistribution in persistent atrial fibrillation enhance calcium release. *Cardiovasc. Res.* **2015**, *108*, 387–398. [CrossRef]

90. Munro, M.L.; van Hout, I.; Aitken-Buck, H.M.; Sugunesegran, R.; Bhagwat, K.; Davis, P.J.; Lamberts, R.R.; Coffey, S.; Soeller, C.; Jones, P.P. Human Atrial Fibrillation Is Not Associated with Remodeling of Ryanodine Receptor Clusters. *Front. Cell Dev. Biol.* **2021**, *9*, 633704. [CrossRef]
91. Franzini-Armstrong, C.; Protasi, F.; Ramesh, V. Shape, size, and distribution of Ca^{2+} release units and couplons in skeletal and cardiac muscles. *Biophys. J.* **1999**, *77*, 1528–1539. [CrossRef]
92. Sobie, E.A.; Guatimosim, S.; Gómez-Viquez, L.; Song, L.-S.; Hartmann, H.; Saleet Jafri, M.; Lederer, W.J. The Ca^{2+} leak paradox and "rogue ryanodine receptors": SR Ca^{2+} efflux theory and practice. *Prog. Biophys. Mol. Biol.* **2006**, *90*, 172–185. [CrossRef]
93. Galice, S.; Xie, Y.; Yang, Y.; Sato, D.; Bers, D.M. Size Matters: Ryanodine Receptor Cluster Size Affects Arrhythmogenic Sarcoplasmic Reticulum Calcium Release. *J. Am. Heart Assoc.* **2018**, *7*, e008724. [CrossRef]
94. Laver, D.R.; Kong, C.H.T.; Imtiaz, M.S.; Cannell, M.B. Termination of calcium-induced calcium release by induction decay: An emergent property of stochastic channel gating and molecular scale architecture. *J. Mol. Cell. Cardiol.* **2013**, *54*, 98–100. [CrossRef]
95. Sheard, T.M.D.; Hurley, M.E.; Colyer, J.; White, E.; Norman, R.; Pervolaraki, E.; Narayanasamy, K.K.; Hou, Y.; Kirton, H.M.; Yang, Z.; et al. Three-Dimensional and Chemical Mapping of Intracellular Signaling Nanodomains in Health and Disease with Enhanced Expansion Microscopy. *ACS Nano* **2019**, *13*, 2143–2157. [CrossRef]
96. Munro, M.L.; Jayasinghe, I.D.; Wang, Q.; Quick, A.; Wang, W.; Baddeley, D.; Wehrens, X.H.T.; Soeller, C. Junctophilin-2 in the nanoscale organisation and functional signalling of ryanodine receptor clusters in cardiomyocytes. *J. Cell Sci.* **2016**, *129*, 4388–4398. [CrossRef]
97. Liu, Y.; Porta, M.; Qin, J.; Ramos, J.; Nani, A.; Shannon, T.R.; Fill, M. Flux regulation of cardiac ryanodine receptor channels. *J. Gen. Physiol.* **2010**, *135*, 15–27. [CrossRef]
98. Zima, A.V.; Picht, E.; Bers, D.M.; Blatter, L.A. Termination of cardiac Ca^{2+} sparks: Role of intra-SR $[Ca^{2+}]$, release flux, and intra-SR Ca^{2+} diffusion. *Circ. Res.* **2008**, *103*, e105–e115. [CrossRef]
99. Gillespie, D.; Fill, M. Pernicious Attrition and Inter-RyR2 CICR Current Control in Cardiac Muscle. *J. Mol. Cell. Cardiol.* **2013**, *58*, 53–58. [CrossRef]
100. Eisner, D.A.; Trafford, A.W.; Díaz, M.E.; Overend, C.L.; O'Neill, S.C. The control of Ca release from the cardiac sarcoplasmic reticulum: Regulation versus autoregulation. *Cardiovasc. Res.* **1998**, *38*, 589–604. [CrossRef]
101. Zima, A.V.; Bovo, E.; Bers, D.M.; Blatter, L.A. Ca^{2+} spark-dependent and -independent sarcoplasmic reticulum Ca^{2+} leak in normal and failing rabbit ventricular myocytes. *J. Physiol.* **2010**, *588*, 4743–4757. [CrossRef]
102. Kashimura, T.; Briston, S.J.; Trafford, A.W.; Napolitano, C.; Priori, S.G.; Eisner, D.A.; Venetucci, L.A. In the $RyR2^{R4496C}$ Mouse Model of CPVT, β-Adrenergic Stimulation Induces Ca Waves by Increasing SR Ca Content and Not by Decreasing the Threshold for Ca Waves. *Circ. Res.* **2010**, *107*, 1483–1489. [CrossRef] [PubMed]
103. Leenhardt, A.; Lucet, V.; Denjoy, I.; Grau, F.; Ngoc, D.D.; Coumel, P. Catecholaminergic polymorphic ventricular tachycardia in children. A 7-year follow-up of 21 patients. *Circulation* **1995**, *91*, 1512–1519. [CrossRef] [PubMed]
104. Sobie, E.A.; Song, L.-S.; Lederer, W.J. Local recovery of Ca^{2+} release in rat ventricular myocytes. *J. Physiol.* **2005**, *565*, 441–447. [CrossRef] [PubMed]
105. Lopez, R.; Janicek, R.; Fernandez-Tenorio, M.; Courtehoux, M.; Matas, L.; Gerbaud, P.; Gomez, A.M.; Egger, M.; Niggli, E. Uptake-leak balance of SR Ca^{2+} determines arrhythmogenic potential of $RyR2^{R420Q+/-}$ cardiomyocytes. *J. Mol. Cell. Cardiol.* **2022**, *170*, 1–14. [CrossRef] [PubMed]
106. Janicek, R.; Agarwal, H.; Gómez, A.M.; Egger, M.; Ellis-Davies, G.C.R.; Niggli, E. Local recovery of cardiac calcium-induced calcium release interrogated by ultra-effective, two-photon uncaging of calcium. *J. Physiol.* **2021**, *599*, 3841–3852. [CrossRef] [PubMed]
107. Brunello, L.; Slabaugh, J.L.; Radwanski, P.B.; Ho, H.-T.; Belevych, A.E.; Lou, Q.; Chen, H.; Napolitano, C.; Lodola, F.; Priori, S.G.; et al. Decreased RyR2 refractoriness determines myocardial synchronization of aberrant Ca^{2+} release in a genetic model of arrhythmia. *Proc. Natl. Acad. Sci. USA* **2013**, *110*, 10312–10317. [CrossRef]
108. Kryshtal, D.O.; Gryshchenko, O.; Gomez-Hurtado, N.; Knollmann, B.C. Impaired calcium-calmodulin-dependent inactivation of Cav1.2 contributes to loss of sarcoplasmic reticulum calcium release refractoriness in mice lacking calsequestrin 2. *J. Mol. Cell. Cardiol.* **2015**, *82*, 75–83. [CrossRef]
109. Nivala, M.; Ko, C.Y.; Nivala, M.; Weiss, J.N.; Qu, Z. Criticality in Intracellular Calcium Signaling in Cardiac Myocytes. *Biophys. J.* **2012**, *102*, 2433–2442. [CrossRef]
110. Izu, L.T.; Means, S.A.; Shadid, J.N.; Chen-Izu, Y.; Balke, C.W. Interplay of ryanodine receptor distribution and calcium dynamics. *Biophys. J.* **2006**, *91*, 95–112. [CrossRef]
111. Chen-Izu, Y.; Ward, C.W.; Stark, W.; Banyasz, T.; Sumandea, M.P.; Balke, C.W.; Izu, L.T.; Wehrens, X.H.T. Phosphorylation of RyR2 and shortening of RyR2 cluster spacing in spontaneously hypertensive rat with heart failure. *Am. J. Physiol.-Heart Circ. Physiol.* **2007**, *293*, H2409–H2417. [CrossRef]
112. Fowler, E.D.; Benoist, D.; Drinkhill, M.J.; Stones, R.; Helmes, M.; Wüst, R.C.I.; Stienen, G.J.M.; Steele, D.S.; White, E. Decreased creatine kinase is linked to diastolic dysfunction in rats with right heart failure induced by pulmonary artery hypertension. *J. Mol. Cell. Cardiol.* **2015**, *86*, 1–8. [CrossRef] [PubMed]
113. Fowler, E.D.; Drinkhill, M.J.; Norman, R.; Pervolaraki, E.; Stones, R.; Steer, E.; Benoist, D.; Steele, D.S.; Calaghan, S.C.; White, E. Beta1-adrenoceptor antagonist, metoprolol attenuates cardiac myocyte Ca^{2+} handling dysfunction in rats with pulmonary artery hypertension. *J. Mol. Cell. Cardiol.* **2018**, *120*, 74–83. [CrossRef] [PubMed]

114. Hiess, F.; Vallmitjana, A.; Wang, R.; Cheng, H.; ter Keurs, H.E.D.J.; Chen, J.; Hove-Madsen, L.; Benitez, R.; Chen, S.R.W. Distribution and Function of Cardiac Ryanodine Receptor Clusters in Live Ventricular Myocytes. *J. Biol. Chem.* **2015**, *290*, 20477–20487. [CrossRef] [PubMed]
115. Santiago, D.J.; Curran, J.W.; Bers, D.M.; Lederer, W.J.; Stern, M.D.; Ríos, E.; Shannon, T.R. Ca Sparks Do Not Explain all Ryanodine Receptor-Mediated SR Ca Leak in Mouse Ventricular Myocytes. *Biophys. J.* **2010**, *98*, 2111–2120. [CrossRef]
116. Al-Khatib, S.M.; Stevenson, W.G.; Ackerman, M.J.; Bryant, W.J.; Callans, D.J.; Curtis, A.B.; Deal, B.J.; Dickfeld, T.; Field, M.E.; Fonarow, G.C.; et al. 2017 AHA/ACC/HRS Guideline for Management of Patients with Ventricular Arrhythmias and the Prevention of Sudden Cardiac Death. *Circulation* **2018**, *138*, e272–e391. [CrossRef]
117. Zaglia, T.; Pianca, N.; Borile, G.; Da Broi, F.; Richter, C.; Campione, M.; Lehnart, S.E.; Luther, S.; Corrado, D.; Miquerol, L.; et al. Optogenetic determination of the myocardial requirements for extrasystoles by cell type-specific targeting of ChannelRhodopsin-2. *Proc. Natl. Acad. Sci. USA* **2015**, *112*, E4495–E4504. [CrossRef]
118. Borile, G.; Zaglia, T.; Lehnart, S.E.; Mongillo, M. Multiphoton Imaging of Ca^{2+} Instability in Acute Myocardial Slices from a $RyR2^{R2474S}$ Murine Model of Catecholaminergic Polymorphic Ventricular Tachycardia. *J. Clin. Med.* **2021**, *10*, 2821. [CrossRef]
119. Wu, J.; Wu, J.; Zipes, D.P. Early Afterdepolarizations, U Waves, and Torsades de Pointes. *Circulation* **2002**, *105*, 675–676. [CrossRef]
120. Volders, P.G.; Vos, M.A.; Szabo, B.; Sipido, K.R.; de Groot, S.H.; Gorgels, A.P.; Wellens, H.J.; Lazzara, R. Progress in the understanding of cardiac early afterdepolarizations and torsades de pointes: Time to revise current concepts. *Cardiovasc. Res.* **2000**, *46*, 376–392. [CrossRef]
121. Kistamás, K.; Veress, R.; Horváth, B.; Bányász, T.; Nánási, P.P.; Eisner, D.A. Calcium Handling Defects and Cardiac Arrhythmia Syndromes. *Front. Pharmacol.* **2020**, *11*, 72. [CrossRef]
122. Pott, C.; Eckardt, L.; Goldhaber, J.I. Triple Threat: The Na^+/Ca^{2+} Exchanger in the Pathophysiology of Cardiac Arrhythmia, Ischemia and Heart Failure. *Curr. Drug Targets* **2011**, *12*, 737–747. [CrossRef]
123. Qu, Z.; Xie, L.-H.; Olcese, R.; Karagueuzian, H.S.; Chen, P.-S.; Garfinkel, A.; Weiss, J.N. Early afterdepolarizations in cardiac myocytes: Beyond reduced repolarization reserve. *Cardiovasc. Res.* **2013**, *99*, 6–15. [CrossRef]
124. Fowler, E.D.; Kong, C.H.T.; Hancox, J.C.; Cannell, M.B. Late Ca^{2+} Sparks and Ripples During the Systolic Ca^{2+} Transient in Heart Muscle Cells. *Circ. Res.* **2018**, *122*, 473–478. [CrossRef]
125. Fowler, E.D.; Wang, N.; Hezzell, M.; Chanoit, G.; Hancox, J.C.; Cannell, M.B. Arrhythmogenic late Ca^{2+} sparks in failing heart cells and their control by action potential configuration. *Proc. Natl. Acad. Sci. USA* **2020**, *117*, 2687–2692. [CrossRef]
126. Litwin, S.E.; Zhang, D.; Bridge, J.H.B. Dyssynchronous Ca^{2+} Sparks in Myocytes from Infarcted Hearts. *Circ. Res.* **2000**, *87*, 1040–1047. [CrossRef]
127. Zhong, M.; Rees, C.M.; Terentyev, D.; Choi, B.-R.; Koren, G.; Karma, A. NCX-Mediated Subcellular Ca^{2+} Dynamics Underlying Early Afterdepolarizations in LQT2 Cardiomyocytes. *Biophys. J.* **2018**, *115*, 1019–1032. [CrossRef]
128. Edwards, A.G.; Mørk, H.; Stokke, M.K.; Lipsett, D.B.; Sjaastad, I.; Richard, S.; Sejersted, O.M.; Louch, W.E. Sarcoplasmic Reticulum Calcium Release Is Required for Arrhythmogenesis in the Mouse. *Front. Physiol.* **2021**, *12*, 744730. [CrossRef]
129. Sankaranarayanan, R.; Li, Y.; Greensmith, D.J.; Eisner, D.A.; Venetucci, L. Biphasic decay of the Ca transient results from increased sarcoplasmic reticulum Ca leak. *J. Physiol.* **2016**, *594*, 611–623. [CrossRef]
130. Benitah, J.-P.; Perrier, R.; Mercadier, J.-J.; Pereira, L.; Gómez, A.M. RyR2 and Calcium Release in Heart Failure. *Front. Physiol.* **2021**, *12*, 734210. [CrossRef]
131. Guo, T.; Zhang, T.; Ginsburg, K.S.; Mishra, S.; Brown, J.H.; Bers, D.M. CaMKIIδ_C Slows $[Ca]_i$ Decline in Cardiac Myocytes by Promoting Ca Sparks. *Biophys. J.* **2012**, *102*, 2461–2470. [CrossRef]
132. Maier, L.; Zhang, T.; Chen, L.; DeSantiago, J.; Brown, J.; Bers, D. Transgenic CaMKIIδ_C overexpression uniquely alters cardiac myocyte Ca^{2+} handling: Reduced SR Ca^{2+} load and activated SR Ca^{2+} release. *Circ. Res.* **2003**, *92*, 904–911. [CrossRef]
133. Liu, N.; Denegri, M.; Dun, W.; Boncompagni, S.; Lodola, F.; Protasi, F.; Napolitano, C.; Boyden, P.A.; Priori, S.G. Abnormal Propagation of Calcium Waves and Ultrastructural Remodeling in Recessive Catecholaminergic Polymorphic Ventricular Tachycardia. *Circ. Res.* **2013**, *113*, 142–152. [CrossRef]
134. Zhao, Y.T.; Valdivia, C.R.; Gurrola, G.B.; Powers, P.P.; Willis, B.C.; Moss, R.L.; Jalife, J.; Valdivia, H.H. Arrhythmogenesis in a catecholaminergic polymorphic ventricular tachycardia mutation that depresses ryanodine receptor function. *Proc. Natl. Acad. Sci. USA* **2015**, *112*, E1669–E1677. [CrossRef]
135. Szentandrássy, N.; Magyar, Z.É.; Hevesi, J.; Bányász, T.; Nánási, P.P.; Almássy, J. Therapeutic Approaches of Ryanodine Receptor-Associated Heart Diseases. *Int. J. Mol. Sci.* **2022**, *23*, 4435. [CrossRef]

biomolecules

Review

Atypically Shaped Cardiomyocytes (ACMs): The Identification, Characterization and New Insights into a Subpopulation of Cardiomyocytes

Mariko Omatsu-Kanbe *, Ryo Fukunaga, Xinya Mi and Hiroshi Matsuura

Department of Physiology, Shiga University of Medical Science, Otsu 520-2192, Shiga, Japan; ryofuku@belle.shiga-med.ac.jp (R.F.); mixinya@phar.kyushu-u.ac.jp (X.M.); matuurah@belle.shiga-med.ac.jp (H.M.)
* Correspondence: m_omatsu@belle.shiga-med.ac.jp; Tel.: +81-77-548-2153

Abstract: In the adult mammalian heart, no data have yet shown the existence of cardiomyocyte-differentiable stem cells that can be used to practically repair the injured myocardium. Atypically shaped cardiomyocytes (ACMs) are found in cultures of the cardiomyocyte-removed fraction obtained from cardiac ventricles from neonatal to aged mice. ACMs are thought to be a subpopulation of cardiomyocytes or immature cardiomyocytes, most closely resembling cardiomyocytes due to their spontaneous beating, well-organized sarcomere and the expression of cardiac-specific proteins, including some fetal cardiac gene proteins. In this review, we focus on the characteristics of ACMs compared with ventricular myocytes and discuss whether these cells can be substitutes for damaged cardiomyocytes. ACMs reside in the interstitial spaces among ventricular myocytes and survive under severely hypoxic conditions fatal to ventricular myocytes. ACMs have not been observed to divide or proliferate, similar to cardiomyocytes, but they maintain their ability to fuse with each other. Thus, it is worthwhile to understand the role of ACMs and especially how these cells perform cell fusion or function independently in vivo. It may aid in the development of new approaches to cell therapy to protect the injured heart or the clarification of the pathogenesis underlying arrhythmia in the injured heart.

Keywords: atypically shaped cardiomyocytes; ACMs; subpopulation of cardiomyocytes; spontaneous beating; fetal cardiac gene proteins; cell fusion; ischemic resistance; cardiomyocyte; cardiac ventricle; heart

Citation: Omatsu-Kanbe, M.; Fukunaga, R.; Mi, X.; Matsuura, H. Atypically Shaped Cardiomyocytes (ACMs): The Identification, Characterization and New Insights into a Subpopulation of Cardiomyocytes. *Biomolecules* **2022**, *12*, 896. https://doi.org/10.3390/biom12070896

Academic Editor: Zoran Ivanovic

Received: 31 May 2022
Accepted: 24 June 2022
Published: 27 June 2022

1. Introduction

The mammalian heart is one of the organs with a low regeneration capacity after birth [1,2]. In the development stage, embryonic cardiomyocytes arise from the early cardiac progenitors or the proliferation of pre-existing cardiomyocytes [3]. Neonatal cardiomyocytes undergo additional rounds of DNA synthesis without cytokinesis, resulting in the binucleation or multinucleation shortly after birth [4], and have transient regeneration potential during this period [5]. In the adult heart, a number of studies have reported evidence to support the notion that the pre-existing cardiomyocytes are capable of re-entering the cell cycle in both the human and mouse heart [6–10], the rate and degree of cardiomyocyte renewal under both physiological and pathophysiological conditions are far too small; thus, dead cells are not substantially replaced by renewed cells [11,12]. In the mouse heart, the turnover of cells through the proliferation of resident cardiomyocytes is estimated to occur at a rate of approximately 1.3–4% per year, whereas in the human heart, the annual turnover rate of cardiomyocytes is observed to gradually decrease from 1% at 25 years of age to 0.45% at 75 years of age; thus, the rate of cardiomyocyte exchange in adulthood is <1% per year [6,13].

Following myocardial infarction, the heart loses approximately 25% of the cardiomyocytes within a few hours [14]; however, it is estimated that <0.1% of cardiomyocytes re-enter the cell cycle [15]. At present, the "proliferative activity" of the postnatal mammalian heart is thought to be limited to the process of multinucleation and polyploidization, which occurs in the early postnatal period and in the failing heart [12]. Under pathophysiological conditions, active cardiomyocyte proliferation to cardiac hypertrophy has been reported based on an autopsy of patients with LEOPARS syndrome [16]. One of the most effective tools for understanding the mechanism of binucleation, polyploidization and cell-cycle arrest of cardiomyocytes is now thought to be a model system using induced pluripotent stem cells (iPSCs) that reproduces the essential factors in the heart [17]. In addition, the upregulation of resident or bone marrow-derived progenitor cells has been reported to give rise to cardiomyocytes for repairing the heart [18–25].

As the regulation of cardiomyocyte regeneration has been one of the most important themes in clinical research, a number of studies have attempted to manipulate endogenous progenitor cells to induce differentiation into functional cardiomyocytes. The first adult resident cardiac stem cells were identified by the expression of stem cell receptor kinase (c-kit) but not blood lineage markers (Lin), c-kit^{+}/Lin^{-} cells, in 2003 [26]. Thereafter, the identification of cardiac stem or progenitor cells based on characteristics such as the expression of stem cell antigen-1 (Sca-1) [27,28] and LIM-homeobox transcription factor (islet 1) [29] and the ability to exclude Hoechst 33342 dye [30], has been reported. During this period, these cells were demonstrated to differentiate into cardiomyocytes in vitro in response to hormones or chemicals, such as 5′-azacytidine [27,28], oxytocin [28,30] and trichostatin A [30]. Adult cardiac stem cells isolated by c-kit^{+}/Lin^{-} selection have been shown to more systematically give rise to cardiomyocytes cultured in leukemia inhibitory factor (LIF)-deprived basic differentiation medium supplemented with several chemicals in sequence, including oxytocin and activin A [31]. Negative views have also been reported, with one review reporting that no reliable data have shown the existence of cardiomyocyte-differentiable stem cells in the adult heart that might be used in practical therapeutics to repair the injured myocardium [32]. Although there are still some limitations that hinder cell therapy using cardiac stem or progenitor cells, the methodology is constantly being innovated, with methods such as proteomic analyses of cardiac progenitor cells, including cells isolated based on surface-marker selection, differentiated human embryonic stem cells (ESCs) and iPSCs [33], and the 3D structure applications composed of cardiac and endothelial progenitor cells [34]. These methods have been accelerated with the aim of translation into cell therapy [1,2].

We detect spontaneously beating cells with peculiar morphologies expressing cardiac-specific proteins but not stem cell markers in the culture of interstitial non-myocyte fraction cells obtained from adult mouse cardiac ventricles and named them "atypically shaped cardiomyocytes (ACMs)" [35]. Similar to cardiomyocytes, ACMs do not actually proliferate. These cells have also been observed mixed in with isolated ventricular myocytes in the studies of other groups [36,37]. Native ACMs are found in the interstitial spaces among ventricular myocytes in both mice and humans [38] and are considered to be a subpopulation of cardiomyocytes or immature cardiomyocytes rather than cardiac stem or progenitor cells. However, although ACMs are able to survive in the myocardium from the neonatal period to the aged period while retaining their ability to develop into a beating cell [39], the fate of ACMs both in vitro and in vivo still remains unclear.

In this review, we focus on the differences in the morphology, protein expression and cellular function between in ACMs and cardiomyocytes to explore the identity of ACMs and consider the physiological function in vivo and their potential utility in treatment.

2. Variety of Heart Cells

The heart is a complex organ comprising multiple cell types, each of which plays an important role under both physiological and pathophysiological conditions [40]. Cardiomyocytes play a major role in maintaining blood circulation via their spontaneous activity and pumping function, but they do not account for the majority of cardiac cells existing in the entire heart. Indeed, cardiomyocytes have been estimated to constitute only 30–40% of total cardiac cells [41–46], indicating that a large number of cells, called "non-myocytes", exist in the interstitial spaces. The non-myocyte group comprises heterogeneous cell lineages, such as endothelial cells, vascular smooth muscle cells, pericytes, fibroblasts, macrophages and other types of cells, such as cardiac stem or progenitor cells. These cells communicate with each other not only by direct physical contact but also by paracrine signaling [40].

2.1. Cardiac Endothelial Cells

Endothelium lines the interior surface of blood vessels and lymphatic vessels, demonstrating heterogeneity in its structure and function [47]. In the heart, endothelial cells can be divided into two types—those localized in the cardiac endothelium and those in the coronary vascular endothelium—that contribute to not only structural roles in cardiovascular homeostasis and angiogenesis but also the regulation of post-infarction remodeling via interaction with cardiomyocytes. Cross-interaction of endothelial cells and cardiomyocytes via paracrine signaling is necessary for cardiac development and regeneration [40,46,48–50]. Representatively, vascular endothelial growth factor (VEGF) is a key factor secreted by cardiomyocytes that can modulate the growth of blood vessels. The deletion of VEGF leads to not only a reduction in the number of coronary microvessels but also thinning of the ventricular wall, contractile dysfunction and an abnormal response to adrenergic stimulation [51].

2.2. Mural Cells

In the vascular system, the major compartment includes endothelial cells, smooth muscle cells and pericytes [40]. Vascular smooth muscle cells maintain the structure of the vessel wall, and their plastic nature enables the regeneration of the injured or diseased heart. Pericytes are vascular mural cells residing in the microvascular basement membrane that enwrap and support the microvessels and play an essential role in vascular remodeling [52–55]. The origin of the mural cells in the heart has been shown to be epicardial cells [56,57], but a recent study demonstrated that endocardial endothelial cells also function as a reservoir providing progenitors for mural cells [58].

2.3. Cardiac Macrophages

Tissue-resident macrophages are phagocytic immune cells that contribute to proper cardiac development and also control local homeostasis in response to diverse changes in microenvironments [59,60]. The heart contains several kinds of macrophages, which can be identified by the expression of specific markers [61]. Cardiac macrophages that originated from embryonic cells are replenished by in situ proliferation, but those derived from bone marrow cells are replenished by monocyte seeding from the circulation, playing diverse roles in the clearance of damaged cells and remodeling after myocardial infarction [61–65].

2.4. Cardiac Fibroblasts

Fibroblasts, classified as a component of connective tissue, synthesize and decompose extracellular matrix collagens and play a critical role in wound healing. Fibroblasts had been believed to constitute between 27 and 50% of the total cells in mouse and rat cardiac ventricles, respectively [41,43]. However, a recent study shows that the reevaluation of the endothelial cell population may reveal that the proportion of fibroblasts among the total cells of the heart is approximately 10%, which is lower than previously thought [45,66]. Under physiological conditions, cardiac fibroblasts are sparsely distributed in the interstitial spaces among cardiomyocytes to maintain the structure of the myocardium and the correct function of the heart via the production and degradation of the extracellular matrix [67]. Under pathophysiological conditions, however, cardiac fibroblasts rapidly proliferate and become activated, and thereafter a portion of these activated fibroblasts further differentiate into myofibroblasts. Myofibroblasts abundantly express α-SMA and other proteins to play a key role in the development of scar tissue after cardiac injury, and in the injury resolution stage, myoblasts gradually become more quiescent, reverting to a resting, senescent or apoptotic fibroblast state [66,68–72].

3. Cardiac Stem or Progenitor Cells

Historically, the types of cell populations have been classified into "static", "transit" and "stem" cell groups [73]. Adult tissue cells are thus generally thought to comprise quiescent stem cells that can proliferate and differentiate into progenitors that possess the ability to differentiate into terminal cell types, and recently, the cellular function of stem cells has been demonstrated to decline with aging [74]. In the heart, the existence and capacity of the regeneration of quiescent "stem" and/or "transit" progenitor cells have been long controversial.

Endogenous adult cardiac stem or progenitor cells were originally identified as interstitial c-kit^{+}/Lin^{-} cells in 2003 [26]. Since then, a number of studies, both in vitro and in vivo, have been undertaken to verify and clarify cardiac stem cells showing self-renewal, clonogenicity and multipotency. Some target antigens, such as Sca-1 [27,28] and Islt-1 [29], have attracted attention in efforts to identify "available" cardiac progenitor cells that might be used for cardiac regeneration, as in studies focusing on c-kit. Most such efforts are based on the immuno-phenotype definition, which is used to sieve and concentrate specifically marked cells. Another approach is to isolate cardiac side population cells identified by their ability to exclude DNA-binding Hoechst33342 dye [30]; this unique ability was found in the side population cells in a variety of organs, including bone marrow and heart [75]. These progenitor cells were demonstrated to differentiate into cardiomyocytes in vitro in response to hormones or chemicals, such as 5′-azacytidine [27,28], oxytocin [28,30] and trichostatin A [30], and also home to the injured heart after in vivo transplantation [27,30]. Consequently, c-kit^{+}/Lin^{-} cells resident in the heart have become one of the most promising targets for cell therapy [31,76–79]. The regenerative activities of these cardiac stem cells have been demonstrated under ischemic injury [80,81], pressure overload [81], and overdose of isoproterenol [82]. In contrast, some studies have reported that cardiac stem cells minimally contribute to cardiomyocytes in the adult heart or have more assertively reported that the adult heart lacks an endogenous functional pool of myogenic precursor cells [83–86]. Further techniques using cardiac stem or progenitor cells, such as proteomic and Glyco(proteo)mic analyses [33] and 3D culture in combination with the niche stem cells [34], are underway with the aim of developing practical cell therapies.

Cardiac fibroblasts are generated from epicardial and endocardial epithelial cells through the epithelial-to-mesenchymal transition and the endothelial-to-mesenchymal transition, respectively [87,88]. Single-cell transcriptional profiling studies reveal that mouse cardiac fibroblasts comprise heterogeneous lineages [89,90]. Colony-forming unit-fibroblast (CFU-F) assay, a method for determining one of the characteristic features of stem cells by examining the colony-forming ability of mesenchymal cells in the culture environment, revealed that CFU-Fs exist in the heart [91]. The cardiac CFU-Fs display similar properties to those of bone marrow mesenchymal stem cells, but lineage tracing studies have demonstrated that cardiac and bone marrow-derived CFU-Fs have different lineage origins. It is reported that cardiac fibroblasts can be directly reprogrammed into functional cardiomyocytes in mice using the developmental transcription factors Gata4, Mef2c and Tbx5 [92], thereby suggesting possible approaches for regeneration after cardiac injury. Subsequently, the direct reprogramming of cardiac fibroblasts in vivo has been reported by the injection of a cocktail of transcription factors, such as GMT and the combination of GMT and Hand2 [93,94]. However, cardiac fibroblasts in the injured heart are thought to originate from diverse cell lineages, and the population of cardiac fibroblasts undergoing successful direct reprogramming into cardiomyocytes in vivo still remains small [95]. Further improvement is considered necessary before the direct cardiac reprogramming method can be applied to human therapeutics [96].

4. Atypically Shaped Cardiomyocytes (ACMs)

We found beating cells in the three-dimensional (3D) culture of cardiomyocyte-removed fraction cells using a methylcellulose-based semi-solid medium. The isolation of cardiomyocytes is performed via coronary perfusion of the whole heart with enzymes using either the retrograde [97] or antegrade [36,98] perfusion method, and the cardiomyocyte-depleted fraction including interstitial heart cells of various cell lineages can then be cultured in semi-solid culture medium. After tightly adhering to the bottom of the appropriate dish, the cells grow over several days and start spontaneously beating. Given its peculiar morphology (Figure 1), we named these cells ACMs [35]. ACMs are also observed in the cell culture with a commonly used liquid culture medium [35], but the number of ACMs, especially beating cells, is extremely low. These beating cells have occasionally been observed as distorted or unusual cells present in culture dishes of isolated cardiomyocytes, where they have not received particular attention [36,37]. Observations indicate that 3D culture is a better condition for the development of these cells in comparison to two-dimensional (2D) culture with a liquid culture medium, which is considered one reason why these cells have not received much attention thus far.

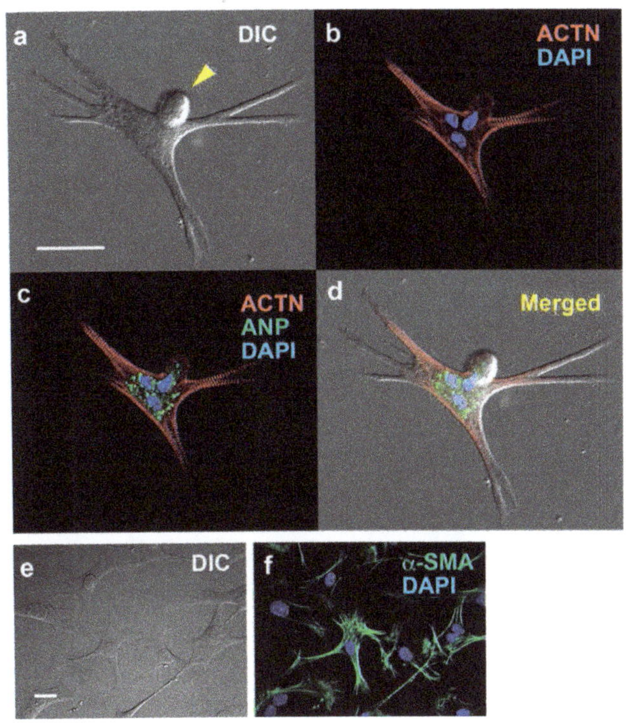

Figure 1. Immunofluorescent microscopy of ACM and cardiac fibroblasts. expressing ACTN and ANP, and cardiac fibroblasts expressing α-SMA. (**a–d**) ACM, bar, 50 μm. (**a**) Differential interference contrast image (DIC). (**b**) Immunostaining for α-actinin (ACTN, red) and DAPI staining for nuclei (blue). (**c**) Merged image for ACTN, DAPI and atrial natriuretic peptide (ANP, green). (**d**) Merged images for (**a–c**). (**e**,**f**) Cardiac fibroblasts, bar, 50 μm. (**e**) DIC. (**f**) Immunostaining for α-smooth muscle actin (α-SMA) and DAPI staining.

ACMs can be detected in cultures prepared from neonatal to aged mice without showing appreciable proliferation during long-term culturing [39]. The data thus indicate that ACMs survive in the heart for a life-long period but lack a self-renewing or clonogenic ability. A few days after plating, ACMs already show the automaticity and express characteristic proteins for ventricular and atrial myocytes, SA nodal pacemaker cells and fetal cardiac cells, indicating that the cells undergo the differentiation process to become cardiomyocytes as opposed to having multipotency. Therefore—at least based on the present data—it is unlikely that these would be classified as progenitor cells; rather, they would be classified as a subpopulation of cardiomyocytes or immature cardiomyocytes.

5. Characteristics of ACMs

5.1. Peculiar Morphology

One of the most obvious characteristics of ACMs, in addition to their spontaneous contracting activity, is that the cell shapes are markedly different from those of normal rod-shaped cardiac ventricular myocytes (Figure 1). ACMs possess a spherical shape at the start of culture, but after a few days of culture, they show peculiar shapes with many branches and/or protrusions, and most possess multiple nuclei with bulge(s) on the cell surface (Figure 1). However, while the cell shapes are complicated, the sarcomere structures are well organized and regularly arranged up to the branched tips [35–37]. The protrusions of ACMs not only come out horizontally, touching the bottom of the culture dish, but also grow three-dimensionally; thus, these cells are thicker than the proliferating fibroblast-like cells observed in the same culture dish, indicating that the cytoplasm is enriched in the contractile proteins. As the tight adherence of the plasma membrane to the culture dish may be a limitation to forming cells, the cell shapes in the heart tissues are not clear.

5.2. Automaticity

In most beating ACMs, the rhythmic action potentials are recorded that are reversibly suppressed by acetylcholine [35], while isoproterenol also shortens the peak intervals of the spontaneous Ca^{2+} transient in ACMs [99]. These findings are similar to the cell responses detected in sino-atrial (SA) node pacemaker cells [100,101], showing that the receptors and signal induction molecules for autonomic nervous systems function properly in ACMs. Unlike SA node pacemaker cells, ACMs often exhibit abnormal electrical automaticity caused by the repeated events of marked hyperpolarization to some extent and the subsequent depolarization. These arrhythmic events occur naturally, and the frequency varies from cell to cell.

It would be interesting to learn whether or not ACMs residing in the heart show automaticity under both physiological and pathophysiological conditions. The heart possesses a highly specialized system for generating rhythmic impulses to cause rhythmic contraction of the cardiomyocytes and for conducting these impulses rapidly throughout the heart. The electrical coupling within cardiomyocytes in ventricles is facilitated by gap junctions mainly composed of connexin 43 (Cx43) localized at the intercalated discs, whereas SA nodal cells do not express this type of gap junction proteins [102,103]. The expression of Cx43 in ACMs is low and particularly detected in the peri-nuclear area at the beginning of the culture before gradually increasing and spreading towards the cell periphery from five to eight days of culture along with the morphological maturation, confirmed by the expression of contractile protein [104]. Similar observations have been reported in which changes in the distribution of Cx43 in cardiomyocytes are observed in both cultured cells [105] and also in the myocardium of individuals with compensated and decompensated cardiac hypertrophy [106]. In addition, ACMs show an average maximal diastolic potential of around −65 mV [35], while the resting membrane potential in the isolated ventricular myocytes obtained from adult mice using our preparation method is approximately −74 mV on average [107], suggesting that the ACMs are not likely to electrically stimulate neighboring cardiomyocytes. These data suggest that the small native ACMs resident in the interstitial spaces have no—or very small—effects on the ventricular myocytes, at least in a healthy heart.

5.3. Protein Expression

While ACMs are small cells, with a diameter of approximately 10 μm immediately after isolation from the adult cardiac ventricular tissues, the large amounts of contractile proteins, such as α-actinin (ACTN) (Figure 1) and cardiac troponin T (cTnT), are synthesized within several days to help them increase their length more than 10-fold. Interestingly, ACMs, isolated from cardiac ventricular tissues, typically express hyperpolarization-activated cyclic nucleotide-gated channel 4 (HCN4), T-type voltage-gated Ca^{2+} channel ($Ca_V3.2$), and atrial natriuretic peptide (ANP) (Table 1). It should be noted that these proteins are not functionally expressed in normal adult ventricles and are referred to as fetal cardiac gene proteins.

Table 1. Protein expression in ACM.

Cardiac Proteins	Protein Expression	Stem Cell Markers	Protein Expression
ACTN	Positive [35]	Sca-1	None [35]
cTnT	Positive [38]	c-kit	None [35]
Cx43	Positive [39]	CD45	None [35]
HCN4	Positive [35]	CD34	None [35]
$Ca_V3.2$	Positive [39]	CD31	None [35]
ANP	Positive [39]	Flk-1	None [35]

Protein expression was determined by immunostaining for the desired protein in ACMs. ACM—atypically shaped cardiomyocyte; ACTN—α-actinin; cTnT—cardiac troponin-T; Cx43—connexin 43; HCN4—hyperpolarization-activated cyclic nucleotide-gated channel 4; $Ca_V3.2$—T-type Ca^{2+} channel; ANP—atrial natriuretic peptide; Sca-1—stem cell antigen-1; c-kit—stem cell factor receptor; CD45—leukocyte common antigen; CD34—muscle stem marker; CD31—platelet-endothelial adhesion molecule; Flk-1—vascular endothelial cell growth factor receptor 2.

Adult cardiac stem cells, which were originally found in the c-kit$^+$/Lin$^-$ cardiac cells subpopulation, possessing the characteristics of self-renewal, clonogenicity and multipotency, can regenerate functional myocardium in vivo [26]. Eliminating cells expressing endothelial and hematopoietic markers, such as CD31 and CD45, has been imperative for the isolation of cardiac stem cells from the heart [31,79]. ACMs do not show appreciable proliferation during long-term culturing and simultaneously express characteristic proteins of ventricular and atrial myocytes, SA nodal pacemaker cells and fetal cardiac cells, suggesting that these cells have different properties from cardiac stem cells or stem cells of other types. As expected, typical stem cell surface markers, Sca-1, c-kit, hematopoietic stem cell marker (CD34), platelet-endothelial adhesion molecule-1 (CD31) and vascular endothelial cell growth factor receptor 2 (Flk-1), are absent in ACMs (Table 1). The evidence suggests that ACMs should not be classified as stem or progenitor cells that can undergo the differentiation process.

5.4. Ischemic Tolerance

Since ACMs are cultured in a methylcellulose-based semi-solid medium, it is more difficult to perform medium exchange than with a liquid medium, so a small droplet of the liquid medium is added from time to time during long-term culture. However, the proliferation of fibroblasts renders the culture condition an acidic one of chronic malnutrition. Surprisingly, ACMs in a medium that has turned a yellowish-orange color continue contracting rhythmically despite the harsh culture environment for over one month. Furthermore, another important feature of ACMs during long-term culture is that they fuse with each other while maintaining spontaneous contraction.

Cardiomyocytes are highly sensitive to hypoxia; for example, in the human heart, ventricular myocytes become irreversibly damaged approximately 30 min after blood flow stops [68,71]. Following the reestablishment of the blood flow, the reperfusion of the coronary artery paradoxically causes further—occasionally lethal—damage to the heart and as well as a wide range of organs [108]. ACMs can survive these lethal ischemic conditions by blocking airflow by covering the cell suspension with oil [109], with half of the cells subsequently able to develop into beating cells, while approximately 85% of ventricular myocytes die within 90 min [99]. ACMs have thus been shown to possess ischemic resistance in addition to resistance to malnutrition and acidic conditions, suggesting that these cells may survive after cardiomyocytes die in injured hearts. The finding that ACMs ($Prp^+/cTnT^+$ cells, as stated in Section 6) survived in the peripheral area of infarction in pathological heart tissue specimens obtained from the patients who had a myocardial infarction [104] supports this view.

5.5. *Constitutively Active Autophagy*

Autophagy is an evolutionally conserved process for the degradation of long-lived and/or damaged proteins and organelles occurring in cells throughout the body [110–112] and during the neonatal period in particular, thus providing a necessary source of energy in various tissues [113]. Under physiological conditions, the autophagy activity in the heart remains low, aiding in the maintenance of the cell components, and is rapidly activated under pathophysiological conditions to support cardio-protection [114–119]. However, during the only early neonatal starvation period, autophagy in the heart is known to be activated [113].

In contrast, autophagy is constitutively activated in ACMs, not only to support cellular functions, but it also plays an essential role in the development of beating cells in the culture [99]. Constitutively active autophagy is thought to enable ACMs to rapidly synthesize proteins to grow into large beating cells and to continue beating in harsh environments by providing a source of energy.

5.6. *Multinucleation and Cell Fusion*

The most obvious characteristic of ACMs is that approximately 76% of these cells have multiple nuclei (Figure 1), sometimes more than four [38]. In mice, the majority of cardiomyocytes undergo additional DNA synthesis without cytokinesis within approximately 14 days after birth, resulting in multinucleation [3,12]. However, we have not observed such nuclear fission in ACMs during culturing; thus, one possible explanation for the multiple nuclei of ACMs is that cells fuse with each other, whether it occurs in vivo or after culturing.

"Native ACMs" that are resident in the heart tissues (as stated in Section 6) have often been observed as clusters [38], potentially resulting in the formation of fused cells with plural nuclei through the cell preparation procedure. Among the various interstitial cells that exist in the culture dish, ACMs are only able to fuse with the same type of cell [38]. When other types of cells are attached, ACMs pull the cell membrane along while beating, but they do not fuse with any cells other than ACMs [99]. These phenomena indicate that ACMs can only fuse with cells that have a cell membrane with the same membrane components, leading to the hypothesis that cardiomyocytes are indeed candidates for fusion with ACMs. Unfortunately, the coculture of ACMs and ventricular myocytes has not succeeded because isolated ventricular myocytes cannot survive in culture while ACMs settle at the bottom of the culture dish and grow. Some ACMs are observed to closely contact with ventricular myocytes but not attach, and then ventricular myocytes shrink and die over time. Therefore, the behavior of native ACMs should be explored in the future.

6. Methods for Identifying Native ACMs

Specifying the origin of ACMs is also an important theme to be resolved. Although not completely elucidated, the facts are that ACMs possess all of the characteristics of ventricular and atrial myocytes and SA node pacemaker cells and survive in neonatal to older hearts with the preserved expression of fetal cardiac gene proteins [35,39] suggest that ACMs originated from immature fetal heart cells. The yield and condition of ACMs obtained from the heart are always much better with the successful isolation of ventricular myocytes, indicating that the separation of native ACMs from the neighboring cells triggers the development of these cells into beating cells. The data may suggest that native ACMs in vivo can survive and grow when these cells become independent from the microenvironment due to the death of adjacent cells in patients who suffer myocardial infarction [104]. Based on the view that ACMs are likely to exist close to ventricular myocytes in the interstitial space in the myocardium, we sought to find native ACMs in the heart.

6.1. Combination of Cellular Prion Protein and Contractile Protein as Markers for ACMs

ACMs can be easily identified in the culture several days after plating based on their peculiar shape and automatic beating activity, but there is no way to identify these cells using specific markers in order to isolate them from a mixture of interstitial cells. In the early developing heart, cardiac transcription factors, such as NK2 transcription factor related, locus 5 (Nkx2.5) and T-box transcription factor (Tbx) 5, are first expressed in the anterior lateral plate mesoderm, and genes encoding cardiac-specific structural proteins are subsequently expressed in the cardiac crescent [120]. However, little is known about proteins that serve as the cell surface markers in the postnatal heart, as cardiomyocytes can be easily identified by their morphological features, such as rod-like shape and sarcomere structure. Identifying specific markers of ACMs, classified as cardiomyocytes, in the myocardium is thus expected to be extremely difficult. However, this problem must be overcome in order to clarify the nature of these cells in vivo.

Cellular prion protein (Prp), also known as CD230, is ubiquitously expressed in mice, except in some specific tissues, such as the liver [121], and can be used to mark nascent cardiomyocytes and cardiac progenitors and cardiogenic populations in differentiating embryonic stem cells (ESCs) [92]. Given that Prp is strongly expressed on the cell membrane of ACMs from just after isolation before development into beating cells, those cells co-expressing Prp and contractile protein cardiac troponin T (cTnT) existing in the interstitial spaces among ventricular myocytes are considered native ACMs [38,104]. $Prp^+/cTnT^+$ cells have been shown to exist sparsely in the interstitial space of myocardial tissues in both mouse [38] and human [104] healthy heart tissue sections. While this approach is a time-consuming way of detecting double-immune-positive cells and an indirect method of identifying ACMs, there was no better way than this, at that time, to identify ACMs, especially in human pathological specimens.

6.2. Fetal Cardiac Specific Gene Proteins for Marking ACMs

ACMs express fetal gene proteins, such as ANP [122,123] and CaV3.2 (Table 1) [124,125], which are potential cell marker candidates. Both ANP and Cav3.2 are preferentially expressed in the fetal heart but not in adult cardiac ventricles and show an increased protein expression in the diseased heart. ANP may have utility as a marker for ACMs since it is absent in the ventricular myocytes in the postnatal healthy heart. However, particular care must be taken, as there is evidence that ANP is activated in heart failure [126,127] and also expressed in macrophages [128]. We are currently conducting analyses involving ACMs marked with these fetal cardiac-specific genes.

7. Future Outlook

Since the discovery of ACMs in the culture, the question arose as to what cell lineage ACMs should be classified into. One possibility was that ACMs are damaged ventricular myocytes. Cardiomyocytes that are damaged during isolation contract irregularly due to the Ca^{2+} overload and eventually die; ACMs develop into complicated and larger beating cells, which indicates that ACMs are not cardiomyocytes with damage. Incomplete or unnecessary cells are usually considered to be removed via the apoptotic pathway in the developing heart [129]. However, ACMs are not thought to be cells that are destined to die, as they can survive for the entire life of the heart or for a long-term period in culture while resisting harsh environmental conditions. The next possibility was that ACMs function as cardiac progenitors, such as satellite cells in skeletal muscle. In skeletal muscles, quiescent myoblasts, known as satellite cells, function as a source of skeletal muscle progenitor cells in the regenerative process of injured muscles, which is a well-confirmed cellular function [130]. However, we did not observe any proliferation of these cells after over a month of culture, leading us to conclude that ACMs actually lack the proliferative ability. ACMs were then determined to be cardiomyocytes that have already undergone differentiation not only due to the absence of stem cell marker proteins but also their spontaneous development into beating cells with the expression of cardiac-specific gene products. The data thus suggest that there may be some as-yet-unknown reason for the existence of these cells.

For several years since the discovery of stem cell marker-positive cells in the heart, a number of studies have reported the development of new types of cell therapy utilizing these cells [11,23,27,30,131]. The purpose of these studies was to regenerate myocardium not only via the differentiation of stem or progenitor cells into new cardiomyocytes but also by the fusion of these cells with damaged cardiomyocytes; their application in medical treatment has not yet been achieved. ACMs are observed to selectively fuse with cells of the same type during culture [38,99], which strongly suggests that these cells can fuse to damaged ventricular myocytes in the injured heart. This fusion ability may be a key clue for clarifying the role of the ACMs in the heart.

A number of therapeutic methods, including cellular therapies using regenerated cells, such as iPSCs [132,133], and acellular therapies using cell-free factors, such as exosomes, microRNAs and antibodies, have been actively studied to overcome heart diseases, and both approaches have their own advantages and disadvantages [134]. It is highly likely that undifferentiated fetal heart cells are the ancestors of ACMs, with the ability to undergo cell fusion and survive for a life-long period. If native ACMs fuse with damaged cardiomyocytes in vivo, it may restore their function. However, conversely, they may also, unfortunately, be the cause of the arrhythmias that often occur after myocardial infarction.

Overall, it would be worthwhile to find a way to consider ACMs as a subpopulation of cardiomyocytes for a new type of cell therapy or to clarify the pathogenesis of ACMs-induced arrhythmias. We hope that this review will help researchers in cardiovascular and other fields to learn about the existence of ACMs and promote interest in starting studies on their cell function and utility.

Funding: This work was supported by a Grant-in-Aid for Scientific Research (C) from the Japan Society for the Promotion of Science (18K06871 and 21K06768 to M.O.-K.).

Institutional Review Board Statement: Not applicable.

Informed Consent Statement: Not applicable.

Data Availability Statement: Not applicable.

Conflicts of Interest: The authors declare no conflict of interest.

References

1. Alonaizan, R.; Carr, C. Cardiac regeneration following myocardial infarction: The need for regeneration and a review of cardiac stromal cell populations used for transplantation. *Biochem. Soc. Trans.* **2022**, *50*, 269–281. [CrossRef] [PubMed]
2. Mehanna, R.A.; Essawy, M.M.; Barkat, M.A.; Awaad, A.K.; Thabet, E.H.; Hamed, H.A.; Elkafrawy, H.; Khalil, N.A.; Sallam, A.; Kholief, M.A.; et al. Cardiac stem cells: Current knowledge and future prospects. *World J. Stem Cells* **2022**, *14*, 1–40. [CrossRef] [PubMed]
3. Zhao, M.T.; Ye, S.; Su, J.; Garg, V. Cardiomyocyte Proliferation and Maturation: Two Sides of the Same Coin for Heart Regeneration. *Front. Cell Dev. Biol.* **2020**, *8*, 594226. [CrossRef] [PubMed]
4. Bishop, S.P.; Zhou, Y.; Nakada, Y.; Zhang, J. Changes in Cardiomyocyte Cell Cycle and Hypertrophic Growth During Fetal to Adult in Mammals. *J. Am. Heart Asoc.* **2021**, *10*, e017839. [CrossRef] [PubMed]
5. Porrello, E.R.; Mahmoud, A.I.; Simpson, E.; Hill, J.A.; Richardson, J.A.; Olson, E.N.; Sadek, H.A. Transient regenerative potential of the neonatal mouse heart. *Science* **2011**, *25*, 1078–1080. [CrossRef]
6. Bergmann, O.; Bhardwaj, R.D.; Bernard, S.; Zdunek, S.; Barnabe-Heider, F.; Walsh, S.; Zupicich, J.; Alkass, K.; Buchholz, B.A.; Druid, H.; et al. Evidence for cardiomyocyte renewal in humans. *Science* **2009**, *324*, 98–102. [CrossRef]
7. Mollova, M.; Bersell, K.; Walsh, S.; Savla, J.; Das, L.T.; Park, S.Y.; Silberstein, L.E.; Dos Remedios, C.G.; Graham, D.; Colan, S.; et al. Cardiomyocyte proliferation contributes to heart growth in young humans. *Proc. Natl. Acad. Sci. USA* **2013**, *110*, 1446–1451. [CrossRef]

8. Senyo, S.E.; Steinhauser, M.L.; Pizzimenti, C.L.; Yang, V.K.; Cai, L.; Wang, M.; Wu, T.D.; Guerquin-Kern, J.L.; Lechene, C.P.; Lee, R.T. Mammalian heart renewal by pre-existing cardiomyocytes. *Nature* **2013**, *493*, 433–436. [CrossRef]
9. Ali, S.R.; Hippenmeyer, S.; Saadat, L.V.; Luo, L.; Weissman, I.L.; Ardehali, R. Existing cardiomyocytes generate cardiomyocytes at a low rate after birth in mice. *Proc. Natl. Acad. Sci. USA* **2014**, *111*, 8850–8855. [CrossRef]
10. Kimura, W.; Xiao, F.; Canseco, D.C.; Muralidhar, S.; Thet, S.; Zhang, H.M.; Abderrahman, Y.; Chen, R.; Garcia, J.A.; Shelton, J.M.; et al. Hypoxia fate mapping identifies cycling cardiomyocytes in the adult heart. *Nature* **2015**, *523*, 226–230. [CrossRef]
11. Laflamme, M.A.; Murry, C.E. Heart regeneration. *Nature* **2011**, *473*, 326–335. [CrossRef] [PubMed]
12. Ponnusamy, M.; Li, P.F.; Wang, K. Understanding cardiomyocyte proliferation: An insight into cell cycle activity. *Cell. Mol. Life Sci.* **2017**, *74*, 1019–1034. [CrossRef] [PubMed]
13. Bergmann, O.; Zdunek, S.; Felker, A.; Salehpour, M.; Alkass, K.; Bernard, S.; Sjostrom, S.L.; Szewczykowska, M.; Jackowska, T.; Dos Remedios, C.; et al. Dynamics of Cell Generation and Turnover in the Human Heart. *Cell* **2015**, *161*, 1566–1575. [CrossRef]
14. Murry, C.E.; Reinecke, H.; Pabon, L.M. Regeneration gaps: Observations on stem cells and cardiac repair. *J. Am. Coll. Cardiol.* **2006**, *47*, 1777–1785. [CrossRef]
15. Beltrami, A.P.; Urbanek, K.; Kajstura, J.; Yan, S.M.; Finato, N.; Bussani, R.; Nadal-Ginard, B.; Silvestri, F.; Leri, A.; Beltrami, C.A.; et al. Evidence that human cardiac myocytes divide after myocardial infarction. *N. Engl. J. Med.* **2001**, *344*, 1750–1757. [CrossRef] [PubMed]
16. Nakagama, Y.; Inuzuka, R.; Ichimura, K.; Hinata, M.; Takehara, H.; Takeda, N.; Kakiuchi, S.; Shiraga, K.; Asakai, H.; Shindo, T.; et al. Accelerated Cardiomyocyte Proliferation in the Heart of a Neonate with LEOPARD Syndrome-Associated Fatal Cardiomyopathy. *Circ. Heart Fail.* **2018**, *11*, e004660. [CrossRef] [PubMed]
17. Milliron, H.Y.; Weiland, M.J.; Kort, E.J.; Jovinge, S. Isolation of Cardiomyocytes Undergoing Mitosis With Complete Cytokinesis. *Circ. Res.* **2019**, *125*, 1070–1086. [CrossRef]
18. Makino, S.; Fukuda, K.; Miyoshi, S.; Konishi, F.; Kodama, H.; Pan, J.; Sano, M.; Takahashi, T.; Hori, S.; Abe, H.; et al. Cardiomyocytes can be generated from marrow stromal cells in vitro. *J. Clin. Investig.* **1999**, *103*, 697–705. [CrossRef]
19. Orlic, D.; Kajstura, J.; Chimenti, S.; Jakoniuk, I.; Anderson, S.M.; Li, B.; Pickel, J.; McKay, R.; Nadal-Ginard, B.; Bodine, D.M.; et al. Bone marrow cells regenerate infarcted myocardium. *Nature* **2001**, *410*, 701–705. [CrossRef]
20. Toma, C.; Pittenger, M.F.; Cahill, K.S.; Byrne, B.J.; Kessler, P.D. Human mesenchymal stem cells differentiate to a cardiomyocyte phenotype in the adult murine heart. *Circulation* **2002**, *105*, 93–98. [CrossRef]
21. Muller, P.; Kazakov, A.; Semenov, A.; Bohm, M.; Laufs, U. Pressure-induced cardiac overload induces upregulation of endothelial and myocardial progenitor cells. *Cardiovasc. Res.* **2008**, *77*, 151–159. [CrossRef] [PubMed]
22. Kubo, H.; Jaleel, N.; Kumarapeli, A.; Berretta, R.M.; Bratinov, G.; Shan, X.; Wang, H.; Houser, S.R.; Margulies, K.B. Increased cardiac myocyte progenitors in failing human hearts. *Circulation* **2008**, *118*, 649–657. [CrossRef] [PubMed]
23. Lutz, M.; Rosenberg, M.; Kiessling, F.; Eckstein, V.; Heger, T.; Krebs, J.; Ho, A.D.; Katus, H.A.; Frey, N. Local injection of stem cell factor (SCF) improves myocardial homing of systemically delivered c-kit$^+$ bone marrow-derived stem cells. *Cardiovasc. Res.* **2008**, *77*, 143–150. [CrossRef] [PubMed]
24. Waring, C.D.; Vicinanza, C.; Papalamprou, A.; Smith, A.J.; Purushothaman, S.; Goldspink, D.F.; Nadal-Ginard, B.; Torella, D.; Ellison, G.M. The adult heart responds to increased workload with physiologic hypertrophy, cardiac stem cell activation, and new myocyte formation. *Eur. Heart J.* **2014**, *35*, 2722–2731. [CrossRef]
25. Marketou, M.E.; Parthenakis, F.; Vardas, P.E. Pathological Left Ventricular Hypertrophy and Stem Cells: Current Evidence and New Perspectives. *Stem Cells Int.* **2016**, *2016*, 5720758. [CrossRef] [PubMed]
26. Beltrami, A.P.; Barlucchi, L.; Torella, D.; Baker, M.; Limana, F.; Chimenti, S.; Kasahara, H.; Rota, M.; Musso, E.; Urbanek, K.; et al. Adult cardiac stem cells are multipotent and support myocardial regeneration. *Cell* **2003**, *114*, 763–776. [CrossRef]
27. Oh, H.; Bradfute, S.B.; Gallardo, T.D.; Nakamura, T.; Gaussin, V.; Mishina, Y.; Pocius, J.; Michael, L.H.; Behringer, R.R.; Garry, D.J.; et al. Cardiac progenitor cells from adult myocardium: Homing, differentiation, and fusion after infarction. *Proc. Natl. Acad. Sci. USA* **2003**, *100*, 12313–12318. [CrossRef]
28. Matsuura, K.; Nagai, T.; Nishigaki, N.; Oyama, T.; Nishi, J.; Wada, H.; Sano, M.; Toko, H.; Akazawa, H.; Sato, T.; et al. Adult cardiac Sca-1-positive cells differentiate into beating cardiomyocytes. *J. Biol. Chem.* **2004**, *279*, 11384–11391. [CrossRef]
29. Laugwitz, K.L.; Moretti, A.; Lam, J.; Gruber, P.; Chen, Y.; Woodard, S.; Lin, L.Z.; Cai, C.L.; Lu, M.M.; Reth, M.; et al. Postnatal isl1+ cardioblasts enter fully differentiated cardiomyocyte lineages. *Nature* **2005**, *433*, 647–653. [CrossRef]
30. Oyama, T.; Nagai, T.; Wada, H.; Naito, A.T.; Matsuura, K.; Iwanaga, K.; Takahashi, T.; Goto, M.; Mikami, Y.; Yasuda, N.; et al. Cardiac side population cells have a potential to migrate and differentiate into cardiomyocytes in vitro and in vivo. *J. Cell Biol.* **2007**, *176*, 329–341. [CrossRef]
31. Vicinanza, C.; Aquila, I.; Scalise, M.; Cristiano, F.; Marino, F.; Cianflone, E.; Mancuso, T.; Marotta, P.; Sacco, W.; Lewis, F.C.; et al. Adult cardiac stem cells are multipotent and robustly myogenic: C-kit expression is necessary but not sufficient for their identification. *Cell Death Differ.* **2017**, *24*, 2101–2116. [CrossRef]
32. He, L.; Nguyen, N.B.; Ardehali, R.; Zhou, B. Heart Regeneration by Endogenous Stem Cells and Cardiomyocyte Proliferation: Controversy, Fallacy, and Progress. *Circulation* **2020**, *142*, 275–291. [CrossRef] [PubMed]
33. Sebastião, M.J.; Marcos-Silva, L.; Gomes-Alves, P.; Alves, P.M. Proteomic and Glyco(proteo)mic tools in the profiling of cardiac progenitors and pluripotent stem cell derived cardiomyocytes: Accelerating translation into therapy. *Biotechnol. Adv.* **2021**, *49*, 107755. [CrossRef] [PubMed]

34. Monsanto, M.M.; Wang, B.J.; Ehrenberg, Z.R.; Echeagaray, O.; White, K.S.; Alvarez, R., Jr.; Fisher, K.; Sengphanith, S.; Muliono, A.; Gude, N.A.; et al. Enhancing myocardial repair with CardioClusters. *Nat. Commun.* **2020**, *11*, 3955. [CrossRef]
35. Omatsu-Kanbe, M.; Matsuura, H. A novel type of self-beating cardiomyocytes in adult mouse ventricles. *Biochem. Biophys. Res. Commun.* **2009**, *381*, 361–366. [CrossRef] [PubMed]
36. Ackers-Johnson, M.; Li, P.Y.; Holmes, A.P.; O'Brien, S.M.; Pavlovic, D.; Foo, R.S. A Simplified, Langendorff-Free Method for Concomitant Isolation of Viable Cardiac Myocytes and Nonmyocytes From the Adult Mouse Heart. *Circ. Res.* **2016**, *119*, 909–920. [CrossRef]
37. Haftbaradaran Esfahani, P.; ElBeck, Z.; Sagasser, S.; Li, X.; Hossain, M.B.; Talukdar, H.A.; Sandberg, R.; Knoll, R. Cell shape determines gene expression: Cardiomyocyte morphotypic transcriptomes. *Basic Res. Cardiol.* **2019**, *115*, 7. [CrossRef]
38. Omatsu-Kanbe, M.; Nishino, Y.; Nozuchi, N.; Sugihara, H.; Matsuura, H. Prion protein- and cardiac troponin T-marked interstitial cells from the adult myocardium spontaneously develop into beating cardiomyocytes. *Sci. Rep.* **2014**, *4*, 7301. [CrossRef]
39. Omatsu-Kanbe, M.; Yamamoto, T.; Mori, Y.; Matsuura, H. Self-beating atypically shaped cardiomyocytes survive a long-term postnatal development while preserving the expression of fetal cardiac genes in mice. *J. Histochem. Cytochem.* **2010**, *58*, 543–551. [CrossRef]
40. Litvinukova, M.; Talavera-Lopez, C.; Maatz, H.; Reichart, D.; Worth, C.L.; Lindberg, E.L.; Kanda, M.; Polanski, K.; Heinig, M.; Lee, M.; et al. Cells of the adult human heart. *Nature* **2020**, *588*, 466–472. [CrossRef]
41. Nag, A.C. Study of non-muscle cells of the adult mammalian heart: A fine structural analysis and distribution. *Cytobios* **1980**, *28*, 41–61.
42. Limana, F.; Urbanek, K.; Chimenti, S.; Quaini, F.; Leri, A.; Kajstura, J.; Nadal-Ginard, B.; Izumo, S.; Anversa, P. bcl-2 overexpression promotes myocyte proliferation. *Proc. Natl. Acad. Sci. USA* **2002**, *99*, 6257–6262. [CrossRef]
43. Banerjee, I.; Fuseler, J.W.; Price, R.L.; Borg, T.K.; Baudino, T.A. Determination of cell types and numbers during cardiac development in the neonatal and adult rat and mouse. *Am. J. Physiol.* **2007**, *293*, H1883–H1891. [CrossRef] [PubMed]
44. Camelliti, P.; Borg, T.K.; Kohl, P. Structural and functional characterisation of cardiac fibroblasts. *Cardiovasc. Res.* **2005**, *65*, 40–51. [CrossRef] [PubMed]
45. Pinto, A.R.; Ilinykh, A.; Ivey, M.J.; Kuwabara, J.T.; D'Antoni, M.L.; Debuque, R.; Chandran, A.; Wang, L.; Arora, K.; Rosenthal, N.A.; et al. Revisiting Cardiac Cellular Composition. *Circ. Res.* **2016**, *118*, 400–409. [CrossRef] [PubMed]
46. Talman, V.; Kivela, R. Cardiomyocyte-Endothelial Cell Interactions in Cardiac Remodeling and Regeneration. *Front. Cardiovasc. Med.* **2018**, *5*, 101. [CrossRef]
47. Aird, W.C. Phenotypic heterogeneity of the endothelium: I. Structure, function, and mechanisms. *Circ. Res.* **2007**, *100*, 158–173. [CrossRef] [PubMed]
48. Brutsaert, D.L. Cardiac endothelial-myocardial signaling: Its role in cardiac growth, contractile performance, and rhythmicity. *Physiol. Rev.* **2003**, *83*, 59–115. [CrossRef] [PubMed]
49. Kivela, R.; Hemanthakumar, K.A.; Vaparanta, K.; Robciuc, M.; Izumiya, Y.; Kidoya, H.; Takakura, N.; Peng, X.; Sawyer, D.B.; Elenius, K.; et al. Endothelial Cells Regulate Physiological Cardiomyocyte Growth via VEGFR2-Mediated Paracrine Signaling. *Circulation* **2019**, *139*, 2570–2584. [CrossRef]
50. Colliva, A.; Braga, L.; Giacca, M.; Zacchigna, S. Endothelial cell-cardiomyocyte crosstalk in heart development and disease. *J. Physiol.* **2020**, *598*, 2923–2939. [CrossRef]
51. Giordano, F.J.; Gerber, H.P.; Williams, S.P.; VanBruggen, N.; Bunting, S.; Ruiz-Lozano, P.; Gu, Y.; Nath, A.K.; Huang, Y.; Hickey, R.; et al. A cardiac myocyte vascular endothelial growth factor paracrine pathway is required to maintain cardiac function. *Proc. Natl. Acad. Sci. USA* **2001**, *98*, 5780–5785. [CrossRef] [PubMed]
52. Armulik, A.; Genove, G.; Betsholtz, C. Pericytes: Developmental, physiological, and pathological perspectives, problems, and promises. *Dev. Cell* **2011**, *21*, 193–215. [CrossRef] [PubMed]
53. Murray, I.R.; Baily, J.E.; Chen, W.C.W.; Dar, A.; Gonzalez, Z.N.; Jensen, A.R.; Petrigliano, F.A.; Deb, A.; Henderson, N.C. Skeletal and cardiac muscle pericytes: Functions and therapeutic potential. *Pharmacol. Ther.* **2017**, *171*, 65–74. [CrossRef] [PubMed]
54. Alex, L.; Frangogiannis, N.G. Pericytes in the infarcted heart. *Vasc Biol.* **2019**, *1*, H23–H31. [CrossRef]
55. Su, H.; Cantrell, A.C.; Zeng, H.; Zhu, S.H.; Chen, J.X. Emerging Role of Pericytes and Their Secretome in the Heart. *Cells* **2021**, *10*, 548. [CrossRef]
56. Olivey, H.E.; Svensson, E.C. Epicardial-myocardial signaling directing coronary vasculogenesis. *Circ. Res.* **2010**, *106*, 818–832. [CrossRef]
57. Wang, G.; Jacquet, L.; Karamariti, E.; Xu, Q. Origin and differentiation of vascular smooth muscle cells. *J. Physiol.* **2015**, *593*, 3013–3030. [CrossRef]
58. Chen, Q.; Zhang, H.; Liu, Y.; Adams, S.; Eilken, H.; Stehling, M.; Corada, M.; Dejana, E.; Zhou, B.; Adams, R.H. Endothelial cells are progenitors of cardiac pericytes and vascular smooth muscle cells. *Nat. Commun.* **2016**, *7*, 12422. [CrossRef]
59. Lavin, Y.; Winter, D.; Blecher-Gonen, R.; David, E.; Keren-Shaul, H.; Merad, M.; Jung, S.; Amit, I. Tissue-resident macrophage enhancer landscapes are shaped by the local microenvironment. *Cell* **2014**, *159*, 1312–1326. [CrossRef]
60. Leid, J.; Carrelha, J.; Boukarabila, H.; Epelman, S.; Jacobsen, S.E.; Lavine, K.J. Primitive Embryonic Macrophages are Required for Coronary Development and Maturation. *Circ. Res.* **2016**, *118*, 1498–1511. [CrossRef]
61. Alvarez-Argote, S.; O'Meara, C.C. The Evolving Roles of Cardiac Macrophages in Homeostasis, Regeneration, and Repair. *Int. J. Mol. Sci.* **2021**, *22*, 7923. [CrossRef] [PubMed]

62. Pinto, A.R.; Paolicelli, R.; Salimova, E.; Gospocic, J.; Slonimsky, E.; Bilbao-Cortes, D.; Godwin, J.W.; Rosenthal, N.A. An abundant tissue macrophage population in the adult murine heart with a distinct alternatively-activated macrophage profile. *PLoS ONE* **2012**, *7*, e36814. [CrossRef] [PubMed]
63. Dick, S.A.; Macklin, J.A.; Nejat, S.; Momen, A.; Clemente-Casares, X.; Althagafi, M.G.; Chen, J.; Kantores, C.; Hosseinzadeh, S.; Aronoff, L.; et al. Self-renewing resident cardiac macrophages limit adverse remodeling following myocardial infarction. *Nat. Immunol.* **2019**, *20*, 29–39. [CrossRef] [PubMed]
64. Deniset, J.F.; Belke, D.; Lee, W.Y.; Jorch, S.K.; Deppermann, C.; Hassanabad, A.F.; Turnbull, J.D.; Teng, G.; Rozich, I.; Hudspeth, K.; et al. Gata6(+) Pericardial Cavity Macrophages Relocate to the Injured Heart and Prevent Cardiac Fibrosis. *Immunity* **2019**, *51*, 131–140.e135. [CrossRef] [PubMed]
65. Gula, G.; Ruminski, S.; Niderla-Bielinska, J.; Jasinska, A.; Kiernozek, E.; Jankowska-Steifer, E.; Flaht-Zabost, A.; Ratajska, A. Potential functions of embryonic cardiac macrophages in angiogenesis, lymphangiogenesis and extracellular matrix remodeling. *Histochem. Cell Biol.* **2021**, *155*, 117–132. [CrossRef] [PubMed]
66. Tallquist, M.D.; Molkentin, J.D. Redefining the identity of cardiac fibroblasts. *Nat. Rev. Cardiol.* **2017**, *14*, 484–491. [CrossRef]
67. Trial, J.; Entman, M.L.; Cieslik, K.A. Mesenchymal stem cell-derived inflammatory fibroblasts mediate interstitial fibrosis in the aging heart. *J. Mol. Cell. Cardiol.* **2016**, *91*, 28–34. [CrossRef]
68. Sheppard, M.N. The coronary arteries—Atherosclerosis and ischaemic heart disease. In *Practical Cardiovascular Pathology*, 2nd ed.; Makepeace, C., Ed.; Hodder Arnold: London, UK, 2011; pp. 24–66.
69. Leask, A. Getting to the heart of the matter: New insights into cardiac fibrosis. *Circ. Res.* **2015**, *116*, 1269–1276. [CrossRef]
70. Rockey, D.C.; Bell, P.D.; Hill, J.A. Fibrosis—A common pathway to organ injury and failure. *N. Engl. J. Med.* **2015**, *372*, 1138–1149. [CrossRef]
71. Schoen, F.J. The Heart. In *Robins and Cotoran Pathologic Basis of Disease*, 9th ed.; Kumar, V., Abbas, A.K., Aster, J.C., Eds.; Elsevier: Philadelphia, PA, USA, 2015; pp. 523–578.
72. Kurose, H. Cardiac Fibrosis and Fibroblasts. *Cells* **2021**, *10*, 1716. [CrossRef]
73. Lajtha, L.G. Stem cell concepts. *Differentiation* **1979**, *14*, 23–34. [CrossRef] [PubMed]
74. Keyes, B.E.; Fuchs, E. Stem cells: Aging and transcriptional fingerprints. *J. Cell Biol.* **2018**, *217*, 79–92. [CrossRef] [PubMed]
75. Goodell, M.A.; Rosenzweig, M.; Kim, H.; Marks, D.F.; DeMaria, M.; Paradis, G.; Grupp, S.A.; Sieff, C.A.; Mulligan, R.C.; Johnson, R.P. Dye efflux studies suggest that hematopoietic stem cells expressing low or undetectable levels of CD34 antigen exist in multiple species. *Nat. Med.* **1997**, *3*, 1337–1345. [CrossRef] [PubMed]
76. He, J.Q.; Vu, D.M.; Hunt, G.; Chugh, A.; Bhatnagar, A.; Bolli, R. Human cardiac stem cells isolated from atrial appendages stably express c-kit. *PLoS ONE* **2011**, *6*, e27719. [CrossRef] [PubMed]
77. Gambini, E.; Pompilio, G.; Biondi, A.; Alamanni, F.; Capogrossi, M.C.; Agrifoglio, M.; Pesce, M. C-kit+ cardiac progenitors exhibit mesenchymal markers and preferential cardiovascular commitment. *Cardiovasc. Res.* **2011**, *89*, 362–373. [CrossRef] [PubMed]
78. Leong, Y.Y.; Ng, W.H.; Ellison-Hughes, G.M.; Tan, J.J. Cardiac Stem Cells for Myocardial Regeneration: They Are Not Alone. *Front. Cardiovasc. Med.* **2017**, *4*, 47. [CrossRef]
79. Cianflone, E.; Cappetta, D.; Mancuso, T.; Sabatino, J.; Marino, F.; Scalise, M.; Albanese, M.; Salatino, A.; Parrotta, E.I.; Cuda, G.; et al. Statins Stimulate New Myocyte Formation After Myocardial Infarction by Activating Growth and Differentiation of the Endogenous Cardiac Stem Cells. *Int. J. Mol. Sci.* **2020**, *21*, 7927. [CrossRef]
80. Linke, A.; Muller, P.; Nurzynska, D.; Casarsa, C.; Torella, D.; Nascimbene, A.; Castaldo, C.; Cascapera, S.; Bohm, M.; Quaini, F.; et al. Stem cells in the dog heart are self-renewing, clonogenic, and multipotent and regenerate infarcted myocardium, improving cardiac function. *Proc. Natl. Acad. Sci. USA* **2005**, *102*, 8966–8971. [CrossRef]
81. Hsieh, P.C.; Segers, V.F.; Davis, M.E.; MacGillivray, C.; Gannon, J.; Molkentin, J.D.; Robbins, J.; Lee, R.T. Evidence from a genetic fate-mapping study that stem cells refresh adult mammalian cardiomyocytes after injury. *Nat. Med.* **2007**, *13*, 970–974. [CrossRef]
82. Aquila, I.; Cianflone, E.; Scalise, M.; Marino, F.; Mancuso, T.; Filardo, A.; Smith, A.J.; Cappetta, D.; De Angelis, A.; Urbanek, K.; et al. c-kit Haploinsufficiency impairs adult cardiac stem cell growth, myogenicity and myocardial regeneration. *Cell Death Dis.* **2019**, *10*, 436. [CrossRef]
83. van Berlo, J.H.; Kanisicak, O.; Maillet, M.; Vagnozzi, R.J.; Karch, J.; Lin, S.C.; Middleton, R.C.; Marban, E.; Molkentin, J.D. c-kit+ cells minimally contribute cardiomyocytes to the heart. *Nature* **2014**, *509*, 337–341. [CrossRef] [PubMed]
84. Nadal-Ginard, B.; Ellison, G.M.; Torella, D. Absence of evidence is not evidence of absence: Pitfalls of cre knock-ins in the c-Kit locus. *Circ. Res.* **2014**, *115*, 415–418. [CrossRef] [PubMed]
85. Sultana, N.; Zhang, L.; Yan, J.; Chen, J.; Cai, W.; Razzaque, S.; Jeong, D.; Sheng, W.; Bu, L.; Xu, M.; et al. Resident c-kit(+) cells in the heart are not cardiac stem cells. *Nat. Commun.* **2015**, *6*, 8701. [CrossRef] [PubMed]
86. Vicinanza, C.; Aquila, I.; Cianflone, E.; Scalise, M.; Marino, F.; Mancuso, T.; Fumagalli, F.; Giovannone, E.D.; Cristiano, F.; Iaccino, E.; et al. Kit(cre) knock-in mice Fail. to fate-map cardiac stem cells. *Nature* **2018**, *555*, E1–E5. [CrossRef] [PubMed]
87. Gittenberger-de Groot, A.C.; Vrancken Peeters, M.P.; Mentink, M.M.; Gourdie, R.G.; Poelmann, R.E. Epicardium-derived cells contribute a novel population to the myocardial wall and the atrioventricular cushions. *Circ. Res.* **1998**, *82*, 1043–1052. [CrossRef]
88. Plikus, M.V.; Wang, X.; Sinha, S.; Forte, E.; Thompson, S.M.; Herzog, E.L.; Driskell, R.R.; Rosenthal, N.; Biernaskie, J.; Horsley, V. Fibroblasts: Origins, definitions, and functions in health and disease. *Cell* **2021**, *184*, 3852–3872. [CrossRef]

89. Muhl, L.; Genove, G.; Leptidis, S.; Liu, J.; He, L.; Mocci, G.; Sun, Y.; Gustafsson, S.; Buyandelger, B.; Chivukula, I.V.; et al. Single-cell analysis uncovers fibroblast heterogeneity and criteria for fibroblast and mural cell identification and discrimination. *Nat. Commun.* **2020**, *11*, 3953. [CrossRef]
90. Skelly, D.A.; Squiers, G.T.; McLellan, M.A.; Bolisetty, M.T.; Robson, P.; Rosenthal, N.A.; Pinto, A.R. Single-Cell Transcriptional Profiling Reveals Cellular Diversity and Intercommunication in the Mouse Heart. *Cell Rep.* **2018**, *22*, 600–610. [CrossRef]
91. Chong, J.J.; Chandrakanthan, V.; Xaymardan, M.; Asli, N.S.; Li, J.; Ahmed, I.; Heffernan, C.; Menon, M.K.; Scarlett, C.J.; Rashidianfar, A.; et al. Adult cardiac-resident MSC-like stem cells with a proepicardial origin. *Cell Stem Cell* **2011**, *9*, 527–540. [CrossRef]
92. Ieda, M.; Fu, J.D.; Delgado-Olguin, P.; Vedantham, V.; Hayashi, Y.; Bruneau, B.G.; Srivastava, D. Direct reprogramming of fibroblasts into functional cardiomyocytes by defined factors. *Cell* **2010**, *142*, 375–386. [CrossRef]
93. Qian, L.; Huang, Y.; Spencer, C.I.; Foley, A.; Vedantham, V.; Liu, L.; Conway, S.J.; Fu, J.D.; Srivastava, D. In vivo reprogramming of murine cardiac fibroblasts into induced cardiomyocytes. *Nature* **2012**, *485*, 593–598. [CrossRef]
94. Song, K.; Nam, Y.J.; Luo, X.; Qi, X.; Tan, W.; Huang, G.N.; Acharya, A.; Smith, C.L.; Tallquist, M.D.; Neilson, E.G.; et al. Heart repair by reprogramming non-myocytes with cardiac transcription factors. *Nature* **2012**, *485*, 599–604. [CrossRef] [PubMed]
95. Deb, A.; Ubil, E. Cardiac fibroblast in development and wound healing. *J. Mol. Cell. Cardiol.* **2014**, *70*, 47–55. [CrossRef] [PubMed]
96. Yamakawa, H.; Ieda, M. Cardiac regeneration by direct reprogramming in this decade and beyond. *Inflamm. Regen.* **2021**, *41*, 20. [CrossRef] [PubMed]
97. Shioya, T. A simple technique for isolating healthy heart cells from mouse models. *J. Physiol. Sci.* **2007**, *57*, 327–335. [CrossRef]
98. Omatsu-Kanbe, M.; Yoshioka, K.; Fukunaga, R.; Sagawa, H.; Matsuura, H. A simple antegrade perfusion method for isolating viable single cardiomyocytes from neonatal to aged mice. *Physiol. Rep.* **2018**, *6*, e13688. [CrossRef]
99. Omatsu-Kanbe, M.; Matsuura, H. Ischemic survival and constitutively active autophagy in self-beating atypically-shaped cardiomyocytes (ACMs): Characterization of a new subpopulation of heart cells. *J. Physiol. Sci.* **2013**, *63*, 17–29. [CrossRef]
100. Honjo, H.; Boyett, M.R.; Kodama, I.; Toyama, J. Correlation between electrical activity and the size of rabbit sino-atrial nodal cells. *J. Physiol.* **1996**, *496*, 795–808. [CrossRef]
101. Irisawa, H.; Brown, H.F.; Giles, W. Cardiac pacemaking in the sinoatrial node. *Physiol. Rev.* **1993**, *73*, 197–227. [CrossRef]
102. Davis, L.M.; Kanter, H.L.; Beyer, E.C.; Saffitz, J.E. Distinct gap junction protein phenotypes in cardiac tissues with disparate conduction properties. *J. Am. Coll. Cardiol.* **1994**, *24*, 1124–1132. [CrossRef]
103. Verheijck, E.E.; van Kempen, M.J.; Veereschild, M.; Lurvink, J.; Jongsma, H.J.; Bouman, L.N. Electrophysiological featuRes. of the mouse sinoatrial node in relation to connexin distribution. *Cardiovasc. Res.* **2001**, *52*, 40–50. [CrossRef]
104. Omatsu-Kanbe, M.; Nozuchi, N.; Nishino, Y.; Mukaisho, K.I.; Sugihara, H.; Matsuura, H. Identification of cardiac progenitors that survive in the ischemic human heart after ventricular myocyte death. *Sci. Rep.* **2017**, *7*, 41318. [CrossRef] [PubMed]
105. Formigli, L.; Francini, F.; Nistri, S.; Margheri, M.; Luciani, G.; Naro, F.; Silvertown, J.D.; Orlandini, S.Z.; Meacci, E.; Bani, D. Skeletal myoblasts overexpressing relaxin improve differentiation and communication of primary murine cardiomyocyte cell cultures. *J. Mol. Cell. Cardiol.* **2009**, *47*, 335–345. [CrossRef] [PubMed]
106. Kostin, S.; Dammer, S.; Hein, S.; Klovekorn, W.P.; Bauer, E.P.; Schaper, J. Connexin 43 expression and distribution in compensated and decompensated cardiac hypertrophy in patients with aortic stenosis. *Cardiovasc. Res.* **2004**, *62*, 426–436. [CrossRef] [PubMed]
107. Hoshino, S.; Omatsu-Kanbe, M.; Nakagawa, M.; Matsuura, H. Postnatal developmental decline in IK1 in mouse ventricular myocytes isolated by the Langendorff perfusion method: Comparison with the chunk method. *Pflugers Arch.* **2012**, *463*, 649–668. [CrossRef] [PubMed]
108. Hausenloy, D.J.; Yellon, D.M. Myocardial ischemia-reperfusion injury: A neglected therapeutic target. *J. Clin. Investig.* **2013**, *123*, 92–100. [CrossRef]
109. Diaz, R.J.; Losito, V.A.; Mao, G.D.; Ford, M.K.; Backx, P.H.; Wilson, G.J. Chloride channel inhibition blocks the protection of ischemic preconditioning and hypo-osmotic stress in rabbit ventricular myocardium. *Circ. Res.* **1999**, *84*, 763–775. [CrossRef]
110. Yamaguchi, O. Autophagy in the Heart. *Circ. J.* **2019**, *83*, 697–704. [CrossRef]
111. Klionsky, D.J.; Emr, S.D. Autophagy as a regulated pathway of cellular degradation. *Science* **2000**, *290*, 1717–1721. [CrossRef]
112. Levine, B.; Klionsky, D.J. Development by self-digestion: Molecular mechanisms and biological functions of autophagy. *Dev. Cell* **2004**, *6*, 463–477. [CrossRef]
113. Kuma, A.; Hatano, M.; Matsui, M.; Yamamoto, A.; Nakaya, H.; Yoshimori, T.; Ohsumi, Y.; Tokuhisa, T.; Mizushima, N. The role of autophagy during the early neonatal starvation period. *Nature* **2004**, *432*, 1032–1036. [CrossRef] [PubMed]
114. Decker, R.S.; Wildenthal, K. Lysosomal alterations in hypoxic and reoxygenated hearts. I. Ultrastructural and cytochemical changes. *Am. J. Pathol.* **1980**, *98*, 425–444. [PubMed]
115. Hamacher-Brady, A.; Brady, N.R.; Gottlieb, R.A. Enhancing macroautophagy protects against ischemia/reperfusion injury in cardiac myocytes. *J. Biol. Chem.* **2006**, *281*, 29776–29787. [CrossRef] [PubMed]
116. Matsui, Y.; Takagi, H.; Qu, X.; Abdellatif, M.; Sakoda, H.; Asano, T.; Levine, B.; Sadoshima, J. Distinct roles of autophagy in the heart during ischemia and reperfusion: Roles of AMP-activated protein kinase and BeClin.1 in mediating autophagy. *Circ. Res.* **2007**, *100*, 914–922. [CrossRef]
117. Nakai, A.; Yamaguchi, O.; Takeda, T.; Higuchi, Y.; Hikoso, S.; Taniike, M.; Omiya, S.; Mizote, I.; Matsumura, Y.; Asahi, M.; et al. The role of autophagy in cardiomyocytes in the basal state and in response to hemodynamic stress. *Nat. Med.* **2007**, *13*, 619–624. [CrossRef]

118. Kuma, A.; Komatsu, M.; Mizushima, N. Autophagy-monitoring and autophagy-deficient mice. *Autophagy* **2017**, *13*, 1619–1628. [CrossRef]
119. Ikeda, S.; Zablocki, D.; Sadoshima, J. The role of autophagy in death of cardiomyocytes. *J. Mol. Cell. Cardiol.* **2021**, *165*, 1–8. [CrossRef]
120. Hochgreb, T.; Linhares, V.L.; Menezes, D.C.; Sampaio, A.C.; Yan, C.Y.; Cardoso, W.V.; Rosenthal, N.; Xavier-Neto, J. A caudorostral wave of RALDH2 conveys anteroposterior information to the cardiac field. *Development* **2003**, *130*, 5363–5374. [CrossRef]
121. Ford, M.J.; Burton, L.J.; Morris, R.J.; Hall, S.M. Selective expression of prion protein in peripheral tissues of the adult mouse. *Neuroscience* **2002**, *113*, 177–192. [CrossRef]
122. Bloch, K.D.; Seidman, J.G.; Naftilan, J.D.; Fallon, J.T.; Seidman, C.E. Neonatal atria and ventricles secrete atrial natriuretic factor via tissue-specific secretory pathways. *Cell* **1986**, *47*, 695–702. [CrossRef]
123. Cameron, V.A.; Ellmers, L.J. Minireview: Natriuretic peptides during development of the fetal heart and circulation. *Endocrinology* **2003**, *144*, 2191–2194. [CrossRef]
124. Perez-Reyes, E. Molecular physiology of low-voltage-activated t-type calcium channels. *Physiol. Rev.* **2003**, *83*, 117–161. [CrossRef] [PubMed]
125. Chiang, C.S.; Huang, C.H.; Chieng, H.; Chang, Y.T.; Chang, D.; Chen, J.J.; Chen, Y.C.; Chen, Y.H.; Shin, H.S.; Campbell, K.P.; et al. The Ca(v)3.2 T-type Ca(2+) channel is required for pressure overload-induced cardiac hypertrophy in mice. *Circ. Res.* **2009**, *104*, 522–530. [CrossRef]
126. Vinnakota, S.; Chen, H.H. The Importance of Natriuretic Peptides in Cardiometabolic Diseases. *J. Endocr. Soc.* **2020**, *4*, bvaa052. [CrossRef] [PubMed]
127. Kuwahara, K. The natriuretic peptide system in heart failure: Diagnostic and therapeutic implications. *Pharmacol. Ther.* **2021**, *227*, 107863. [CrossRef] [PubMed]
128. Vollmar, A.M.; Schulz, R. Gene expression and secretion of atrial natriuretic peptide by murine macrophages. *J. Clin. Invest.* **1994**, *94*, 539–545. [CrossRef] [PubMed]
129. Barbosky, L.; Lawrence, D.K.; Karunamuni, G.; Wikenheiser, J.C.; Doughman, Y.Q.; Visconti, R.P.; Burch, J.B.; Watanabe, M. Apoptosis in the developing mouse heart. *Dev. Dyn.* **2006**, *235*, 2592–2602. [CrossRef]
130. Yin, H.; Price, F.; Rudnicki, M.A. Satellite cells and the muscle stem cell niche. *Physiol. Rev.* **2013**, *93*, 23–67. [CrossRef]
131. Barile, L.; Messina, E.; Giacomello, A.; Marban, E. Endogenous cardiac stem cells. *Prog. Cardiovasc. Dis.* **2007**, *50*, 31–48. [CrossRef]
132. Masumoto, H.; Ikuno, T.; Takeda, M.; Fukushima, H.; Marui, A.; Katayama, S.; Shimizu, T.; Ikeda, T.; Okano, T.; Sakata, R.; et al. Human iPS cell-engineered cardiac tissue sheets with cardiomyocytes and vascular cells for cardiac regeneration. *Sci. Rep.* **2014**, *4*, 6716. [CrossRef]
133. Martinez-Falguera, D.; Iborra-Egea, O.; Galvez-Monton, C. iPSC Therapy for Myocardial Infarction in Large Animal Models: Land of Hope and Dreams. *Biomedicines* **2021**, *9*, 1836. [CrossRef] [PubMed]
134. Zhu, D.; Cheng, K. Cardiac Cell Therapy for Heart Repair: Should the Cells Be Left Out? *Cells* **2021**, *10*, 641. [CrossRef] [PubMed]

biomolecules

MDPI

Article

Preferential Expression of Ca^{2+}-Stimulable Adenylyl Cyclase III in the Supraventricular Area, including Arrhythmogenic Pulmonary Vein of the Rat Heart

Yosuke Okamoto [1,*], Naing Ye Aung [2], Masahiro Tanaka [3], Yuji Takeda [4], Daichi Takagi [5], Wataru Igarashi [5], Kuniaki Ishii [3], Mitsunori Yamakawa [2] and Kyoichi Ono [1,*]

1 Department of Cell Physiology, Akita Graduate School of Medicine, Hondo 010-8543, Japan
2 Pathological and Image Analysis Center, Cancer Research Center, Faculty of Medicine, Yamagata University, Iida-Nishi 990-9585, Japan; a.naing@med.id.yamagata-u.ac.jp (N.Y.A.); myamakawa236.kamihozawa@gmail.com (M.Y.)
3 Department of Pharmacology, Faculty of Medicine, Yamagata University, Iida-Nishi 990-9585, Japan; breathless.06@icloud.com (M.T.); kuishii@med.id.yamagata-u.ac.jp (K.I.)
4 Department of Immunology, Faculty of Medicine, Yamagata University, Iida-Nishi 990-9585, Japan; yu-takeda@med.id.yamagata-u.ac.jp
5 Department of Cardiovascular Surgery, Akita Graduate School of Medicine, Hondo 010-8543, Japan; takagi@med.akita-u.ac.jp (D.T.); vvataruig@gmail.com (W.I.)
* Correspondence: okamoto@med.akita-u.ac.jp (Y.O.); onok@med.akita-u.ac.jp (K.O.)

Abstract: Ectopic excitability in pulmonary veins (PVs) is the major cause of atrial fibrillation. We previously reported that the inositol trisphosphate receptor in rat PV cardiomyocytes cooperates with the Na^+-Ca^{2+} exchanger to provoke ectopic automaticity in response to norepinephrine. Here, we focused on adenylyl cyclase (AC) as another effector of norepinephrine stimulation. RT-PCR, immunohistochemistry, and Western blotting revealed that the abundant expression of Ca^{2+}-stimulable AC3 was restricted to the supraventricular area, including the PVs. All the other AC isotypes hardly displayed any region-specific expressions. Immunostaining of isolated cardiomyocytes showed an enriched expression of AC3 along the t-tubules in PV myocytes. The cAMP-dependent response of L-type Ca^{2+} currents in the PV and LA cells is strengthened by the 0.1 mM intracellular Ca^{2+} condition, unlike in the ventricular cells. The norepinephrine-induced automaticity of PV cardiomyocytes was reversibly suppressed by 100 μM SQ22536, an adenine-like AC inhibitor. These findings suggest that the specific expression of AC3 along t-tubules may contribute to arrhythmogenic automaticity in rat PV cardiomyocytes.

Keywords: adenylyl cyclase; supraventricular area; pulmonary vein; arrhythmia; atrial fibrillation; t-tubule

Citation: Okamoto, Y.; Aung, N.Y.; Tanaka, M.; Takeda, Y.; Takagi, D.; Igarashi, W.; Ishii, K.; Yamakawa, M.; Ono, K. Preferential Expression of Ca^{2+}-Stimulable Adenylyl Cyclase III in the Supraventricular Area, including Arrhythmogenic Pulmonary Vein of the Rat Heart. *Biomolecules* **2022**, *12*, 724. https://doi.org/10.3390/biom12050724

Academic Editor: Pietro Scicchitano

Received: 5 April 2022
Accepted: 19 May 2022
Published: 20 May 2022

1. Introduction

Ectopic excitability originating in pulmonary veins (PVs) leads to symptomatic atrial fibrillation, the most frequently observed arrhythmia [1]. Several studies have investigated the arrhythmogenic properties of PVs [2–6], and various types of spontaneous activities have been recorded in PV cardiomyocytes, including the sinoatrial node (SAN)-like spontaneous action potentials [2,4], digitalis-induced arrhythmia [7], pacing-induced spontaneous activity after ryanodine treatment [8], and burst-pacing-induced atrial fibrillation after genetic deletion of ryanodine receptor-regulating subunit FKBP12.6 [9]. This arrhythmogenic activity partly depends on the distinct electrophysiological characteristics of cells and involves intracellular Ca^{2+} dynamics [10]. Previous studies have demonstrated the specific mechanisms by which Ca^{2+} contributes to the firing of action potentials under α- and β-adrenoceptor stimulation [3,5,11]. In response to norepinephrine application, local diastolic Ca^{2+} release via the inositol 1,4,5-triphosphate receptor 2 (IP_3R_2) activates the Na^+-Ca^{2+} exchanger (NCX), which accelerates diastolic depolarization. The functional coupling between IP_3R_2 and NCX is facilitated by both the structural characteristics of PV cardiomyocytes and the localization of the two molecules, in that PV cardiomyocytes possess enriched t-tubules compared with atrial myocytes, and IP_3R_2 co-localizes with NCX in t-tubule microdomains [5,11]. The potential automaticity of PV cardiomyocytes is manifested by the cooperation of spontaneous Ca^{2+} release and an ensemble of sarcolemmal electrogenic molecules, including hyperpolarization-activated K^+ current (I_{KH}) in dog [12] and hyperpolarization-activated Cl^- current ($I_{Cl,h}$) in rat [13].

Numerous studies have shown that the cardiac autonomic nervous system is involved in the occurrence and persistence of atrial fibrillation [14]. Indeed, abundant sympathetic nerves innervate human PVs [15]. Norepinephrine released from sympathetic nerve terminals stimulates both α- and β-adrenergic receptors, which in turn activate G_q- and G_s-protein-coupled signaling pathways, respectively. The former pathway activates IP_3R_2 and thus induces sporadic $[Ca^{2+}]_i$ elevations [16,17], while the latter pathway involves cyclic-adenosine monophosphate (cAMP) and acts on a variety of downstream effectors, including protein kinase A. Protein kinase A phosphorylation of intracellular targets coordinates a number of physiological outputs, such as myocyte contraction and relaxation and pacemaker activity. The synthesis of cAMP from ATP is controlled by the enzyme adenylyl cyclase (AC). The AC family is composed of nine membrane-bound isoforms (AC1–9) and a soluble form (AC10). Most of the AC isoforms can be found in the heart [18]. While AC5 and AC6 are major isoforms in adult cardiomyocytes [19], AC1 is proposed to function as a Ca^{2+}-stimulable AC in the SAN and to contribute to normal pacemaker rhythm [20]. In fact, cAMP activity in SAN cells is enriched compared with that in ventricular cardiomyocytes [21], enabling spontaneous, rhythmic, local diastolic Ca^{2+} releases from storage sites and subsequent activation of the NCX inward current that accelerates diastolic depolarization. It is now believed that the release and uptake of Ca^{2+} via the ryanodine receptor (Ca^{2+} clock) facilitates SAN automaticity with an ensemble of several ion currents on the membrane (membrane clock) [22].

In the present study, we found a unique expression pattern of another Ca^{2+}-stimulable AC, AC3, in the supraventricular area and investigated the impacts of its activity on the potential automaticity of PV cardiomyocytes using the patch-clamp technique.

2. Materials and Methods

2.1. Tissue Preparation

For RNA extraction from different regions of the cardiopulmonary tissues, male Wistar rats (10–15 weeks of age) were euthanized. The rats were deeply anesthetized by intraperitoneal injection of 0.15 mg/kg of medetomidine, 2.0 mg/kg of midazolam, and 2.5 mg/kg of butorphanol. After checking the suppression of the nociceptive reflex, the chest cavity was opened under artificial ventilation, and the aorta was cannulated in situ to perfuse the coronary arteries. The heart and lungs were excised in a block, mounted on a Langendorff apparatus, and sequentially perfused with ice-cold, Ca^{2+}-free, and heparinized Tyrode solution for 5 min. The composition of Tyrode solution (in mM) was: NaCl 136.9, KCl 5.4, NaH_2PO_4 0.33, HEPES 5.0, $MgCl_2$ 0.5, glucose 5.5; pH 7.4 adjusted with NaOH. The heart-lung block was pinned to a tissue bath. Soft tissue containing the vagus nerve and adipose tissue was trimmed off under a stereomicroscope (Figure 1A). The PV was exposed after excision of the pulmonary trunk; PV samples were collected from the proximal and peripheral left PVs, while right PVs were routinely excluded. The left atrium (LA) adjacent to the PV, the left ventricle (LV) mass, and the right atrium were isolated. Each LV mass was dissected into three pieces (i.e., samples). All cardiopulmonary tissues were isolated within 10 min after removal of the heart from the body. During tissue isolation, the external solution was perfused at a rate of 10 mL/min on ice. All tissue samples were fresh-frozen in liquid nitrogen and stored at −80 °C for later use.

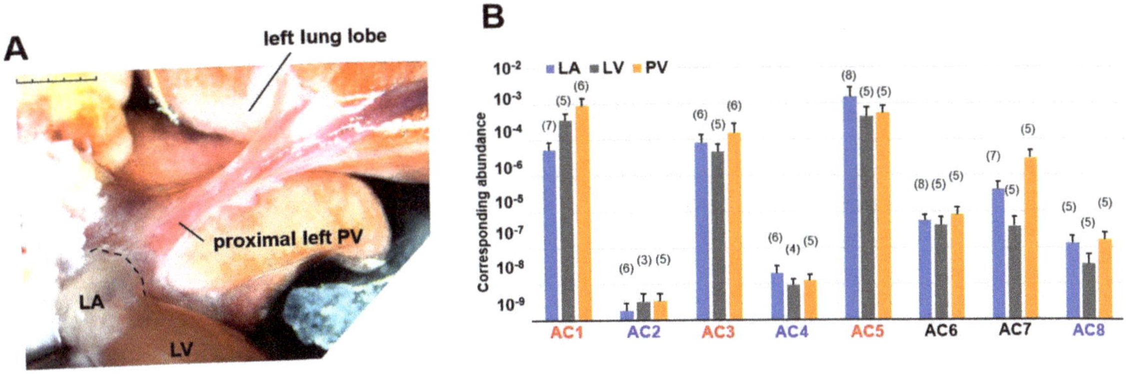

Figure 1. Macroscopic view of the preparation and genetic RT-PCR screening for adenylyl cyclase (AC) superfamily members. (**A**) A stereomicroscopic snapshot of the cardiopulmonary preparation focusing on a left pulmonary vein (PV). To visualize the veins, fresh blood was left inside. The PV extends from the left atrium (LA) and visibly branches into the left superior lung lobe. The post caval lobe of the lung is behind the left PV. The branches from the right PV are out of focus and cannot be fully seen in this view. A part of the left ventricular (LV) mass is observed at the bottom. Scale bar: 5 mm with 1 mm increments. (**B**) Quantitative RT-PCR for AC1–8. By fitting each quantitative cycle to a standard linear curve, the abundance of each reverse-transcribed cDNA is measured quantitatively. The exogenous standard curves are prepared by RT-PCR of diluted RNA extracted from the right atrium. Accordingly, the vertical axis corresponds to the same amount of standard cDNA. The ordinary one-way analysis of variance indicates that expression levels significantly vary from isotype to isotype ($p < 0.0001$). Tukey's multiple comparisons test identified statistically significant differences. AC1, AC3, and AC5 (in red) are all expressed more strongly than AC2, AC4, and AC8 (in blue). To simplify our PCR data, bar graphs with means ± standard deviations instead of dot plots. The number of experiments is provided in parentheses.

2.2. *Reverse Transcription-Polymerase Chain Reaction (RT-PCR)*

RNA was extracted using RNeasy Mini Columns (Qiagen, Chatsworth, CA, USA) and then reverse-transcribed using the High-Capacity cDNA Reverse Transcription Kit (Applied Biosystems™, Waltham, MA, USA). The transcribed cDNA was amplified by the LightCycler Nano System (Roche Holdings AG, Basel, Switzerland) by monitoring SYBR Green fluorescence. The AC expression levels in the LA, LV, and PV were analyzed by both simplified-absolute quantifications. For absolute quantification with standard linearity, cDNA from the right atrium was diluted 1 to 10^{-7} times in 10× decrements. The dilutions were amplified by the PCR primer for glyceraldehyde-3-phosphate dehydrogenase (GAPDH), and their quantification cycles (Cq) were adopted as an exogenous standard. The accuracies of all the PCR primers designed for the current study (Table 1) were evaluated by a singular peak of the melting curves and the specific bands on the Tris-borate-EDTA gel (1% agar). The specificity of all the primers was confirmed on the NCBI database (https://blast.ncbi.nlm.nih.gov/Blast.cgi (accessed on 1 March 2017).

Table 1. List of PCR primers for rat AC1–8 and GAPDH. Specificities were checked in the database NCBI-BLAST (https://blast.ncbi.nlm.nih.gov/Blast.cgi (accessed on 1 January 2020), and accuracies were checked by the singular melting peaks using LightCycler. RT-PCR products on the Tris-borate-EDTA gel. The rat GAPDH primer is routinely used in our experiments [23].

	Forward Sequences	Reverse Sequences
rAdcy1	5′-TATCCTGCTGTTCTCATGCACG-3′	5′-TTGGACATGAGGAAGTGCTGTG-3′
rAdcy2	5′-ATCATCAGGCATCATCGCCAAC-3′	5′-ATTCACTCTGTCTGCCCAGAAC-3′
rAdcy3	5′-ATGGTGAAGCTGACACTCATGC-3′	5′-AAATGGCGTGCAACATGCTC-3′
rAdcy4	5′-TGCCAAGTTCTTCCAGGTCATC-3′	5′-TGGGAGTGCAGAAATAGGGAAC-3′
rAdy5	5′-ATTGACAGACTGCGATCCGAAC-3′	5′-ATGACATGCGTTCACCTGGTTC-3′
rAdcy6	5′-ATGGCAGTTTGATGTCTGGTCC-3′	5′-ATCGGACGGTGTTAAGTTCAGC-3′
rAdcy7	5′-ATGAACAGCCACACCCTTGCTC-3′	5′-GCTCCTTCTTAAACTTCTTCT-3′
rAdcy8	5′-AGAATACTCTGGCTGCCCTAAC-3′	5′-AGCCTCGAAGAAGGAAGCAAAC-3′
rGapdh	5′-ACCACAGTCCATGCCATCAC-3′	5′-TCCACCACCCTGTTGCTGTA-3′

2.3. *Immunohistochemistry*

For heart fixation, a similar procedure as for tissue preparation was applied. After the blood was washed out from the heart with Ca^{2+}-free Tyrode solution, the heart-lung block was detached from the Langendorff apparatus. To expand the PVs, 10% neutral-buffered formalin solution was injected from the aortic arch and LV free wall into the LV cavity with a 20G needle. The block was immersed in a 10% neutral-buffered formalin solution for 24–48 h. Immunohistochemistry of the rat heart was performed with rabbit polyclonal antibodies against AC1 (Antibodies-Online, Atlanta, GA, USA; cat. no. ABIN702565), AC2 (Abcam, Cambridge, UK; cat. no. ab151470), AC3 (Antibodies-Online; cat. no. ABIN2736239), AC6 (Antibodies-Online; cat. no. ABIN751198), and AC8 (Antibodies-Online; cat. no. ABIN751228) at the manufacturer-recommended concentrations. Next, 3 µm thick sections were cut and deparaffinized. Endogenous peroxidase activity was blocked with methanol containing 0.3% hydrogen peroxide for 30 min on ice. Antigen retrieval was performed using EDTA (Antigen Retrieval Solution pH 9; Nichirei Biosciences, Tokyo, Japan) or citric acid (Antigen Retrieval Solution pH 6; Iatron Laboratories Inc., Tokyo, Japan) in an autoclave (2 atmospheres, 121 °C, 20 min). An EnVision+ System HRP-labeled polymer (anti-rabbit; DAKO, Carpinteria, CA, USA) was used. Positive reactions were detected as brown coloration with 3,3-diaminobenzidine tetrahydrochloride (Dojindo, Kumamoto, Japan). The sections were then counterstained with hematoxylin. The brain cortex was used as a positive control for AC1 and AC3. Lung epithelium cells were used as positive controls for AC2. AC6 was positively stained in the entire heart regions.

2.4. *Cell Isolation*

To isolate cardiomyocytes from the rat heart, a similar procedure as for tissue preparation was applied. The heart-lung block was mounted on a Langendorff apparatus and sequentially perfused with the following buffers at 37 °C: (1) normal Tyrode solution for approximately 3 min; (2) Ca^{2+}-free Tyrode solution for 5 min; and (3) Ca^{2+}-free Tyrode solution containing 0.05% collagenase and 0.005% elastase (Wako Pure Chemicals, Osaka, Japan) for 30 min. The composition of the normal Tyrode solution (in mM) was: NaCl 136.9, KCl 5.4, $CaCl_2$ 1.8, $MgCl_2$ 0.5, NaH_2PO_4 0.33, HEPES 5.0, glucose 5.5; pH 7.4 adjusted with NaOH. After digestion, the heart-lung block was perfused with 50 mL of a high-K^+ and low-Cl^- solution. The composition of the solution (in mM) was: L-glutamic acid 70, KOH 70, KCl 30, KH_2PO_4 10, $MgCl_2$ 1, taurine 20, glucose 10, EGTA 0.3, HEPES 10; pH 7.4 adjusted with KOH. The soft tissue containing the vagus nerves, adipose tissue, and PA was trimmed off under a stereomicroscope, and the left PV was excised from the digested block. The PV was minced gently in the high-K^+ and low-Cl^- solution, and the pieces were gently agitated to dissociate the cells. The cell suspension was stored at 4 °C for later use.

2.5. Immunocytochemistry

Freshly isolated cardiomyocytes were incubated for 30 min on poly-D-lysine-coated coverslips and subsequently fixed with 4% paraformaldehyde for 30 min at room temperature. The fixed cells were permeabilized with phosphate-buffered saline (PBS) containing 0.2% Triton X-100 for 30 min, blocked with blocking buffer (PBS containing 5% bovine serum albumin), and incubated with primary antibodies against AC3 (Proteintech, Chicago, IL, USA; cat. no. 19492-1-AP) and α-actinin (Abcam; cat. no. ab9465) at 1:500 dilution in PBS containing 1% bovine serum albumin. The cells were then incubated with appropriate secondary antibodies conjugated with either Alexa Fluor 488 or Alexa Fluor 594 (at 1:1000 dilution) and mounted on glass slides with glycerol with 10% 1,4-diazobicyclo-(2,2,2)-octane. Confocal images were obtained using a Zeiss 780 confocal microscope at ×100 magnification (pinhole size of an arbitrary unit). The captured images were processed using ZEN Image software (Zeiss, Oberkochen, Germany). The fluorescence intensity profiles of α-actinin and AC3 for each cell type were analyzed by ImageJ software, and the cross-correlations were calculated in IGOR software ver. 6.36 (Wavemetrics, Portland, OR, USA).

2.6. Patch-Clamp Electrophysiology

The conventional and perforated whole-cell patch-clamp configurations were used to record membrane currents with a patch-clamp amplifier (Axopatch 200B; Molecular Devices, Chicago, IL, USA). The composition of normal Tyrode solution (in mM) was: NaCl 136.9, KCl 5.4, NaH_2PO_4 0.33, $CaCl_2$ 1.8, HEPES 5.0, $MgCl_2$ 0.5, glucose 5.5; pH 7.4 adjusted with NaOH. During the patch-clamp experiments, the normal or Cs-Tyrode were perfused as the external solution. The Cs-Tyrode solution was prepared by replacing KCl with equimolar CsCl for L-type Ca^{2+} recording. The composition of the pipette solution (in mM) for L-type Ca^{2+} recording was: CsOH 120, aspartic acid 80, CsCl 20, Na_2-ATP 5, $MgCl_2$ 5, HEPES 5, Na_2-GTP 0.1 and EGTA 5 or $CaCl_2$ 0.1; pH 7.2 adjusted with aspartic acid. Borosilicate glass electrodes with tip resistances of 2.0–5.0 MΩ, when filled with internal solution, were used. The cell membrane capacitance (Cm) was determined by applying a 30 ms hyperpolarizing voltage-clamp step from a holding potential of −40 mV to −50 mV and integrating the area under the capacitive transient. The Cm of LA, LV, and PV cardiomyocytes were 107.25 ± 20.23 (n = 13), 238.41 ± 52.50 (n = 9), and 210.91 ± 61.93 (n = 28) pF, respectively. L-type Ca^2+ current was recorded every 10 s, where a cell was held at a potential of −80 mV. The Ca^{2+} current was induced by a depolarizing pulse to 0 mV for 500 ms, followed by a depolarizing pulse to −40 mV for 80 ms to inactivate the Na^+ current. The series resistance estimated from the Cm value and the time constant of the capacitive transient was 6.05 ± 2.14 MΩ (n = 20), selecting 5, 5, and 10 from the LA, LV, and PV cardiomyocyte experiments, respectively. The composition of the pipette solution (in mM) for norepinephrine-induced automaticity was: KOH 110, aspartic acid 110, KCl 30, NaCl 10, HEPES 5, EGTA 10; pH 7.2 adjusted with KOH. Amphotericin B at 0.3 mg/mL was also added. Membrane currents and potentials were recorded under voltage- and current-clamp conditions, respectively, at 37 ± 1 °C. Data acquisition and storage were accomplished with CLAMPEX software ver. 10.7 (Molecular Devices) running on a personal computer. The sampling frequency was 1 kHz, and low-pass filtering was performed at 500 Hz. All patch-clamp data were analyzed using IGOR software ver. 6.36 (Wavemetrics).

2.7. Drugs

Chemicals and reagents were purchased from Sigma-Aldrich (St. Louis, MO, USA) or Fujifilm Wako Pure Chemical Corporation. Stock solutions of dl-norepinephrine and dl-isoproterenol were prepared in standard and Cs-Tyrode solution, respectively. Amphotericin

B and SQ22536 were dissolved in dimethyl sulfoxide under light-free conditions and diluted 1000-fold for experiments.

2.8. Statistics

Statistical significance was determined by an ordinary one-way analysis of variance (ANOVA) with Tukey's multiple comparisons test for comparing more than three groups or by the Welch's *t*-test to compare two groups, using GraphPad Prism ver. 9.0.0. As the higher mean value of the expression levels measured by RT-PCR clearly indicates a higher variance (Figure 1B), we tested the expression levels by a parametric analysis with log transformation. To avoid negative values, a constant ($10^9 + 2$) was added to the raw values, logtransformed with a base of two, and then evaluated with a one-way ANOVA. Data were expressed as mean ± standard deviation. Values of $p < 0.05$ (*) and $p < 0.01$ (**) indicated statistical significance.

3. Results

3.1. Genetic Screening of AC Family Members in Cardiopulmonary Organs by RT-PCR

We performed multiplex PCR analysis to characterize the expression patterns of AC isoforms (AC1–8) in the heart. The activity of AC9 was estimated to be less than 3% of the total AC activity by the previous report [24], and it was not initially considered as a major AC candidate. This is why we did not include AC9 for this screening. AC10 is a soluble AC, not a membrane enzyme. In this study, we focused on membrane enzymes that contribute to arrhythmias and action potentials, and AC10 was not included. In our PCR experiments, total RNAs extracted from three different regions of cardiopulmonary tissues, left atrium (LA), left ventricle (LV), and pulmonary vein (PV) (Figure 1A), were reverse-transcribed and amplified by the LightCycler. We evaluated absolute gene expression levels in each cardiopulmonary region (Figure 1B). In addition to Ca^{2+}-inhibitable AC5, known as the major heart isotype [25], Ca^{2+}-stimulable AC1 and AC3 were predominantly expressed across the entire heart. AC1 expression was significantly higher than AC2, AC4, AC6, and AC7 expressions, AC3 expression was significantly higher than AC2, AC4, and AC8 expressions, and AC5 expression was significantly higher than AC2, AC4, AC6, AC7, and AC8 expressions ($p < 0.05$, see Figure 1B, and Table 2).

Table 2. Results of simplified-absolute PCR quantification. Statistical *p*-values for each isotype are given below the table. The number of experiments is provided in parentheses.

	$\times 10^9$	LA		LV		PV	
corresponding abundance	AC1 *	53,200.08 ± 78,097.71	(7)	340,281.06 ± 402,755.99	(5)	876,466.31 ±1,272,990.19	(6)
	AC2 #	1.95 ± 2.79	(6)	3.38 ± 3.79	(3)	3.52 ± 4.59	(5)
	AC3 $	80,931.17 ±126,819.78	(6)	46,348.88 ± 57,065.14	(5)	150,449.16 ± 301,138.14	(6)
	AC4 †	20.37 ±28.00	(6)	9.43 ± 8.09	(4)	12.32 ± 12.63	(5)
	AC5 &	1,451,129.43 ±3,541,141.54	(8)	408,442.32 ± 785,280.03	(5)	536,885.37 ± 661,839.27	(5)
	AC6	549.27 ±730.34	(8)	417.79 ± 681.51	(5)	810.37 ± 1038.31	(5)
	AC7	4051.16 ± 6371.10	(7)	393.87 ± 725.97	(5)	29,443.41 ± 39,503.56	(5)
	AC8	130.21 ± 168.44	(5)	35.74 ± 67.52	(5)	167.89 ± 205.04	(5)

* $p < 0.0001$, vs. AC2; $p < 0.0001$, vs. AC4; $p = 0.0029$, vs. AC6; $p = 0.0026$, vs. AC7; $p < 0.0001$, vs. AC8. # $p = 0.0002$, vs. AC3; $p < 0.0001$, vs. AC5. $ $p = 0.0011$, vs. AC4; $p = 0.0048$, vs. AC8. † $p < 0.0001$, vs. AC5. & $p = 0.0014$, vs. AC6; $p = 0.0012$, vs. AC7; $p < 0.0001$, vs. AC8.

3.2. Immunohistochemistry of AC Family Members in Cardiopulmonary Regions

Immunohistochemistry of cardiopulmonary regions was performed to characterize protein expression patterns. To outline the heart structure, modified Masson's trichrome (M-T) staining of a left-sided heart preparation was performed (Figure 2A). In the M-T-stained preparation, muscular areas (purple) were abundantly observed in not only the LV and LA but also in the left and right PVs. The LV and ventricular septum were separated from the LA by collagen-enriched mitral valves (turquoise blue). The left and right PVs were both identified as tubular structures extending from the LA. All of the structures appeared to consist of striated muscle, consistent with previous reports [26,27]. Immunostaining of AC6 was used as a positive control [28] and showed intense brown staining that was detected throughout the heart and in the PVs (Figure 2B). AC1 was moderately positive (Figure 2C) and distributed throughout the entire heart. AC2, a Ca^{2+}-insensitive AC isoform, was almost negative (Figure 2D). Meanwhile, AC3-positive regions were specifically restricted to the LA and PVs. AC8, another Ca^{2+}-stimulable AC, was mildly stainable through the entire heart (Figure 2F). The unique expression of AC3 in the supraventricular area was confirmed in the right-sided preparation, except for the sinoatrial node (SAN) (Figure 3B). Classically, SAN is defined as an intercaval heart area surrounded by connective tissues and containing the SAN artery. In the expanded view of Figure 4Ab, a high AC3 expression was uniformly observed in the LA, while the AC3 expression in the PV was enriched on the inner side of the vessel. There was an unstained wall between the visibly stained LA and PV. Peripheral portions of the PV were also unstained. Figure 4B shows the intercaval area in a right-sided rat heart with M-T staining (upper panel) and immunostaining of AC3 (lower panel). The M-T staining identified the SAN as a muscle region containing cross-sections of two arteries, which were surrounded by connective tissues (Figure 4Ba). The corresponding SAN segment was not stained by the anti-AC3 antibody (Figure 4Bb).

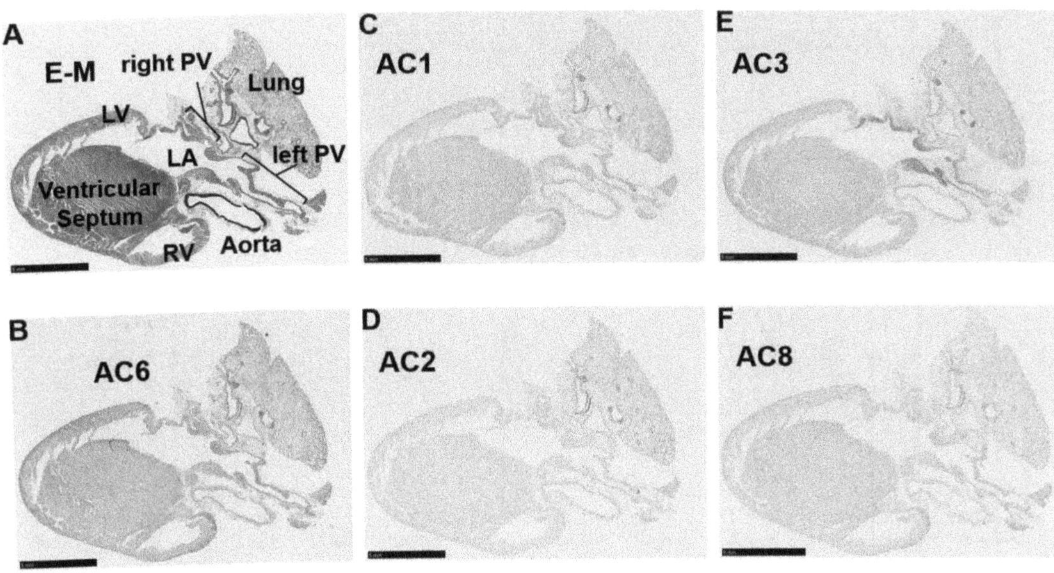

Figure 2. Immunohistochemistry of a left-sided rat heart for adenylyl cyclase AC1, AC2, AC3, AC6, and AC8. (**A**) Masson's trichrome staining in a left-sided heart from a male Wistar rat. Muscle, collagen-enriched tissue, and elastin-enriched fibers are stained purple, turquoise, and dark blue, respectively. The cardiopulmonary regions are indicated by bold letters. LA, left atrium; LV, left ventricle; PV, pulmonary vein; RV, right ventricle. (**B**) Immunohistochemistry for AC6. AC6 is known as a heart subtype of AC and presents a positive control with solid staining (brown). (**C–F**) Immunohistochemistry for AC1–3 and AC8. AC1 and AC8 are both moderately positive in (**C**,**F**). (**D**) AC2 is almost negative in the heart, while the lung epithelium is positively stained. (**E**) AC3 expression is restricted to the LA and PV. Scale bars: 5 mm.

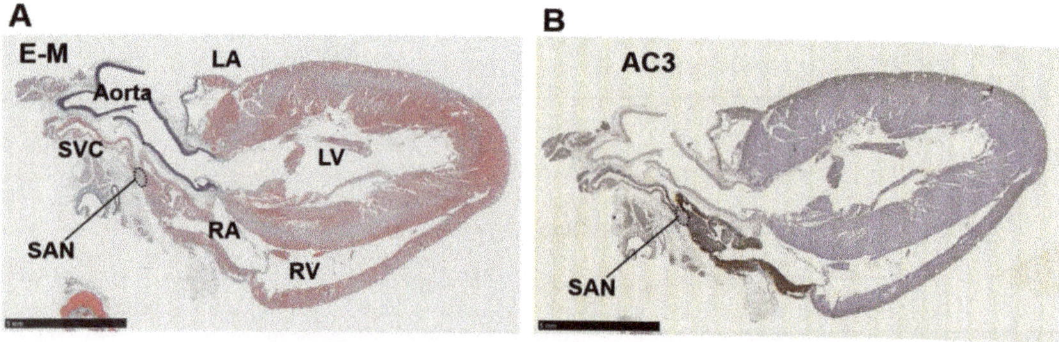

Figure 3. Immunohistochemistry of a right-sided rat heart for adenylyl cyclase AC3. (**A**) Masson's trichrome staining. LA, left atrium; LV, left ventricle; RA, right atrium; RV, right ventricle, SAN, sinoatrial node; SVC, supra vena cava. (**B**) Immunohistochemistry for AC3. The supraventricular area except the SAN is stained positive. Scale bars: 5 mm.

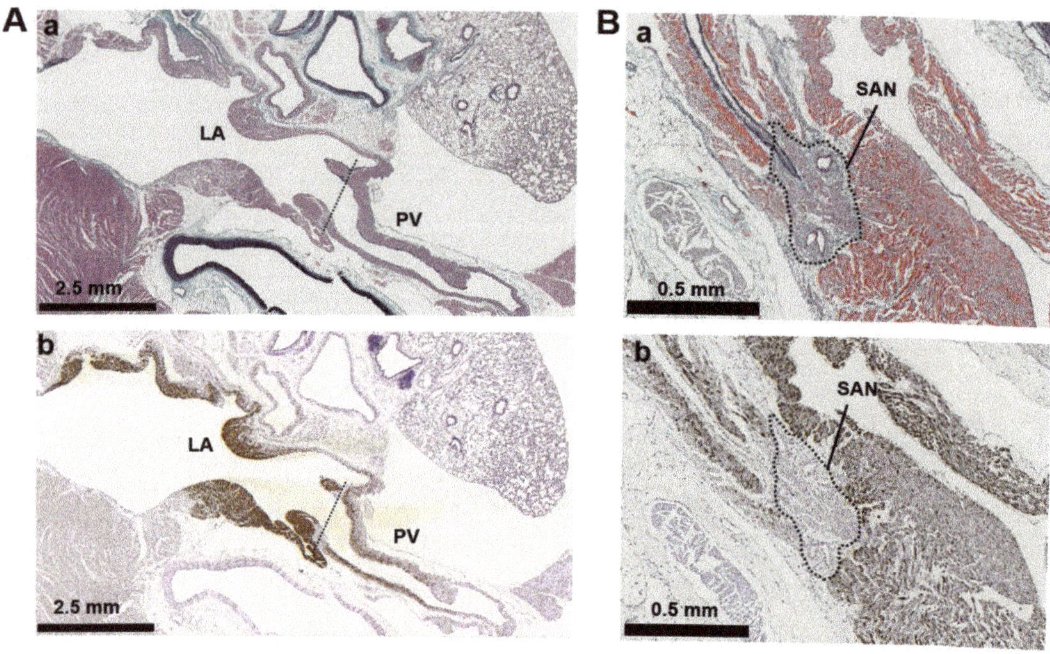

Figure 4. Immunostaining of adenylyl cyclase AC3. (A) Supraventricular regions with Masson's trichrome staining (Aa) and immunostaining for AC3 (Ab). The ventricular septum is visible beneath the mitral valve on the lower left, a connective tissue-enriched thin structure. The lung is on the upper right. The tentative boundary line between the left atrium (LA) and pulmonary vein (PV) is indicated by a dotted line. (B) Intercaval regions with Masson's trichrome staining (Ba) and AC3 (Bb) immunostaining. The sinoatrial node (SAN) is defined as a myocardial region surrounded by collagen-enriched connective tissues and contains the SAN artery. The artery is observed as circular cavities. The area indicated by the dotted line is the SAN. The superior vena cava is located immediately above the SAN to the right.

3.3. AC3 Is Localized in the t-Tubule Microdomains of PV Cardiomyocytes

Figure 5 shows the cellular locations of α-actinin (green) and AC3 (red) on the freshly isolated LA, LV, and PV cardiomyocytes. α-actinin is localized at the cardiac z-line in each myocyte. LA myocytes display weak transverse striations of AC3, while the AC3 expression along the cell contour was relatively clear (Figure 5A, middle panel). The cross-correlation between α-actinin and the AC3 signal is relatively small (Figure 5D). A previous study revealed that transverse tubules (t-tubules) of the LA were less enriched [5]. Accordingly, blunt striations in AC3 staining and a weak cross-correlation between α-actinin and the AC3 signal are supposed to reflect immature t-tubules of LA cells. Similarly, as with AC5, AC6 [29], and AC9 [30], AC3 was suggested to co-localize with α-actinin in both LV and PV cells (Figure 5B,C,E,F). As the z-line is known to overlap with t-tubules in ventricular myocytes [31], the co-localization of α-actinin and AC3 in these cells indicates that in PV cells with a well transversally organized tubular system, AC3 assembles near t-tubules. It should be noted that the nuclei of LA and LV are stained with AC3 in Figure 5A and B. It is unlikely that LV was captured under overexposed conditions compared to PV for the following reasons: (1) Neither the laser intensity of the microscope nor the gain during imaging is increased in these images compared to PV imaging in Figure 5C. (2) Microstructures in the transverse stripes were captured with similar resolution in LV and PV.

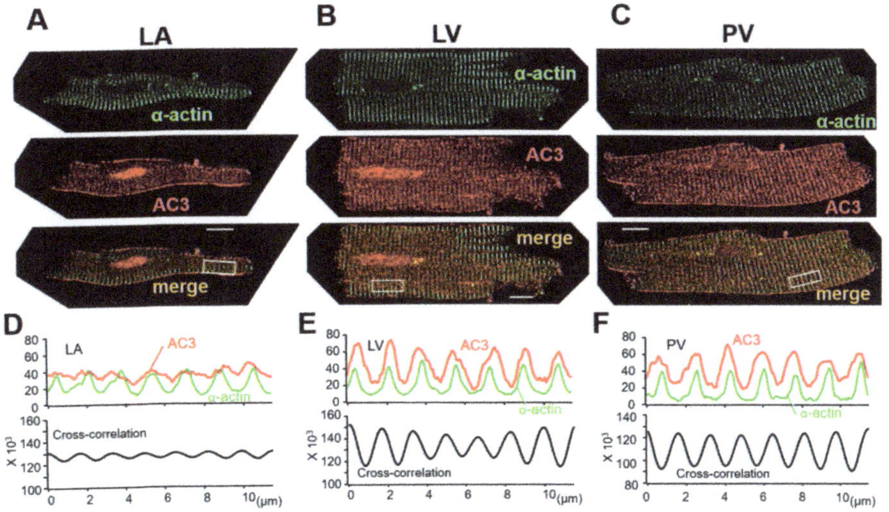

Figure 5. Immunocytochemistry for adenylyl cyclase AC3 in the myocytes of (**A**) isolated left atrium (LA), (**B**) left ventricle (LV), and (**C**) pulmonary vein (PV). α-actinin (green, all top panels) localized to the z-line of cardiomyocytes, indicating the position of the t-tubules. AC3 staining (red, all middle panels) revealed transverse striations in LV and PV myocytes, unlike in LA myocytes. AC3 is strongly expressed along the cell contour in LA myocytes. The bottom panels display the overlay of AC3 on α-actinin (yellow-orange). The white bars in the bottom panel indicate 10 μm. (**D**,**E**) and (**F**) display the profiles of α-actin and AC3 staining intensities in (**A**–**C**), and the cross-correlation between both intensity profiles, respectively. The area to be intensity-profiled is chosen from the rectangular box in the bottom panel of (**A**–**C**), all of the same size. The cross-correlation waveforms indicate that the intensities of actin and AC3 are synchronized, suggesting that they are localized near each other [11,32]. It should be noted that the nuclei of LA and LV are stained with AC3. Neither the laser intensity of the microscope nor the gain during imaging is increased in these imaging compared to PV imaging. Considering that the transverse microstructures can be captured, it is more likely that the nuclei were stained by anti-AC3 antibody rather than overexposure.

3.4. Full Activation of L-Type Ca^{2+} Current by β-Adrenergic Receptor Agonist Requires Intracellular Ca^{2+} in the Supraventricular Area, Unlike in the Ventricles

Isolated PV, LV, and LA cardiomyocytes were voltage-clamped by the patch-clamp technique. In L-type Ca^{2+} current recording, the extracellular solution was the K^+-free Cs-Tyrode solution, and the K^+-free internal solution was utilized. The Na^+ current was inactivated by the depolarization step from a holding potential of −80 mV to −40 mV for 80 ms; subsequent depolarization to 0 mV activated the Ca^{2+} current. After confirming that the amplitude of the L-type Ca^{2+} current in our system hardly ran down within 10 min, all the recordings were completed within 10 min in this study. In Figure 6A,C,E, three stable sweeps are superimposed, one under control conditions (black) and one under isoproterenol-administrated conditions (red). Every Ca^{2+} current was enhanced by the β-adrenergic stimulation of 1 μM isoproterenol, and the intracellular Ca^{2+} concentration ($[Ca^{2+}]_i$) also affected the magnitude of the stimulant effect in PV myocytes. Under the $[Ca^{2+}]_i$-free condition with 5 mM EGTA, the β-adrenergic stimulation increased the Ca^{2+} currents amplitude of the PV cell by 1.50 ± 0.18 times. At $[Ca^{2+}]_i$ of 0.1 mM, the currents were further increased by 2.59 ± 0.58 times (Figure 6A,B). On the other hand, the currents of LV cells were magnified by 2.05 ± 0.43 and 1.91 ± 0.55 times under $[Ca^{2+}]_i$-free and -rich conditions, respectively (Figure 6C,D). There is no statistically significant difference between these values. In the LA cardiomyocytes, the Ca^{2+} currents were amplified by isoproterenol by 1.54 ± 0.12-fold in the chelated state of $[Ca^{2+}]_i$, while the currents were boosted to 1.97 ± 0.30-fold by the 0.1 mM $[Ca^{2+}]_i$ (Figure 6E,F). These data are summarized in Table 3 as well as Figure 6B,D,F. Of note, cells slowly repeated contractions and relaxations during the recording under $[Ca^{2+}]_i$-rich conditions. Table 3 reveals that the amplitude of the Ca^{2+} current is smaller under $[Ca^{2+}]_i$-rich conditions. This is presumably attributed to the characteristics of the L-type Ca^{2+} current, which is Ca^{2+}-dependent inactivation [33]. We assessed the influence of AC3 activity on the property of PV cardiomyocytes by pharmacological experiments using an AC inhibitor. Because an AC3-selective reagent was not available, the nonspecific adenine-like inhibitor SQ22536 was employed for the pharmacological characterization. According to a previously published report, the IC_{50} of the inhibitor on AC3, AC5, and AC6 are 110, 3.5, and 5.8 μM, respectively, while AC9 is resistant to the drug [34]. In rat PV cardiomyocytes, the increase in the L-type Ca^{2+} current induced by isoproterenol is reduced by 5 μM of the inhibitor under the $[Ca^{2+}]_i$-free condition (Figure 7). The isoproterenol-induced augmentation of the Ca^{2+} current decreased from 1.50 ± 0.18 to 1.18 ± 0.23 (p = 0.023) in the absence of $[Ca^{2+}]_i$ (Figure 8A,B).

Table 3. Comparison of L-type Ca^{2+} current amplitudes under different conditions. The current amplitudes are an average of three stably recorded sweeps. *Baseline* and *Isoproterenol* conditions indicate the current amplitudes in the absence and presence of 1 μM isoproterenol, respectively. The amplitude in the presence of isoproterenol is divided by the baseline to calculate the effect. Statistical p-values for the differences between the values are given below the table. PV, pulmonary vein; LA, left ventricle; LA, left atrium; Cm, membrane cell capacitance.

Cell Type	*$[Ca^{2+}]_i$*	*Baseline (pA/pF)*	*Isopreterenol (pA/pF)*	*Effect (-Fold)*	*Cm (pF)*	*n =*
PV	$[EGTA]_i$ = 5 mM	9.45 ± 4.18	14.27 ± 6.89	* 1.50 ± 0.18	194.80 ± 76.97	5
PV	$[Ca^{2+}]_i$ = 0.1 mM	7.07 ± 1.92	17.67 ± 3.24	* 2.59 ± 0.58	230.28 ± 40.64	5
LV	$[EGTA]_i$ = 5 mM	11.0 ± 3.14	21.66 ± 3.97	2.05 ± 0.43	223.07 ± 51.45	5
LV	$[Ca^{2+}]_i$ = 0.1 mM	6.86 ± 2.76	12.62 ± 4.96	1.91 ± 0.55	246.05 ± 40.62	5
LA	$[EGTA]_i$ = 5 mM	14.1 ± 2.54	21.55 ± 3.12	# 1.54 ± 0.12	117.13 ± 41.96	5
LA	$[Ca^{2+}]_i$ = 0.1 mM	9.47 ± 4.49	17.74 ± 5.47	# 1.97 ± 0.30	141.76 ± 38.50	6

* p = 0.0109; # p = 0.0087.

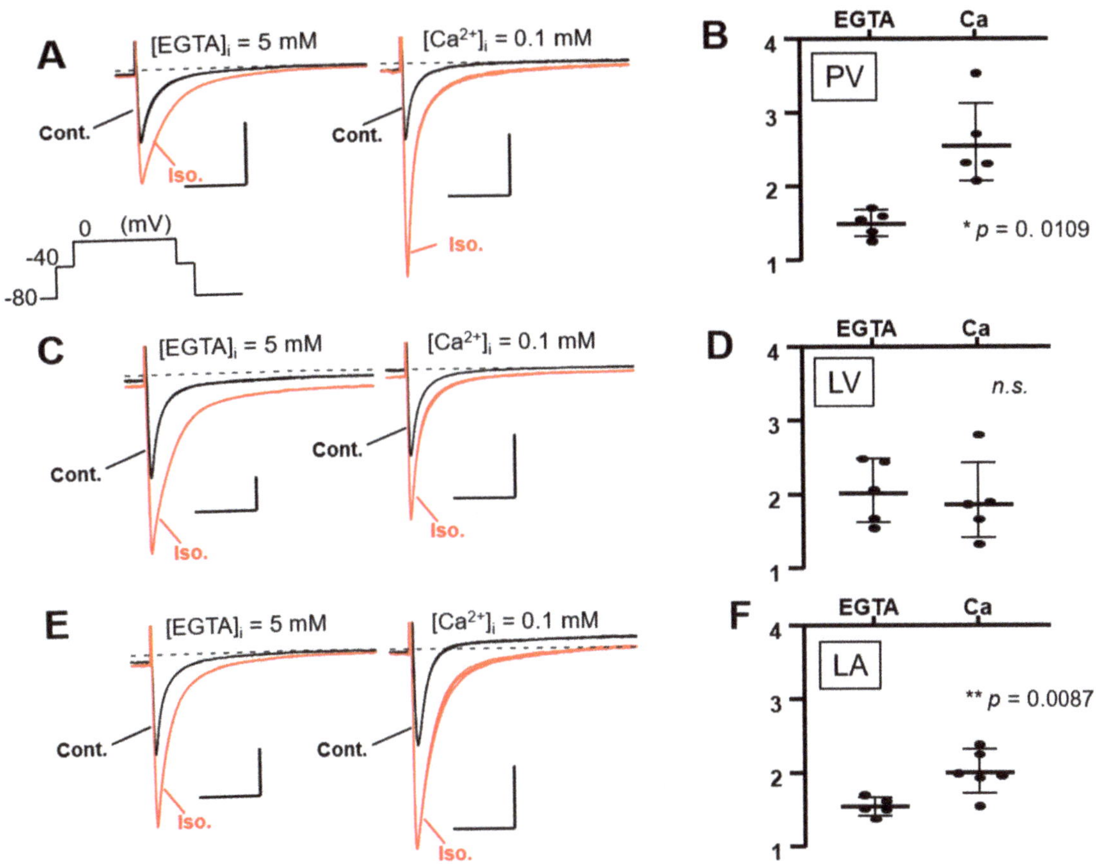

Figure 6. Ca^{2+}-dependent effects of β-adrenergic stimulation on L-type Ca^{2+} currents in three types of cardiomyocytes. As shown in the inset of (**A**), the Ca^{2+} current is induced by a depolarizing pulse to 0 mV for 500 ms, followed by a depolarizing pulse to −40 mV for 80 ms to inactivate the Na^+ current. The L-type Ca^{2+} current is amplified by 1 μM isoproterenol (Iso.), a β-adrenergic receptor-selective agonist. The strength of the β-adrenergic stimulation is evaluated by the rate of amplification of the current amplitudes. It was observed whether the absence or presence of intracellular Ca^{2+} changed the strength of the β-adrenergic reaction. In Figure 8A,C,E, three stable sweeps are superimposed under control conditions (Cont., black) or Iso. conditions (red). Inserted time and current scales indicate 50 ms and 5 pA/pF, respectively. Dashed lines indicate zero current level. Intracellular Ca^{2+} is chelated by 5 mM EGTA under the Ca^{2+}-free condition, while the internal solution contains 0.1 mM Ca^{2+} in the other. (**A**) Representative data from pulmonary vein (PV) cardiomyocytes. A statistical summary of an increase in calcium current by Iso. for each condition is shown in (**B**). (**C**) Representative data from left ventricular (LV) cardiomyocytes and summarized in (**D**). (**E**) Representative data from left atrial (LA) cardiomyocytes and summarized in (**F**). *n.s.*, not significant.

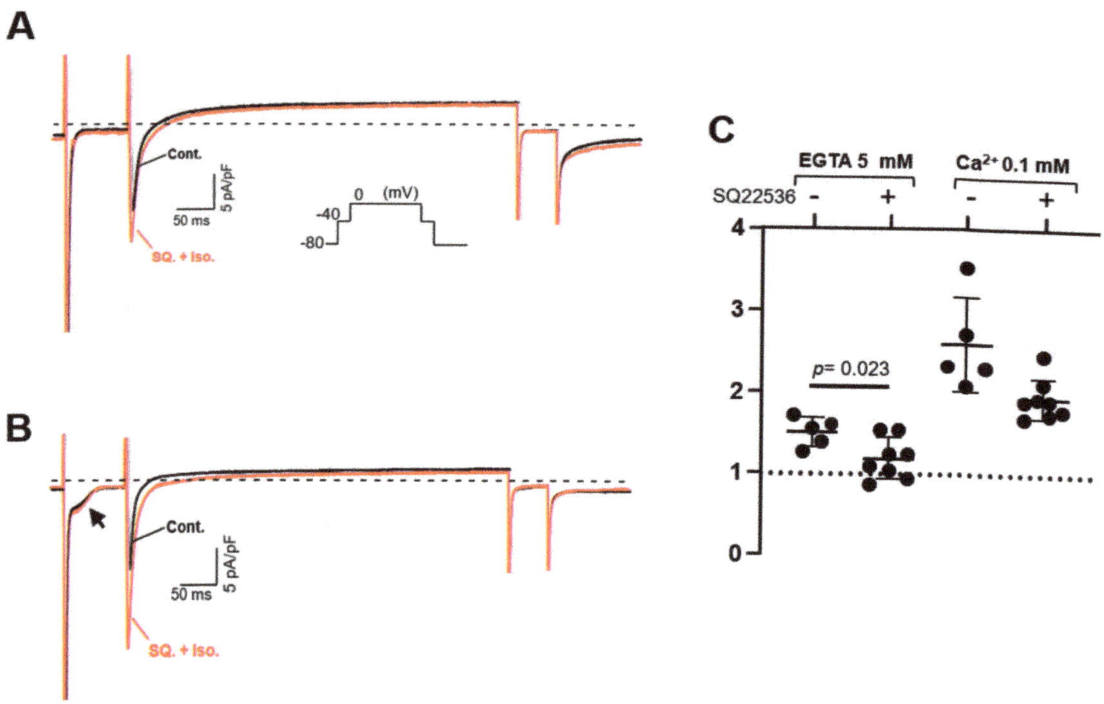

Figure 7. Effect of SQ22536 (SQ) on L-type Ca^{2+} current in pulmonary vein (PV) cardiomyocytes. The response of the Ca^{2+} current was recorded after 5 μM SQ application for approximately 2 min. The pulse protocol is inserted in (**A**). Intracellular Ca^{2+} is chelated by 5 mM EGTA in (**A**), while the internal solution contains 0.1 mM Ca^{2+} in (**B**). Dashed lines indicate zero current level. The rate of change in the current amplitudes for each condition is summarized in (**C**). Of note, under Ca^{2+}-rich conditions, a Ca^{2+}-dependent inward current indicated by an arrow in (**B**) is always recorded. This current is estimated to be of Na^+-Ca^{2+} exchanger origin.

3.5. Suppression of Norepinephrine-Induced PV Automaticity by AC Inhibitor Treatment

The spontaneous automaticity evoked by 10 μM norepinephrine (NE) [5] could not be suppressed by 5 μM SQ22536 (n = 4, Figure 8A). However, the PV excitability was reversibly arrested by applying 100 μM (Figure 8B). The cessation of action potentials and gradual attenuation of transient depolarization was observed before it stopped completely (Figure 8C). These sporadic depolarizations originate from inward currents via NCX, which is evoked by transient Ca^{2+} elevation [5]. The NE-induced automaticity was always suppressed by 100 μM SQ22536 (n = 4).

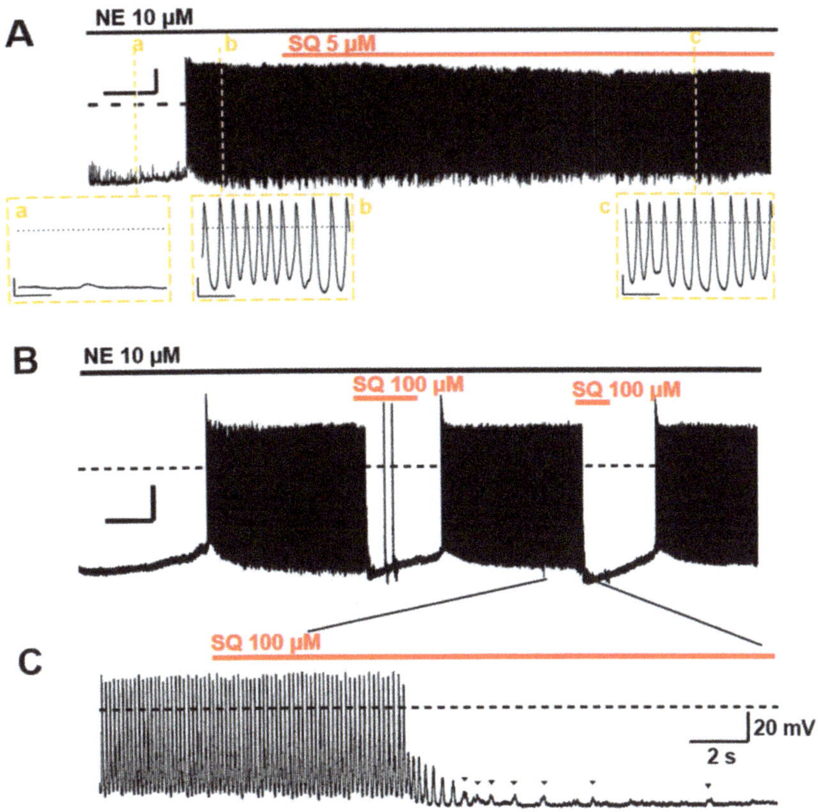

Figure 8. Effects of SQ22536 (SQ) on the norepinephrine (NE)-induced automaticity of pulmonary vein (PV). (**A**) An example of NE-induced spontaneous automaticity. NE 10 μM evoked a series of action potentials in isolated rat PV cardiomyocytes. The action potentials recorded at a, b, and c are displayed in an expanded time scale below. Inserted time and voltage scales indicate 50 s and 20 mV, respectively. The time scale in the expanded view indicates 500 ms. The induced automaticity was uninhibited by 5 μM SQ. (**B**) The induced automaticity was reversibly suppressed by 100 μM SQ. Inserted time and voltage scales indicate 50 s and 20 mV, respectively. The time course of the inhibition is expanded in (**C**). The sporadic depolarizations corresponding to Ca^{2+} release were gradually attenuated (arrowheads). Similar results were confirmed four times ($n = 4$, for each). The timing and duration of drug applications are indicated by horizontal bars. A dashed line indicates the baseline.

4. Discussion

4.1. Major Findings

The identity of the AC isoforms responsible for cAMP production in PV cardiomyocytes has not been previously elucidated. As a first step toward clarification of this critical issue, we employed RT-PCR, immunohistochemistry, and immunocytochemistry to determine the expression profiles in the rat heart and PVs. The gene expression profile of the AC isoforms by RT-PCR (Figure 1B) confirmed that all AC isoforms were expressed in the heart [35–37]; these results are consistent with previous studies that showed AC5 as the major isoform in cardiomyocytes. The most striking finding of the current study is that AC3 is specifically expressed in the supraventricular region, including PVs. In contrast, AC1, AC6, and AC8 were expressed throughout the heart and did not show reigion-specific expression patterns (Figures 2 and 4A). AC2 protein was hardly detected in cardiomyocytes. In cell staining, AC3 was apparently localized along the t-tubule microdomain in PV cardiomyocytes with a well transversally organised tubular system (Figure 5). However, pulmonary vein cardiomyocytes showed a wide range of transverse tubular incidence and organization, going from their virtual absence, as described in atrial myocytes, to well transversally organized tubular systems, as in ventricular myocytes [11]. Therefore, the variability in the tubular system could also be associated with the cell-to-cell heterogeneity of AC3 localization in the PV myocardium compared with either the atrium or the ventricule. The t-tubule is an $[Ca^{2+}]_i$-concentrating apparatus. For example, during an action potential, the $[Ca^{2+}]_i$ below the t-tubule is estimated to exceed 200 μM [38,39]. Because AC3 is a Ca^{2+}-stimulable isotype, there may be a connection between its location along the t-tubule and the reinforcement of the cAMP-dependent response by $[Ca^{2+}]_i$. In the patch-clamp experiments, the cAMP-dependent response of L-type Ca^{2+} currents in supraventricular cells, unlike ventricular cells, is enhanced when $[Ca^{2+}]_i$ is abundant (~100 μM, Figure 6). To evaluate the impact of ACs and their Ca^{2+} dependency on the arrhythmogenic property of PV, adenine-like AC inhibitor SQ22536 is administrated during the automaticity. An amount of 5 μM of this reagent suppressed the β-adrenegic stimulant effect on the Ca^{2+} current of PV under $[Ca^{2+}]_i$-free conditions, while the effects of the β-adrenergic response remain to a certain extent under $[Ca^{2+}]_i$-rich conditions (Figure 7). This drug fails to abolish the NE-induced automaticity at the same concentration (Figure 8A). By contrast, 100 μM of SQ22536 reproductively arrested the excitability (Figure 8B,C). Accordingly, these results suggest that Ca^{2+}-stimulable AC3 plays the primary role in the NE-induced PV automaticity as a part of β-adrenergic signaling. To summarize the reasons: (1) Highly expressed Ca^{2+}-stimulable AC3 localizes at the t-tubule at least in PV cardiomyocytes with a well transversally organised tubular system, where it is strongly influenced by $[Ca^{2+}]_i$. (2) The β-adrenergic effect evaluated by the Ca^{2+} current is enhanced by $[Ca^{2+}]_i$ in PV. Finally, (3) $[Ca^{2+}]_i$ is supposedly at a high level during the PV automaticity, and the inhibitory effect of SQ22536 on the automaticity is demonstrated at an adequate concentration of 100 μM against AC3, rather than AC5, AC6, or AC9.

4.2. Potential Role of AC3 in the Rat Heart, among Other ACs

Among the nine membrane-bound AC isoforms, AC1, AC3, and AC8 are Ca^{2+}-stimulable, while AC5 and AC6 are Ca^{2+}-inhibitable [40]. Residual AC2, AC4, AC7, and AC9 are Ca^{2+}-insensitive [41]. AC3 was first discovered in the olfactory epithelium [42] and was originally reported to be directly stimulated by Ca^{2+} and calmodulin [43]. In AC3 knockout mice ($AC3^{-/-}$ mice), cAMP- and IP_3-induced responses to odorants are completely abolished as measured by electro-olfactograms [44], urine and pheromones are not detected, and male sexual behavior is no longer observed [45]. However, as an inhibition of AC3 by the calmodulin-dependent protein kinase, which Ca^{2+} also activates, was recorded in vitro [46], the Ca^{2+}-dependent activity of AC3 was controversial. Dr. Cooper's team proved that AC3 is activated by submillimolar $[Ca^{2+}]_i$ in cellular experiments [47], whereas AC5 and AC6 were entirely inhibited by similar conditions [48,49]. In the current study, we estimated AC activities by the β-adrenergic response of L-type Ca^{2+} currents on a cell-by-cell basis. Consistent with LA and PV, where AC3 expression was regionally up-regulated, the β-adrenergic effect was enhanced by 0.1 mM $[Ca^{2+}]_i$ (Figure 6, Table 3). During sustained arrhythmias, such as the PV automaticity, $[Ca^{2+}]_i$ is kept high, and the activity of AC3 could be enhanced. In contrast, the activity of the major cardiac types, i.e., AC5 and 6, is declined. It should be considered that these AC types are apparently along the t-tubule (Figure 5 and [29]). Simulation studies predict that Ca^{2+} release from cardiac ryanodine receptors can increase $[Ca^{2+}]_i$ to >200 μM at the t-tubules [38,39]. As pharmacological charasteristics, SQ22536-sensitive isotypes are AC5 and 6 (IC_{50}; 3.5 and 5.8 μM, respectively), moderately sensitive isotypes are AC1 and AC3 (IC_{50}; 54 and 110 μM, respectively), and less sensitive are AC2, 4, and 7 (IC_{50}; 210, 280, and 440 μM, respectively), while AC8 and AC9 are resistant to the drug [34]. Accordingly, the inhibitory effect of 100 μM SQ22536 in Figure 8B,C targets AC1 and AC3 rather than AC5, 6, and 9, indicating that Ca^{2+}-stimulable AC may be involved in the catecholamine-dependent arrhythmogenic property of PV by producing cAMP under the high-$[Ca^{2+}]_i$ conditions.

On the other hand, AC1 is believed to underlie normal automaticity in the sinoatrial node (SAN) [21]. AC1 is more highly expressed in the SAN than in the ventricles that lack automaticity. The basis of the coupled-clock theory of SAN beating is that a strong expression of AC1 in the localized region produces the unique calcium dynamics via cAMP-dependent phosphorylation of ryanodine receptors and phospholamban [50]. In our experiments, AC1 is not more highly expressed in the pulmonary veins than in the ventricles, so we propose that the pulmonary venous automaticity is derived from AC3, not AC1, as an analogy to the coupled-clock theory. As for cardiac majors, AC5 and AC6 are very closely related isotypes and are believed to be dominantly expressed in the heart. In rats, AC5 expression becomes more dominant with advancing age [51], and $AC5^{-/-}$ mice have been found to have impaired responses to sympathetic stimulation as well as parasympathetic responses such as baroreceptor reflexes [52]. In addition, $AC6^{-/-}$ mice dramatically reduce not only AC6 but also AC5 protein expression. Therefore, the phenotype of $AC6^{-/-}$ mice is stronger than that of $AC5^{-/-}$ mice, and the positive inotropic effects induced by dobutamine is greatly impaired. This is because Ca^{2+}-related proteins such as phospholamban are less likely to be phosphorylated by cAMP-dependent kinase [53]. In recent years, AC9 has been extensively studied [54–56]. Although cAMP production by AC9 is less than 3% of the total cardiac cAMP production, $AC9^{-/-}$ mice show impaired cardiac diastolic capacity and reduced heart rate [24]. It is thought that AC9 binds to KCNQ1 channels via A-kinase anchor protein and promotes the repolarization phase of the action potential during sympathetic nerve stimulation, thereby increasing heart rate [57].

As described above, while AC1, AC5, AC6, and AC9 have been reported as the main physiological cAMP producers in the heart, AC3 is distinguished by its pathophysiological function.

4.3. Interpretation of Region-Dependent AC3 Expression

Surprisingly, immunohistochemistry with anti-AC3 revealed four heart regions. The AC3-negative ventricular region was separated from the AC3-positive supraventricular area. The PV was heterogeneously stained, whereas the atrium was uniformly stained. The high AC3 expression on the inner side of the vein was isolated from the unstained zone in the proximal and peripheral portions of the vessel. The SAN was an unstained island in the supraventricular area. To our knowledge, this is the first case in which simple staining has discerned four heart regions. Indeed, these regions are known to undergo different developmental processes. For example, several transcription factors appear in heart region-specific manners during early developmental stages [58]. *Irx4* [59] and *Hrt2* [60] are strongly stained by in situ hybridization in the ventricular precursor region during embryonic day 9.5–10.5, whereas *Hrt1* expression is restricted to the atrial precursor region. *Pitx2c* is required for cardiomyocyte proliferation in the PV after developing the atrial and ventricular precursors during embryonic days 11.5–12.5 [61]. *Shox2* directs the expression of *Isl1* in the SAN, which establishes the pacemaker activity by suppressing needless genes before the embryonic heart starts to beat regularly [62,63]. These transcription factors are core molecules that organize the region-dependent expression of functional proteins such as hormones, enzymes, and ion channels. AC3 can thus be considered as one region-specific enzyme.

4.4. Limitations

The main limitation of our study is the absence of direct experiments that specifically target AC3, such as selective inhibitors of AC3, genetic recombination experiments with AC3, or knockout mice for AC3. In addition, we have not been able to establish a method to measure cAMP at the cellular level, so we can only indirectly analogize the function of AC by evaluating the amplitude of the Ca^{2+} current. Thus, even though it is certain that AC3 is strongly expressed in the supraventricular region of the heart, direct experiments on the function of AC3 in the heart are still insufficient. We hope that future advances in materials will reproduce and strengthen our conclusion that AC3 is involved in PV arrhythmias.

5. Conclusions

This study identified specific AC3 expression in the rat supraventricular area, including arrhythmogenic PV cardiomyocytes. In the t-tubule membrane structure of PV cells, Ca^{2+}-stimulable AC3 may play an arrhythmogenic role in norepinephrine-induced automaticity under high-$[Ca^{2+}]_i$ conditions, rather than other Ca^{2+}-inhibitable and -insensitive ACs. Accordingly, SQ22536 suppressed the automaticity potently and reversibly at the concentration appropriately against AC3. These findings suggest that the blockade of AC3 can suppress the ectopic excitability of PVs. Our data may provide valuable information on how ectopic excitability occurs in PVs and may contribute to the development of drugs that can pharmacologically suppress the Ca^{2+}-dependent arrhythmogenic property of PVs because AC3-selective reagents are currently unavailable.

Author Contributions: Y.O. wrote the manuscript and designed and performed experiments and funding. N.Y.A. wrote the manuscript and performed immunohistochemistry. M.T., Y.T., D.T. and W.I. performed experiments. M.Y. and K.I. supervised experiments. K.O. wrote the manuscript and supervised the study. All authors have read and agreed to the published version of the manuscript.

Funding: This research was funded by JSPS KAKENHI (Grant-in-Aid for Young Scientists (B)), grant number 17H15554 and 20K16116 to Y.O., and by collaboration with Ono Pharmaceutical Cooperation, Ltd. (Osaka, Japan).

Institutional Review Board Statement: The protocols used in this study were approved by the Animal Ethics Committees of the Akita University School of Medicine (a-1-2953; December 2017, a-1-0212; May 2020).

Informed Consent Statement: Not applicable.

Data Availability Statement: The data in this study are available on a request basis.

Acknowledgments: We thank Eva Lasic, from Edanz Group (https://en-author-services.edanz.com/ac (accessed on 1 January 2021), for editing a draft of this manuscript.

Conflicts of Interest: The authors declare no conflict of interest.

References

1. Chugh, S.S.; Havmoeller, R.; Narayanan, K.; Singh, D.; Rienstra, M.; Benjamin, E.J.; Gillum, R.F.; Kim, Y.H.; McAnulty, J.H.; Zheng, Z.J.; et al. Worldwide epidemiology of atrial fibrillation: A global burden of disease 2010 study. *Circulation* **2014**, *129*, 837–847. [CrossRef] [PubMed]
2. Cheung, D.W. Electrical activity of the pulmonary vein and its interaction with the right atrium in the guinea-pig. *J. Physiol.* **1981**, *314*, 445–456. [CrossRef] [PubMed]
3. Doisne, N.; Maupoil, V.; Cosnay, P.; Findlay, I. Catecholaminergic automatic activity in the rat pulmonary vein: Electrophysiological differences between cardiac muscle in the left atrium and pulmonary vein. *Am. J. Physiol. Heart Circ. Physiol.* **2009**, *297*, H102–H108. [CrossRef] [PubMed]
4. Namekata, I.; Tsuneoka, Y.; Takahara, A.; Shimada, H.; Sugimoto, T. Short Communication Involvement of the Na+/Ca 2+ Exchanger in the Automaticity of Guinea-Pig Pulmonary Vein Myocardium as Revealed by SEA0400. *J. Pharmacol. Sci.* **2009**, *116*, 111–116. [CrossRef]
5. Okamoto, Y.; Takano, M.; Ohba, T.; Ono, K. Arrhythmogenic coupling between the Na+–Ca2+ exchanger and inositol 1,4,5-triphosphate receptor in rat pulmonary vein cardiomyocytes. *J. Mol. Cell. Cardiol.* **2012**, *52*, 988–997. [CrossRef]
6. Bredeloux, P.; Pasqualin, C.; Bordy, R.; Maupoil, V.; Findlay, I. Automatic activity arising in cardiac muscle sleeves of the pulmonary vein. *Biomolecules* **2022**, *12*, 23. [CrossRef]
7. Cheung, D.W. Pulmonary vein as an ectopic focus in digitalis-induced arrhythmia. *Nature* **1981**, *294*, 582–584. [CrossRef]
8. Honjo, H.; Boyett, M.R.; Niwa, R.; Inada, S.; Yamamoto, M.; Mitsui, K.; Horiuchi, T.; Shibata, N.; Kamiya, K.; Kodama, I. Pacing-induced spontaneous activity in myocardial sleeves of pulmonary veins after treatment with ryanodine. *Circulation* **2003**, *107*, 1937–1943. [CrossRef]
9. Sood, S.; Chelu, M.G.; Van Oort, R.J.; Skapura, D.G.; Santonastasi, M.; Dobrev, D.; Wehrens, X.H.T. Intracellular calcium leak due to FKBP12.6 deficiency in mice facilitates the inducibility of atrial fibrillation. *Heart Rhythm* **2009**, *1*, 1047–1054. [CrossRef]

10. Landstrom, A.P.; Dobrev, D.; Wehrens, X.H.T. Calcium Signaling and Cardiac Arrhythmias. *Circ. Res.* **2017**, *120*, 1969–1993. [CrossRef]
11. Maupoil, V.; Bronquard, C.; Freslon, J.-L.; Cosnay, P.; Findlay, I. Ectopic activity in the rat pulmonary vein can arise from simultaneous activation of alpha1- and beta1-adrenoceptors. *Br. J. Pharmacol.* **2007**, *150*, 899–905. [CrossRef]
12. Pasqualin, C.; Yu, A.; Malécot, C.O.; Gannier, F.; Cognard, C.; Godin-Ribuot, D.; Morand, J.; Bredeloux, P.; Maupoil, V. Structural heterogeneity of the rat pulmonary vein myocardium: Consequences on intracellular calcium dynamics and arrhythmogenic potential. *Sci. Rep.* **2018**, *8*, 3244. [CrossRef]
13. Ehrlich, J.R.; Cha, T.-J.; Zhang, L.; Chartier, D.; Villeneuve, L.; Hébert, T.E.; Nattel, S. Characterization of a hyperpolarization-activated time-dependent potassium current in canine cardiomyocytes from pulmonary vein myocardial sleeves and left atrium. *J. Physiol.* **2004**, *557*, 583–597. [CrossRef]
14. Okamoto, Y.; Kawamura, K.; Nakamura, Y.; Ono, K. Pathological impact of hyperpolarization-activated chloride current peculiar to rat pulmonary vein cardiomyocytes. *J. Mol. Cell. Cardiol.* **2014**, *66*, 53–62. [CrossRef]
15. Chen, P.S.; Chen, L.S.; Fishbein, M.C.; Lin, S.F.; Nattel, S. Role of the autonomic nervous system in atrial fibrillation: Pathophysiology and therapy. *Circ. Res.* **2014**, *114*, 1500–1515. [CrossRef]
16. Tan, A.Y.; Chen, P.-S.; Chen, L.S.; Fishbein, M.C. Autonomic nerves in pulmonary veins. *Heart Rythm.* **2007**, *4*, 57–60. [CrossRef]
17. Miyakawa, T.; Maeda, A.; Yamazawa, T.; Hirose, K.; Kurosaki, T.; Iino, M. Encoding of Ca2+ signals by differential expression of IP3 receptor subtypes. *EMBO J.* **1999**, *18*, 1303–1308. [CrossRef]
18. Okubo, Y.; Kanemaru, K.; Suzuki, J.; Kobayashi, K.; Hirose, K.; Iino, M. IP 3 R2-independent Ca 2+ release from the endoplasmic reticulum in astrocytes. *Glia* **2018**, *14*, 321042. [CrossRef]
19. Sadana, R.; Dessauer, C.W. Physiological roles for G protein-regulated adenylyl cyclase isoforms: Insights from knockout and overexpression studies. *NeuroSignals* **2009**, *17*, 5–22. [CrossRef]
20. Efendiev, R.; Dessauer, C.W. AKAPs and Adenylyl Cyclase in Cardiovascular Physiology and Pathology. *Acc. Chem. Res.* **2008**, *45*, 788–802. [CrossRef]
21. Vinogradova, T.M.; Lyashkov, A.E.; Zhu, W.; Ruknudin, A.M.; Sirenko, S.; Yang, D.; Deo, S.; Barlow, M.; Johnson, S.; Caffrey, J.L.; et al. High basal protein kinase A-dependent phosphorylation drives rhythmic internal Ca2+ store oscillations and spontaneous beating of cardiac pacemaker cells. *Circ. Res.* **2006**, *98*, 505–514. [CrossRef]
22. Mattick, P.; Parrington, J.; Odia, E.; Simpson, A.; Collins, T.; Terrar, D. Ca2+-stimulated adenylyl cyclase isoform AC1 is preferentially expressed in guinea-pig sino-atrial node cells and modulates the I(f) pacemaker current. *J. Physiol.* **2007**, *582*, 1195–1203. [CrossRef]
23. Lakatta, E.G.; Maltsev, V.A.; Vinogradova, T.M. A coupled SYSTEM of intracellular Ca2+ clocks and surface membrane voltage clocks controls the timekeeping mechanism of the heart's pacemaker. *Circ. Res.* **2010**, *106*, 659–673. [CrossRef]
24. Li, Y.; Baldwin, T.A.; Wang, Y.; Subramaniam, J.; Carbajal, A.G.; Brand, C.S.; Cunha, S.R.; Dessauer, C.W. Loss of type 9 adenylyl cyclase triggers reduced phosphorylation of Hsp20 and diastolic dysfunction. *Sci. Rep.* **2017**, *7*, 5522. [CrossRef]
25. Baldwin, T.A. Function of Adenylyl Cyclase in Heart: The AKAP Connection. *J. Cardiovasc. Dev. Dis.* **2018**, *5*, 2. [CrossRef]
26. Henry, A.D.; MacQuaide, N.; Burton, F.L.; Rankin, A.C.; Rowan, E.G.; Drummond, R.M. Spontaneous Ca2+transients in rat pulmonary vein cardiomyocytes are increased in frequency and become more synchronous following electrical stimulation. *Cell Calcium* **2018**, *76*, 36–47. [CrossRef]
27. Rietdorf, K.; Bootman, M.D.; Sanderson, M.J. Spontaneous, pro-arrhythmic calcium signals disrupt electrical pacing in mouse pulmonary vein sleeve cells. *PLoS ONE* **2014**, *9*, e88649. [CrossRef] [PubMed]
28. Katsushika, S.; Chen, L.; Kawabe, J.; Nilakantan, R.; Halnon, N.J.; Homcy, C.J.; Ishikawa, Y. Cloning and characterization of a sixth adenylyl cyclase isoform: Types V and VI constitute a subgroup within the mammalian adenylyl cyclase family. *Proc. Natl. Acad. Sci. USA* **1992**, *89*, 8774–8778. [CrossRef] [PubMed]
29. Gao, T.; Puri, T.S.; Gerhardstein, B.L.; Chien, A.J.; Green, R.D.; Hosey, M.M. Identification and subcellular localization of the subunits of L-type calcium channels and adenylyl cyclase in cardiac myocytes. *J. Biol. Chem.* **1997**, *272*, 19401–19407. [CrossRef] [PubMed]
30. Wilson, Z.T.; Jiang, M.; Geng, J.; Kaur, S.; Workman, S.W.; Hao, J.; Bernas, T.; Tseng, G.N. Delayed KCNQ1/KCNE1 assembly on the cell surface helps IKs fulfil its function as a repolarization reserve in the heart. *J. Physiol.* **2021**, *599*, 3337–3361. [CrossRef]
31. Crossman, D.J.; Young, A.A.; Ruygrok, P.N.; Nason, G.P.; Baddelely, D.; Soeller, C.; Cannell, M.B. T-tubule disease: Relationship between t-tubule organization and regional contractile performance in human dilated cardiomyopathy. *J. Mol. Cell. Cardiol.* **2015**, *84*, 170–178. [CrossRef]
32. Kashino, Y.; Obara, Y.; Okamoto, Y.; Saneyoshi, T.; Hayashi, Y.; Ishii, K. ERK5 phosphorylates Kv4.2 and inhibits inactivation of the A-type current in PC12 cells. *Int. J. Mol. Sci.* **2018**, *19*, 2008. [CrossRef]
33. Findlay, I. Voltage- and cation-dependent inactivation of L-type Ca2+ channel currents in guinea-pig ventricular myocytes. *J. Physiol.* **2002**, *541*, 731–740. [CrossRef]
34. Brand, C.S.; Hocker, H.J.; Gorfe, A.A.; Cavasotto, C.N.; Dessauer, C.W. Isoform selectivity of adenylyl cyclase inhibitors: Characterization of known and novel compounds. *J. Pharmacol. Exp. Ther.* **2013**, *347*, 265–275. [CrossRef]
35. Gottle, M.; Geduhn, J.; Ko, B.; Gille, A.; Ho, K.; Seifert, R. Characterization of Mouse Heart Adenylyl Cyclase. *Jpet* **2009**, *329*, 1156–1165. [CrossRef]

36. Timofeyev, V.; Myers, R.E.; Kim, H.J.; Woltz, R.L.; Sirish, P.; Heiserman, J.P.; Li, N.; Singapuri, A.; Tang, T.; Yarov-Yarovoy, V.; et al. Adenylyl cyclase subtype-specific compartmentalization: Differential regulation of L-Type Ca2+current in ventricular myocytes. *Circ. Res.* **2013**, *112*, 1567–1576. [CrossRef]
37. Wang, T.; Brown, M.J. Differential expression of adenylyl cyclase subtypes in human cardiovascular system. *Mol. Cell. Endocrinol.* **2004**, *223*, 55–62. [CrossRef]
38. Stern, M.D. Theory of excitation-contraction coupling in cardiac muscle. *Biophys. J.* **1992**, *63*, 497–517. [CrossRef]
39. Laver, D.R.; Kong, C.H.T.; Imtiaz, M.S.; Cannell, M.B. Termination of calcium-induced calcium release by induction decay: An emergent property of stochastic channel gating and molecular scale architecture. *J. Mol. Cell. Cardiol.* **2013**, *54*, 98–100. [CrossRef]
40. Ishikawa, Y.; Homcy, C.J. The adenylyl cyclases as integrators of transmembrane signal transduction. *Circ. Res.* **1997**, *80*, 297–304. [CrossRef]
41. Cooper, D.M. Regulation and organization of adenylyl cyclases and {cAMP.}. *Biochem. J.* **2003**, *375*, 517–529. [CrossRef]
42. Bakalyar, H.; Randall, R. Identification of a Specialized Adenylyl Cyclase That May Mediate Odorant Detection. *Science* **1990**, *250*, 1403–1406. [CrossRef]
43. Choi, E.J.; Xia, Z.; Storm, D.R. Stimulation of the Type III Olfactory Adenylyl Cyclase by Calciumand Calmodulin. *Biochemistry* **1992**, *31*, 6492–6498. [CrossRef]
44. Trinh, K.; Storm, D.R. Vomeronasal organ detects odorants in absence of signaling through main olfactory epithelium. *Nat. Neurosci.* **2003**, *6*, 519–525. [CrossRef]
45. Wang, Z.; Sindreu, C.B.; Li, V.; Nudelman, A.; Chan, G.C.K.; Storm, D.R. Pheromone detection in male mice depends on signaling through the type 3 adenylyl cyclase in the main olfactory epithelium. *J. Neurosci.* **2006**, *26*, 7375–7379. [CrossRef]
46. Wei, J.; Wayman, G.; Storm, D.R. Phosphorylation and inhibition of type III adenylyl cyclase by calmodulin-dependent protein kinase II in vivo. *J. Biol. Chem.* **1996**, *271*, 24231–24235. [CrossRef]
47. Cooper, D.M.F.; Mons, N.; Karpen, J.W. Adenylyl cyclases and the interaction between calcium and cAMP signaling. *Nature* **1995**, *374*, 421–424. [CrossRef]
48. Guillou, J.L.; Nakata, H.; Cooper, D.M.F. Inhibition by calcium of mammalian adenylyl cyclases. *J. Biol. Chem.* **1999**, *274*, 35539–35545. [CrossRef]
49. Hu, B.; Nakata, H.; Gu, C.; De Beer, T.; Cooper, D.M.F. A critical interplay between Ca2+ inhibition and activation by Mg2+ of AC5 revealed by mutants and chimeric constructs. *J. Biol. Chem.* **2002**, *277*, 33139–33147. [CrossRef]
50. Vinogradova, T.M.; Lakatta, E.G. Dual activation of phosphodiesterase 3 and 4 regulates basal cardiac pacemaker function and beyond. *Int. J. Mol. Sci.* **2021**, *22*, 8414. [CrossRef]
51. Tobise, K.; Ishikawa, Y.; Holmer, S.R.; Im, M.J.; Newell, J.B.; Yoshie, H.; Fujita, M.; Susannie, E.E.; Homcy, C.J. Changes in type VI adenylyl cyclase isoform expression correlate with a decreased capacity for cAMP generation in the aging ventricle. *Circ. Res.* **1994**, *74*, 596–603. [CrossRef] [PubMed]
52. Okumura, S.; Kawabe, J.I.; Yatani, A.; Takagi, G.; Lee, M.C.; Hong, C.; Liu, J.; Takagi, I.; Sadoshima, J.; Vatner, D.E.; et al. Type 5 adenylyl cyclase disruption alters not only sympathetic but also parasympathetic and calcium-mediated cardiac regulation. *Circ. Res.* **2003**, *93*, 364–371. [CrossRef] [PubMed]
53. Tang, T.; Gao, M.H.; Lai, N.C.; Firth, A.L.; Takahashi, T.; Guo, T.; Yuan, J.X.J.; Roth, D.M.; Hammond, H.K. Adenylyl cyclase type 6 deletion decreases left ventricular function via impaired calcium handling. *Circulation* **2008**, *117*, 61–69. [CrossRef] [PubMed]
54. Cumbay, M.G.; Watts, V.J. Novel regulatory properties of human type 9 adenylate cyclase. *J. Pharmacol. Exp. Ther.* **2004**, *310*, 108–115. [CrossRef]
55. Pálvölgyi, A.; Simpson, J.; Bodnár, I.; Bíró, J.; Palkovits, M.; Radovits, T.; Skehel, P.; Antoni, F.A. Auto-inhibition of adenylyl cyclase 9 (AC9) by an isoform-specific motif in the carboxyl-terminal region. *Cell. Signal.* **2018**, *51*, 266–275. [CrossRef]
56. Hacker, B.M.; Tomlinson, J.E.; Wayman, G.A.; Sultana, R.; Chan, G.; Villacres, E.; Disteche, C.; Storm, D.R. Cloning, chromosomal mapping, and regulatory properties of the human type 9 adenylyl cyclase (ADCY9). *Genomics* **1998**, *50*, 97–104. [CrossRef]
57. Li, Y.; Chen, L.; Kass, R.S.; Dessauer, C.W. The A-kinase anchoring protein yotiao facilitates complex formation between adenylyl cyclase type 9 and the IKs potassium channel in heart. *J. Biol. Chem.* **2012**, *287*, 29815–29824. [CrossRef]
58. Franco, D.; Domínguez, J.; Del Pilar de Castro, M.; Aránega, A. Regulation of myocardial gene expression during heart development. *Rev. Esp. Cardiol.* **2002**, *55*, 167–184. [CrossRef]
59. Bruneau, B.G.; Bao, Z.Z.; Tanaka, M.; Schott, J.J.; Izumo, S.; Cepko, C.L.; Seidman, J.G.; Seidman, C.E. Cardiac expression of the ventricle-specific homeobox gene Irx4 is modulated by Nkx2-5 and dHand. *Dev. Biol.* **2000**, *217*, 266–277. [CrossRef]
60. Nakagawa, O.; Nakagawa, M.; Richardson, J.A.; Olson, E.N.; Srivastava, D. HRT1, HRT2, and HRT3: A new subclass of bHLH transcription factors marking specific cardiac, somitic, and pharyngeal arch segments. *Dev. Biol.* **1999**, *216*, 72–84. [CrossRef]
61. Mommersteeg, M.T.M.; Brown, N.A.; Prall, O.W.J.; De Gier-de Vries, C.; Harvey, R.P.; Moorman, A.F.M.; Christoffels, V.M. Pitx2c and Nkx2-5 are required for the formation and identity of the pulmonary myocardium. *Circ. Res.* **2007**, *101*, 902–909. [CrossRef]
62. Espinoza-lewis, R.A.; Yu, L.; He, F.; Liu, H.; Tang, R.; Shi, J.; Sun, X.; Martin, J.F.; Wang, D.; Yang, J.; et al. Shox2 is essential for the differentiation of cardiac pacemaker cells by repressing Nkx2-5. *Dev. Biol.* **2009**, *327*, 376–385. [CrossRef]
63. Hoffmann, S.; Berger, I.M.; Glaser, A.; Bacon, C.; Li, L.; Gretz, N.; Steinbeisser, H.; Rottbauer, W.; Just, S.; Rappold, G. Islet1 is a direct transcriptional target of the homeodomain transcription factor Shox2 and rescues the Shox2-mediated bradycardia. *Basic Res. Cardiol.* **2013**, *108*, 339. [CrossRef]

biomolecules

MDPI

Article

Propofol, an Anesthetic Agent, Inhibits HCN Channels through the Allosteric Modulation of the cAMP-Dependent Gating Mechanism

Morihiro Shimizu [1], Xinya Mi [2], Futoshi Toyoda [2], Akiko Kojima [1], Wei-Guang Ding [2,*], Yutaka Fukushima [1], Mariko Omatsu-Kanbe [2], Hirotoshi Kitagawa [1] and Hiroshi Matsuura [2]

1 Department of Anesthesiology, Shiga University of Medical Science, Otsu 520-2192, Japan; mshimizu@belle.shiga-med.ac.jp (M.S.); akiko77@belle.shiga-med.ac.jp (A.K.); yf@belle.shiga-med.ac.jp (Y.F.); hirotosi@belle.shiga-med.ac.jp (H.K.)

2 Department of Physiology, Shiga University of Medical Science, Otsu 520-2192, Japan; mi1990@belle.shiga-med.ac.jp (X.M.); toyoda@belle.shiga-med.ac.jp (F.T.); m_omatsu@belle.shiga-med.ac.jp (M.O.-K.); matuurah@belle.shiga-med.ac.jp (H.M.)

* Correspondence: ding@belle.shiga-med.ac.jp; Tel.: +81-77-548-2152; Fax: +81-77-548-2348

Abstract: Propofol is a broadly used intravenous anesthetic agent that can cause cardiovascular effects, including bradycardia and asystole. A possible mechanism for these effects is slowing cardiac pacemaker activity due to inhibition of the hyperpolarization-activated, cyclic nucleotide-gated (HCN) channels. However, it remains unclear how propofol affects the allosteric nature of the voltage- and cAMP-dependent gating mechanism in HCN channels. To address this aim, we investigated the effect of propofol on HCN channels (HCN4 and HCN2) in heterologous expression systems using a whole-cell patch clamp technique. The extracellular application of propofol substantially suppressed the maximum current at clinical concentrations. This was accompanied by a hyperpolarizing shift in the voltage dependence of channel opening. These effects were significantly attenuated by intracellular loading of cAMP, even after considering the current modification by cAMP in opposite directions. The differential degree of propofol effects in the presence and absence of cAMP was rationalized by an allosteric gating model for HCN channels, where we assumed that propofol affects allosteric couplings between the pore, voltage-sensor, and cyclic nucleotide-binding domain (CNBD). The model predicted that propofol enhanced autoinhibition of pore opening by unliganded CNBD, which was relieved by the activation of CNBD by cAMP. Taken together, these findings reveal that propofol acts as an allosteric modulator of cAMP-dependent gating in HCN channels, which may help us to better understand the clinical action of this anesthetic drug.

Keywords: hyperpolarization-activated cyclic-nucleotide gated channel; cyclic nucleotide binding domain; I_f; propofol; cAMP

Citation: Shimizu, M.; Mi, X.; Toyoda, F.; Kojima, A.; Ding, W.-G.; Fukushima, Y.; Omatsu-Kanbe, M.; Kitagawa, H.; Matsuura, H. Propofol, an Anesthetic Agent, Inhibits HCN Channels through the Allosteric Modulation of the cAMP-Dependent Gating Mechanism. *Biomolecules* **2022**, *12*, 570. https://doi.org/10.3390/biom12040570

Academic Editors: Yosuke Okamoto and Kyoichi Ono

Received: 24 March 2022
Accepted: 10 April 2022
Published: 12 April 2022

1. Introduction

Propofol (2,6-diisopropylphenol) is an intravenous anesthetic drug that is widely used in the induction and maintenance of general anesthesia and sedation because of its advantages, which include a controllable state and fast awakening [1]. The potentiation of $GABA_A$ receptor Cl^- channels has long been considered a major mechanism for the anesthetic action of propofol [2–4]. This drug also modulates the function of several voltage gated ion channels, including hyperpolarization-activated, cyclic nucleotide-gated (HCN) channels [5–7], which may also contribute to its principal anesthetic effects. It is also known that propofol infusion can cause several side effects (propofol infusion syndrome) in several organs, including myocardial cells, such as metabolic acidosis, hyperkalaemia, rhabdomyolysis, or signs of heart failure. A dosage reduction or discontinuation is recommended when these conditions developed [8]. On the other hand, propofol provides cardioprotection of cardiomyocyte function under ischemic condition e.g., in animal model of heart failure, propofol was reported to protect against peroxidative and functional damages induce by exogenous H_2O_2 [9].

HCN channels control the electrical excitability and rhythmicity of neuronal and cardiac pacemaker cells by producing hyperpolarization-activated cationic inward current (I_h in neurons or I_f in the heart) [10]. In fact, mutation in hHCN4 is known to result in sinus bradycardia [11]. HCN channels allow Na^+ influx in the cells under physiological situation, and an increase in channel activation might elevate intracellular Ca^{2+} transiently by a reverse mode of Na^+/Ca^{2+} exchange current [12]. Therefore, in certain pathophysiological situations, the inhibition of HCN channels might protect against several damages from cardiac remodeling after myocardial infraction, such as cardiomyopathy and arrhythmia.

The blocking of I_h in central neurons may be associated with the general anesthetic action of propofol [5,13,14], whereas the I_f inhibition in cardiac pacemaker cells has been implicated in propofol-induced bradycardia [15]. Despite its fundamental importance, the molecular mechanism underlying the action of propofol on the HCN channels is poorly understood. The HCN channels (HCN1–4) are members of the voltage-gated cation channel superfamily that share a common six transmembrane-tetrameric architecture, consisting of one central pore domain (PD) surrounded by four peripheral voltage-sensor domains (VSDs). Unlike most other voltage-gated channels that are opened by membrane depolarization, the HCN channels are activated by membrane hyperpolarization. In addition, a distinguishing feature of HCN channels is the presence of cyclic nucleotide-binding domains (CNBDs) at their intracellular C-termini. Previous functional and structural analyses have suggested that CNBDs act as an inhibitory module of the inner pore gate [16–20]. Binding of cAMP to CNBDs can relieve the autoinhibition and thereby facilitate HCN channel opening. Thus, the opening of HCN channels is operated by both membrane voltage and cAMP, with distinct but allosterically coupled gating pathways [21,22]. Recent cryo-electron microscopy (EM) studies revealed molecular details of the structural elements of the PD, VSD, and CNBD, opening the door for the realization of the mechanics and dynamics of HCN channel gating [17,23]. This advanced knowledge on the structure and function provide an attractive framework for understanding the HCN channel behaviors as well as drug actions.

The present study investigated the effect of propofol on HCN channels in terms of their gating mechanism. For this purpose, we performed whole-cell patch clamp recordings to assess the drug responses of HCN4 and HCN2 channels in a heterologous expression system. In addition, to better interpret the drug effect, our experimental data were further analyzed using an allosteric gate model for HCN channels [22]. With these approaches, our results provide a better understanding of how propofol affects the gating properties of HCN channels.

2. Materials and Methods

2.1. Complementary DNA, Cell Culture, and Transfection

The full-length cDNA encoding human HCN4 (GenBank accession number NM_005477) and HCN2 (GenBank accession number NM_001194) channels, subcloned into the mammalian expression vector pcDNA3, ere a kind gift from Dr. Juliane Stieber (Lehrstuhl Pharmakologie und Toxikologie, Munich, Germany).

Chinese hamster ovary (CHO; RRID: CVCL_0213) cells were maintained in DMEM/Ham's F-12 medium supplemented with 10% fetal bovine serum (Sigma Chemical Company, St Louis, MO, USA) and antibiotics (100 IU/mL penicillin and 100 μ/mL streptomycin) in a humidified atmosphere under 5% CO_2 at 37 °C. The cells were passaged twice a week and a fraction of the cells was plated onto glass coverslips (3×5 mm^2) in 35-mm culture dishes. Each HCN4 (0.5 μg) or HCN2 DNA (0.5 μg) was transiently transfected into CHO cells together with green fluorescent protein (GFP) cDNA (0.5 μg) using Lipofectamine (Invitrogen Life Technologies, Carlsbad, CA, USA). After transfection for approximately 48 h, the GFP-positive cells were used in the patch-clamp study.

2.2. Whole-Cell Patch-Clamp Recording and Data Analysis

Whole-cell membrane currents were recorded in voltage-clamp models with an EPC-8 patch-clamp amplifier (HEKA Elektronik, Lambrecht, Germany). Patch electrodes were fabricated from glass capillaries (outer diameter, 1.5 mm; inner diameter, 0.9 mm; Narishige Scientific Instrument Laboratory, Tokyo, Japan) using a Brown-Flamming microelectrode puller (model P-97, Sutter Instrument, Novato, CA, USA), and the tip was then fire-polished using an MF-830 microforge (Narishige, Tokyo, Japan). The electrodes had a resistance of 2.5–4.0 MΩ when filled with pipette solution. The liquid junction potential (approximately −10 mV) was compensated for. A glass coverslip with adherent CHO cells was placed in the chamber mounted on the stage of an inverted microscope (ECLIPSE TE2000-U, Nikon, Tokyo, Japan), and continuously perfused at a rate of 2 mL/min with normal Tyrode solution at 37 °C. Signals were low-pass filtered at 1 kHz, and acquired at 5 kHz through an LIH 1600 analogue-to-digital converter (Instrutech, NY, USA). The data were stored and analyzed using the Patchmaster software program (HEKA Elektronik).

The bath solution for whole cell recording was normal Tyrode solution containing (in mM) 140 NaCl, 5.4 KCl, 1.8 $CaCl_2$, 0.5 $MgCl_2$, 0.33 NaH_2PO_4, 5.5 glucose, and 5.0 HEPES (pH was adjusted to 7.4 with NaOH). The pipette solution contained (in mM) 70 potassium aspartate, 50 KCl, 10 KH_2PO_4, 1 $MgSO_4$, 3 Na_2-ATP (Sigma Chemical Company), 5 HEPES, 5 EGTA, 0.1 Li_2-GTP (Sigma), and 2 $CaCl_2$ (pH was adjusted to 7.2 with KOH). Propofol (Sigma) was dissolved in DMSO and then added to the bath solution at concentrations of 0.3, 1, 3, and 10 μM. Our previous studies have confirmed that DMSO as a solvent had no influence on the membrane ion currents in cell line and native cardiac cells when its concentration was between 0.04~0.1% (v/v) [24,25]. In the present study, the concentration of DMSO in the final solution was ≤0.002%. Cyclic adenosine 3′, 5′-monophosphate (cAMP, Sigma) was dissolved in pipette solution at a concentration of 50 μM.

The HCN2 and HCN4 currents were activated by 2-s hyperpolarizing voltage-clamp steps applied from a holding potential of −30 mV to test potentials of −40 to −150 mV in 10-mV increments, with each subsequent depolarizing step to 0 mV to record the current of deactivation. In each cell, the tail current amplitudes were measured and normalized to cell membrane capacitance to yield current density (I_{tail}). The current density–voltage (I_{tail}, density-V) relationships were fitted to a Boltzmann equation: $I_{tail} = I_{tail,max}/(1 + \exp((V_t - V_h)/k))$, where $I_{tail,max}$ is the maximal tail current density, V_h is the voltage at half-maximal activation, V_t is the test potential, and k is the slope factor. For the analysis of voltage dependence for the channel activation, the I_{tail}, density-V relationships were all normalized to the maximum current density.

2.3. Model Simulation of Electrophysiological Data

The allosteric gate model for HCN channel activation [22] was used to analyze our patch clamp data. This model was developed by parameterizing the HA model [26] for the BK channel with the assumption that the allosteric coupling between the PD, VSD and CNBD was autoinhibitory. The two-state equilibrium behaviors of three functional modules of the PD, VSD, and CNBD were intrinsically characterized by distinct equilibrium constants, L, $H(V)$, and K, respectively. In the present study, these parameters were fixed as follows: (1) the PD close–open equilibrium constant, L = 85.3, in terms of an intrinsic bias towards to the open state, (2) the VSD resting-activated equilibrium constant, $H = H_0 e^{(V_t/V_h)}$, where the zero voltage value $H_0 = 0.003$ and the half-activation voltage $V_h = -11.3$ mV, and (3) the CNBD cAMP binding constant $K = [\text{cAMP}]/K_D$, where the dissociation constant $K_D = 75$ μM. The autoinhibitory coupling strengths of the PD with each of the four VSDs and four CNBDs were represented by allosteric factors F and C, respectively, such that the PD equilibrium constant decreased to a minimum of $L/(F^4C^4)$ when both VSDs and CNBDs were resting. However, the PD equilibrium constant increased F-fold with the activation of each VSD and also C-fold with the cAMP binding to each CNBD to a maximum of L. Reciprocally, intrinsic equilibrium constants in the VSD and CNBD decreased to H/F and K/C, respectively, when the PD is closed. This autoinhibitory constraint was again removed by the opening of the PD. Similarly, the autoinhibitory mechanism was also adopted for functional interaction between the VSD and CNBD, with an allosteric factor E. This model consists of 70 states (Horrigan and Aldrich, 2002) and the steady state open probability (P_O) was calculated using the following equation:

$$P_O = \frac{L/(CF)^4 \times (1+H/E+K/E+HK/E)^4}{L/(CF)^4 \times (1+H/E+K/E+HK/E)^4 + (1+H/(EF)+K/(CE)+HK/(CEF))^4} \quad (1)$$

The model was fitted to the experimental P_O-V data in the presence and absence (control) of propofol, in which values were normalized to the maximum available current density in control to yield the relative fraction of the open channel (i.e., relative P_O). The model fitting was performed using the Microsoft Excel Solver and parameter sets of allosteric factors, F, C, and E, were determined. The obtained parameters were employed in a simulation study. Given that stepwise activation of CNBDs occurs by the binding of cAMP to each of four subunits, the P_O was simplified for individual states of CNBD activation as follows:

$$P_O = \frac{L/(CF)^4 \times (1+H/E)^4}{L/(CF)^4 \times (1+H/E)^4 + (1+H/(EF))^4} \quad (2)$$

when none of the four CNBDs are activated.

$$P_O = \frac{L/(CF)^4 \times (1+H)(1+H/E)^3}{L/(CF)^4 \times (1+H) \times (1+H/E)^3 + (1/C+H/(CF)) \times (1+H/(EF))^3} \quad (3)$$

when one of four CNBDs is activated.

$$P_O = \frac{L/(CF)^4 \times (1+H)^2 \times (1+H/E)^2}{L/(CF)^4 \times (1+H)^2 \times (1+H/E)^2 + (1/C+H/(CF))^2 \times (1+H/(EF))^2} \quad (4)$$

when two of four CNBDs are activated.

$$P_O = \frac{L/(CF)^4 \times (1+H)^3 \times (1+H/E)}{L/(CF)^4 \times (1+H)^3 \times (1+H/E) + (1/C+H/(CF))^3 \times (1+H/(EF))} \quad (5)$$

when three of four CNBDs are activated.

$$P_O = \frac{L/(CF)^4 \times (1+H)^4}{L/(CF)^4 \times (1+H)^4 + (1/C+H/(CF))^4} \quad (6)$$

when all four CNBDs are activated.

2.4. Statistical Analyses

All data are expressed as the mean ± S.E.M. and the number of cells is indicated by n. Statistical comparisons were performed using the Student's two-tailed paired or unpaired t-test or using ANOVA with the Tukey's post hoc test (Prism Version 5.0), as appropriate. p values of <0.05 were considered to indicate statistical significance.

3. Results

3.1. Inhibitory Effects of Propofol on HCN4 Channels Expressed in CHO Cells

The effect of propofol on the current property of HCN4 channels was examined using the whole-cell patch-clamp technique (Figure 1). Propofol was administered for 2–3 min, until it was ensured that the current had approached a steady state. As shown in a representative experiment in Figure 1A, propofol caused substantial inhibition of the HCN4 current; the effect was characterized by a reduction in the tail currents at 0 mV after hyperpolarizing voltage steps to different membrane potentials. The I_{tail}, density-V

relationships for tail currents revealed that the inhibition is at least due to a decrease in the saturating tail current density (Figure 1C). The inhibitory effect of propofol was dose-dependent (Figure 1E) and was even observed at clinically relevant plasma concentrations (~1.0 μM) [27]. The tissue concentration is estimated as 1.0 μM is approximately 27 μg/g in brain, 39 μg/g in liver, and 18 μg/g in kidney [27,28]. On average, 10 μM propofol produced a ~35% decrease in the current density determined after hyperpolarization to −120 mV.

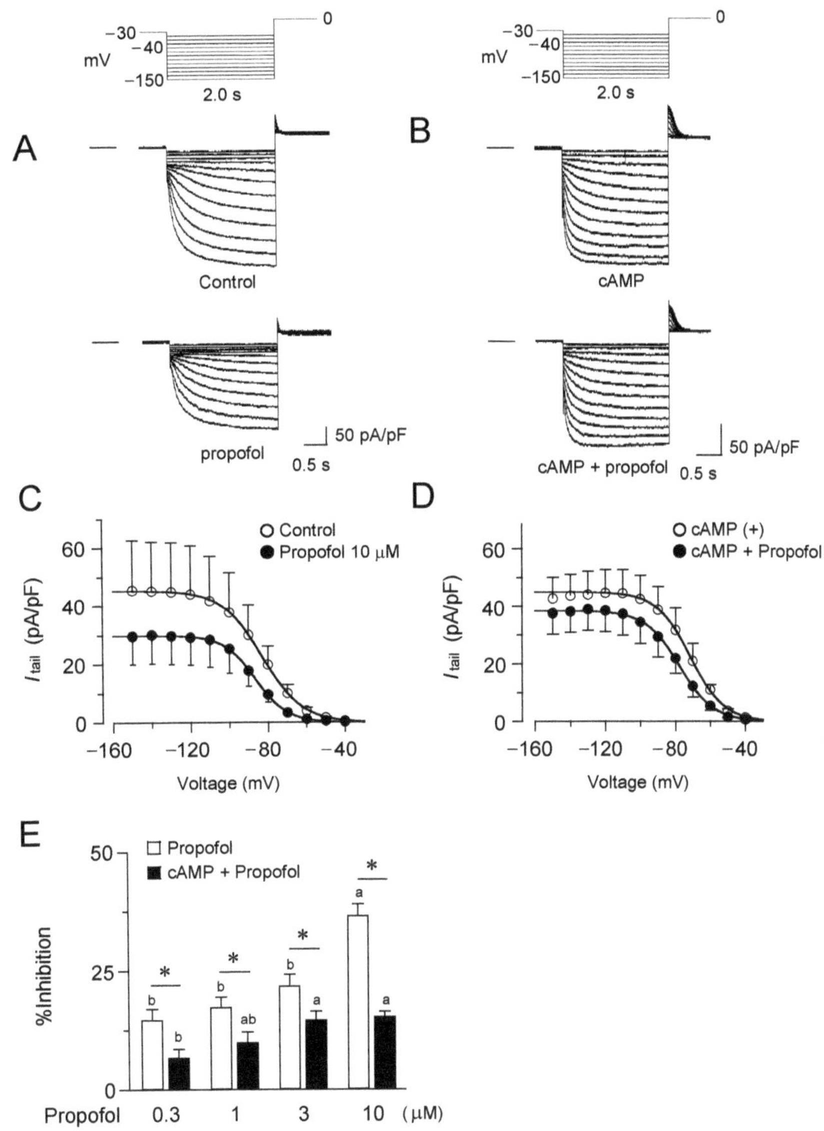

Figure 1. Inhibitory effects of propofol on HCN4 channels. (**A**) Superimposed current traces from

HCN4-expressing CHO cells, recorded using the pulse protocol indicated above, before (upper, control) and 2–3 min after exposure to 10 μM propofol (lower). The horizontal line to the left of the current traces indicates zero level. (**B**) Current recordings in the cell preloaded with cAMP (50 μM) via pipette. The concentration of propofol is the same as A. (**C**) The mean I_{tail}, density-V relationships for the current density in the presence and absence of propofol (n = 6). (**D**) The mean I_{tail}, density-V relationships obtained in the presence of cAMP in the pipette (n = 6). (**E**) The percentage inhibition of the saturating tail current density at various concentrations of propofol in the presence and absence of cAMP. The data represent the mean ± S.E.M. (n = 6), and those followed by different letters indicate a significant difference at $p < 0.05$ according to the Tukey's multiple range test and t-test (* indicates the difference in the presence and absence of cAMP; a, b, indicate different propofol concentration.).

To examine the involvement of cAMP-dependent gating in the current inhibition by propofol, we recorded HCN4 currents in the presence of cAMP at a concentration of 50 μM in the pipette (Figure 1B). Measurements were taken >15 min after the start of cAMP loading to achieve the steady state. In agreement with previous reports on the HCN4 channels [29], cAMP caused a depolarizing shift in V_h (see also Figure 2A,B) with a slight but non-significant increase in the maximum current density (Figure 1D). As is evident in Figure 1B, in the presence of cAMP, propofol had sparing effect on the HCN4 currents. The inhibition of the tail current by propofol was significantly attenuated, and this attenuation was similar for all concentrations examined (Figure 1E), suggesting that cAMP influenced the propofol effect in a noncompetitive manner.

In Figure 2, the effect of propofol on the voltage dependence of HCN4 current activation was investigated. As illustrated in Figure 2A, propofol shifted the voltage-dependent opening of HCN4 channels to more hyperpolarized membrane potentials. This effect is important, as the hyperpolarizing shift in the activation threshold can profoundly reduce the current within the physiological range of membrane potentials [5]. The propofol-induced shift in the voltage dependence was evaluated with the changes in V_h (ΔV_h) at various concentrations, which expanded in a dose-dependent manner. We also examined these effects in the presence of cAMP (Figure 2B). Intracellular loadings of cAMP led to a shift in the voltage range of HCN4 channel activation toward more depolarized potentials (V_h, −71 ± 0.99 mV and −82 ± 0.73 mV in the presence and absence of cAMP, respectively; $p < 0.05$). Subsequent application of propofol caused a slight shift in the activation curve to the opposite direction, but to a much lesser extent than that expected based on the observation without cAMP. In the presence of cAMP, propofol yielded significantly smaller ΔV_h values at all tested concentrations than that in the absence of cAMP ($p < 0.05$, Figure 2C).

3.2. Inhibitory Effects of Propofol on HCN2 Channels Expressed in CHO Cells

It has been reported that propofol modulation varies among different isoforms of HCN channels [30]. Therefore, the same sets of experiments were carried out for the HCN2 channels (Figure 3). Consistent with previous reports [31], the HCN2 channels were activated more rapidly and at less negative potential in comparison to HCN4 channels. However, propofol exerted nearly identical effects on HCN2 channels in terms of the impact on the current density, as well as the voltage dependence for activation. Furthermore, in the presence of cAMP, propofol was less effective for modulating the functions of HCN2 channels, which was again similar to the observations for the HCN4 channels.

3.3. Computational Simulation of Propofol Effects on HCN4 Channels Using an Allosteric Gate Model

Our patch clamp data showed that the inhibition of the HCN channels by propofol was attenuated in the presence of cAMP, raising the question of whether and how cAMP-dependent gating is involved in the drug effect. To address this point, we employed an allosteric gating model for HCN channels, which was developed by Flynn and Zagotta [22]. This model includes autoinhibitory interactions between the PD, VSDs, and CNBDs, and the coupling strengths are represented by distinct parameters of F, C, and E (Figure 4A).

The model fits the experimentally obtained plots of the relative tail current amplitudes (P_O-V curves) for HCN4 channels in the presence and absence of propofol (Figure 4B). The best-fit parameters revealed that propofol significantly increased the coupling factor between the PD and CNBDs, without changing other coupling strengths (Figure 4C). Results were validated by evaluating the sensitivity and specificity for each parameter (Figure S1, Supplementary Materials). Using the obtained parameter sets, the P_O-V curves were simulated with cAMP at 50 μM, where the equilibrium constant for the CNBD activation (K was assumed to be 0.667). As shown in Figure 4D, cAMP caused a depolarizing shift in the P_O-V curve. Furthermore, the model successfully reproduced the lesser effect of propofol on the P_O-V curve in the presence of cAMP. To gain further insight into these results, the P_O-V curves were generated by a simplified model, in which the activation of CNBD was fixed in a given state (Figure 5). The simulation clearly illustrated that the effects of propofol—both of the reduced maximum P_O and the hyperpolarizing shift in the P_O-V curve—were gradually attenuated by the stepwise activation of four CNBDs and eventually abolished when the CNBDs were fully activated.

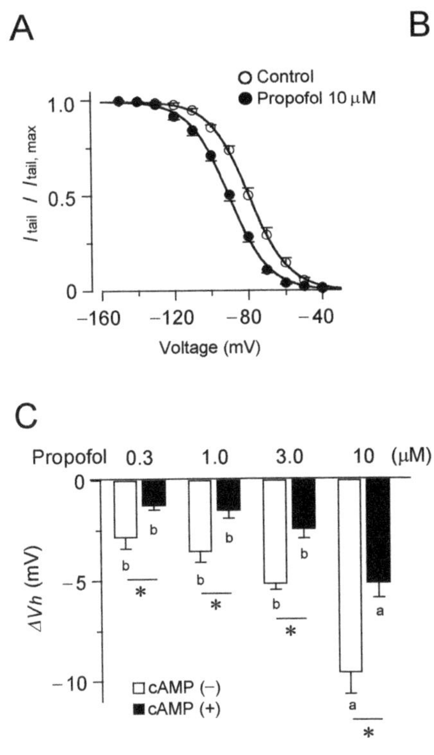

Figure 2. Propofol causes a hyperpolarizing shift in voltage dependence for HCN4 channel activation. The voltage dependence of HCN4 channel activation, in the presence and absence of propofol, was obtained from the cells using a pipette without (**A**) and with (**B**) cAMP. Tail current densities were normalized to the maximum value. The data points are mean ± S.E.M. (n = 6) and the smooth curves are the fit to the data using the Boltzmann equation. (**C**) Shifts in V_h caused by propofol at various concentrations in the presence and absence of cAMP. Bars represent the mean ± S.E.M. (n = 6), and those followed by different letters indicate a significant difference at $p < 0.05$ according to the Tukey's multiple range test and t-test (* indicates the difference in the presence and absence of cAMP; a, b, indicate different propofol concentration.).

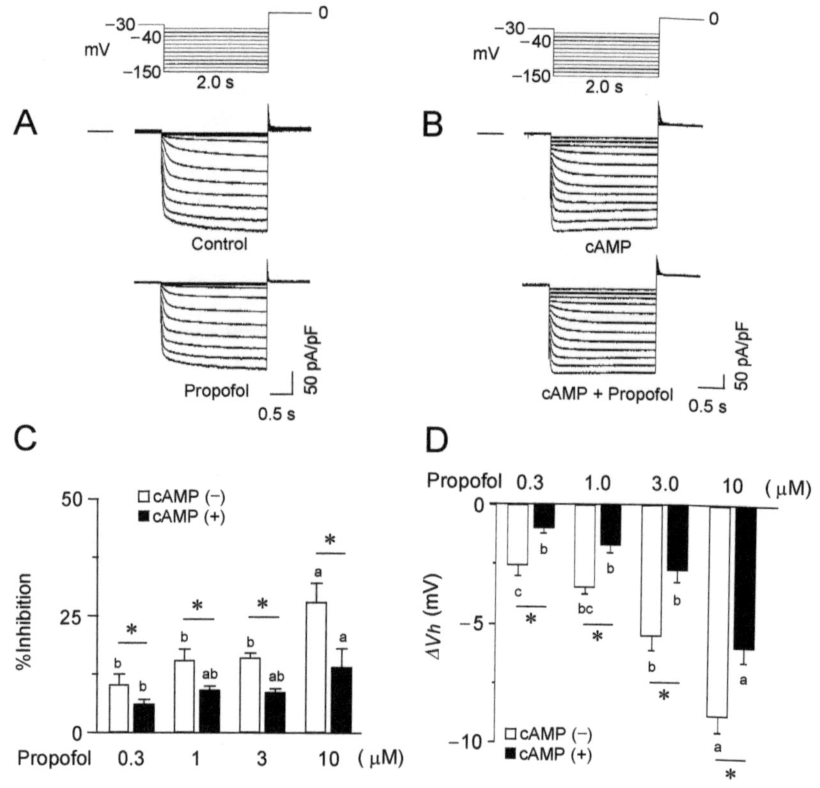

Figure 3. Inhibitory effects of propofol on HCN2 channels. (**A**) Superimposed current traces from HCN2-expressing CHO cells, recorded using the above-mentioned pulse protocol, before (upper, control) and 2–3 min after exposure to propofol at 10 μM (lower). The horizontal line to the left of the current traces indicates zero level. (**B**) Current recordings in a cell preloaded with cAMP (50 μM) via pipette. The concentration of propofol is the same as A. (**C**) The percentage inhibition of the saturating current density at various concentrations of propofol in the presence and absence of cAMP. Data represent the mean ± S.E.M. *, $p < 0.05$. (**D**) Shifts in V_h caused by propofol at various concentrations in the presence and absence of cAMP. Bars represent the mean ± S.E.M. ($n = 6$). The different letters in (**C**,**D**) indicate a significant difference at $p < 0.05$ according to the Tukey's multiple range test and t-test (* indicates the difference in the presence and absence of cAMP; a, b, c, indicate different propofol concentration.).

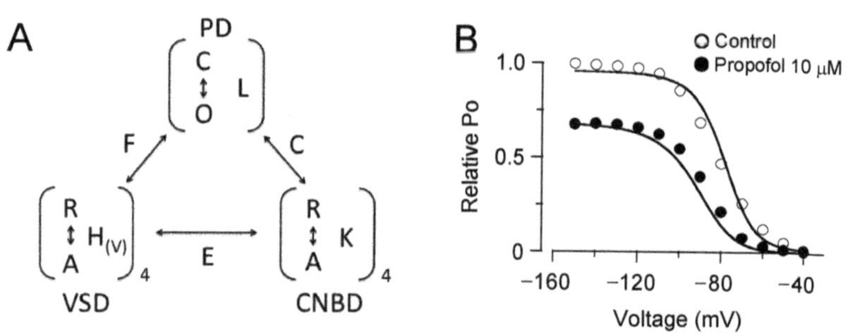

Figure 4. *Cont.*

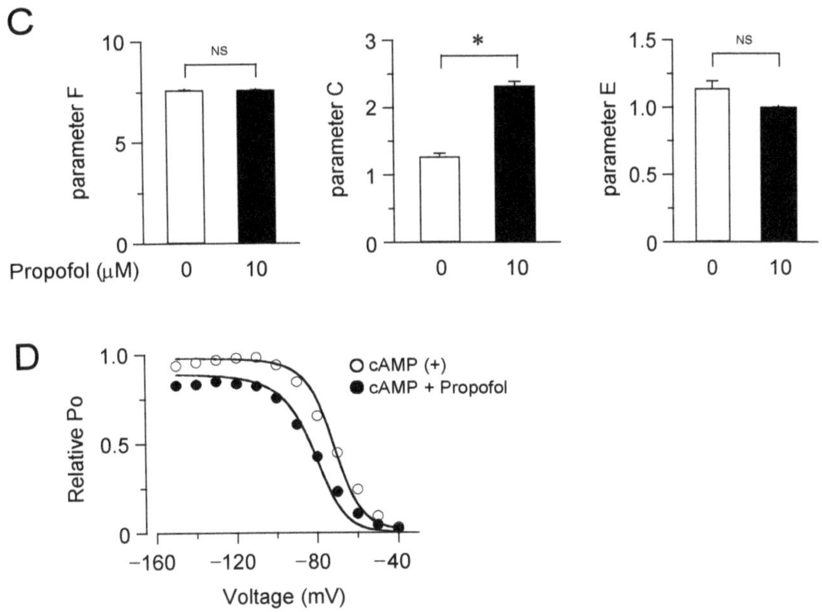

Figure 4. Allosteric gate model simulation of the inhibition of HCN4 channels by propofol. (**A**) The allosteric gate model for HCN channel activation, where *L, H(V), K* are the equilibrium constants for the transitions between close and open (C–O) or rest and active (R–A) states in the PD, VSD, and CNBD, respectively. *F, C,* and *E* are the allosteric coupling factors between the PD and the VSD, between the PD and CNBD, and between the VSD and CNBD, respectively. (**B**) Model fitting to the P_O-V data in the presence and absence (control) of propofol, obtained by normalizing the current density (relative P_O) to the maximum value in control. The best-fit parameters of *F, C,* and *E* are 8.2, 1.6, and 1.6, respectively, in controls, and became 7.8, 2.9, and 1.3, respectively, in the presence of propofol. (**C**) Parameters of model fitting to the P_O -V data in the presence and absence (control) of propofol. Bars represent the mean ± S.E.M. ($n = 6$). *, $p < 0.05$, NS, not significant. (**D**) Simulated P_O -V curves with cAMP (50 μM) as determined by the model with the same parameter sets used in (**B**), which closely overlay the experimental data points.

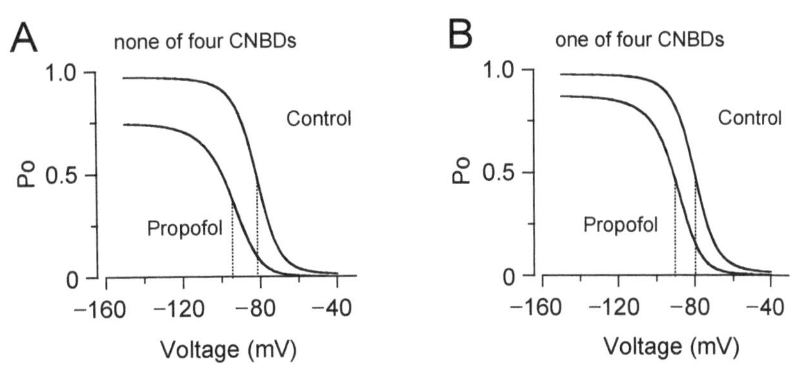

Figure 5. *Cont.*

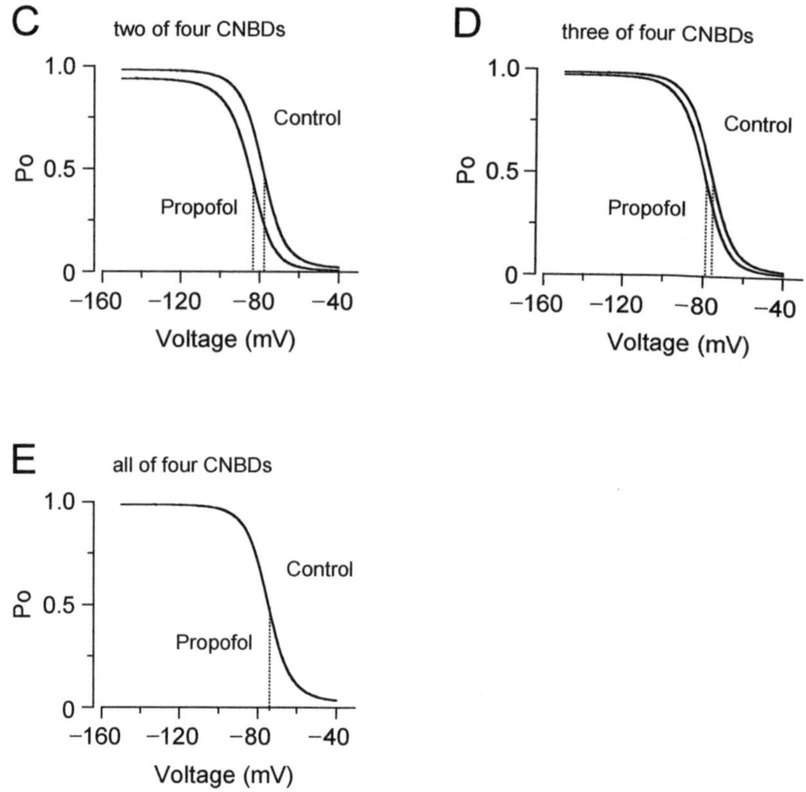

Figure 5. Simulation of the attenuated effects of propofol by activation of the CNBDs. The P_O-V curves in the presence and absence (control) of propofol were simulated depending on the activation state of CNBDs: none (**A**) and one to all (**B–E**) of four CNBDs activated. The vertical dashed line indicates the V_h.

4. Discussion

Several anesthetic agents inhibit the HCN channel function [15,16,32–34]. However, their mechanism of action remains largely elusive, which is at least due to the complex nature of HCN channel gating. In our patch clamp study, inhibition of the HCN channels by propofol was characterized by a decrease in the maximum current density and a hyperpolarizing shift of the voltage dependence of channel opening, in accordance with previous studies [30,35]. Most importantly, we found that these effects were considerably attenuated in the presence of cAMP, suggesting the involvement of cAMP-dependent gating in the action of propofol. Indeed, the model fitting analysis showed that propofol primarily affected allosteric coupling between the PD and CNBDs to exert its inhibitory effect, providing a key to understanding the complicated effects of propofol on the HCN channels. The relief of autoinhibitory interactions between the PD and CNBDs has been suggested to be a mechanism for cAMP-dependent gating in the HCN channels. In contrast to CNG channels that typically require the cyclic nucleotides for their opening [36], HCN channels are principally opened by the voltage, even without cAMP. Thus, the autoinhibition is weak and incomplete in the HCN channels. In the present study, we found that propofol increased the coupling strength between the PD and CNBDs, indicating the enhanced tonic inhibition of the pore opening by unliganded CNBDs. This would account for the reduction of the maximum current by propofol. In addition, the enhanced autoinhibition can allosterically reduce the open fraction via the voltage-dependent gating, resulting in a hyperpolarizing shift in the activation curve. Considering that the autoinhibition is relieved by the activation of the CNBD, the effect of propofol could be attenuated in a graded manner with increased cAMP levels, while the increased coupling strength would greatly accentuate the cAMP-dependent facilitation of the HCN channels. Collectively, our findings illustrate how propofol acts as an allosteric modulator of cAMP-dependent gating in the HCN channels.

The site of action for propofol is still unidentified at a structural level. However, given its effects on cAMP-dependent gating, we hypothesize that the relevant structure resides between the PD and CNBDs, such as the C-linker. Previous functional and structural studies have revealed that the C-linker is critical for transmitting cAMP-induced conformational changes in the CNBD to the PD. Recently, a novel ligand-binding pocket was identified at the boundary between the C-linker and the CNBD [37]. The occupancy of this pocket by exogenously applied compounds such as cyclic dinucleotides [38] and $TRIP8b_{nano}$ [39] interferes with the transmission of movement from the CNBD to the C-linker, thereby suppressing the responsiveness of HCN channels to cAMP. These findings provide the example of the allosteric regulation of the cAMP-dependent gate via the C-linker. Of note, the C-linker region is involved in the inhibitory effects of various drugs, including anesthetics [40]. Chen et al. [40] reported that HCN isoform-specific sensitivity to inhibition by halothane is imposed by differences in their structural arrangements of the C-linker domains. It is worth mentioning that the effect of halothane can be relieved by cAMP, similar to that observed for propofol in the present study. Besides, there is a growing body of structural evidence for a direct interaction of the CNBD with VSD though the HCN domain [41]. In this regard, we consider that the coupling factor E in the model may reflect a functional role of the HCN domain. Although disabled in the earlier study by Flynn and Zagotta [22], this parameter was included as a variable in the present study. However, in our model analysis, E was nearly one and was hardly affected by propofol. Further studies may illustrate the significance of functional coupling between CNBD and VSD in the HCN channels.

Given the fundamental role of the HCN channels in cardiac pacemaking, the inhibition of the HCN channel activity has been implicated as a cause of propofol-associated bradycardia [16]. According to the ACC/AHA/HRS guidelines [42], parasympathetic blockade with intravenous atropine is recommended for reducing bradycardia during the induction of total intravenous anesthesia with propofol. In light of recent experimental evidence from genetic mouse studies that investigated the HCN channels, this treatment seems to be reasonable. HCN4 knockout mice showed recurrent sinus pause and bradycardia [43–45], which typically manifested following vagal stimulation [46]. Interestingly, this pacemaker dysfunction was almost totally rescued by genetic ablation of the G-protein-coupled inward rectifier K^+ channel, a primary contributor to the negative chronotropic response to muscarinic acetylcholine receptor stimulation [47]. Meanwhile, in the clinical setting, β-adrenergic agonists, such as isoproterenol, are also administered for severe bradycardia after propofol induction. This would also make sense because cAMP may exceedingly facilitate HCN channel opening by the activation of CNBDs, relieving the inhibition by propofol, as suggested by our observations. Accordingly, a clinical study reported that propofol anesthesia enhanced the heart rate increase in response to isoproterenol infusion [48]. In addition, cAMP may also cause cAMP-induced arrhythmia when it is accumulated [49,50]. Caution should however be exercised in interpreting such results in terms of the altered ionic mechanism underlying pacemaker activity, as well as the diverse pharmacological activity of propofol. β-adrenergic stimulation potentiates not only HCN-mediated I_f but also other ionic currents, including $Ca_V1.3$-mediated $I_{Ca,L}$ [51] and I_{st} [52–54], and $K_V7.1$-mediated I_{Ks} [25], which are also potential targets of propofol [16]. Moreover, propofol significantly reduces sympathetic nerve activity [55].

5. Conclusions

Our patch clamp recordings and model fitting analyses of allosteric gating revealed that the anesthetic agent propofol inhibits HCN channels by functionally interacting with the cAMP-dependent gate. The model predicted that propofol facilitated the autoinhibition of pore opening by unliganded CNBDs, which could be relieved by cAMP. The model-based approaches provide a perspective for examinations and understanding the functional interactions of HCN channels with allosteric modulators.

Supplementary Materials: The following supporting information can be downloaded at: https://www.mdpi.com/article/10.3390/biom12040570/s1, Figure S1: Effects of changing each parameter on the P_O-V relationship.

Author Contributions: Conceptualization, W.-G.D., F.T., M.S. and H.M.; Methodology, M.S., X.M. and M.O.-K.; Simulation, F.T. and M.S.; Validation, M.S., X.M., A.K., W.-G.D., F.T., Y.F., M.O.-K., H.K. and H.M.; Data Analysis, M.S., W.-G.D., M.O.-K., Y.F. and A.K.; Writing-Original Draft Preparation, M.S., X.M., F.T. and W.-G.D.; Writing-Review and Editing, F.T. and W.-G.D.; Supervision, W.-G.D., H.K. and H.M.; Funding Acquisition, F.T. All authors have read and agreed to the published version of the manuscript.

Funding: This study was supported by a Grant-in-Aid for Scientific Research (No. P17K08537A to F.T.) from Japan Society for the Promotion of Science (Tokyo, Japan).

Institutional Review Board Statement: Not applicable.

Informed Consent Statement: Not applicable.

Data Availability Statement: The data that support the findings of this study are available from the corresponding author upon reasonable request.

Conflicts of Interest: The authors declare no conflict of interest.

References

1. Smith, I.; White, P.F.; Nathanson, M.; Gouldson, R. Propofol: An update on its clinical use. *Anesthesiology* **1994**, *81*, 1005–1043. [PubMed]
2. Franks, N.P. General anaesthesia: From molecular targets to neuronal pathways of sleep and arousal. *Nat. Rev. Neurosci.* **2008**, *9*, 370–386. [CrossRef] [PubMed]
3. Hemmings, H.C., Jr.; Akabas, M.H.; Goldstein, P.A.; Trudell, J.R.; Orser, B.A.; Harrison, N.L. Emerging molecular mechanisms of general anesthetic action. *Trends Pharmacol. Sci.* **2005**, *26*, 503–510. [CrossRef] [PubMed]
4. Rudolph, U.; Antkowiak, B. Molecular and neuronal substrates for general anaesthetics. *Nat. Rev. Neurosci.* **2004**, *5*, 709–720. [CrossRef]
5. Chen, X.; Shu, S.; Bayliss, D.A. Suppression of ih contributes to propofol-induced inhibition of mouse cortical pyramidal neurons. *J. Neurophysiol.* **2005**, *94*, 3872–3883. [CrossRef]
6. Higuchi, H.; Funahashi, M.; Miyawaki, T.; Mitoh, Y.; Kohjitani, A.; Shimada, M.; Matsuo, R. Suppression of the hyperpolarization-activated inward current contributes to the inhibitory actions of propofol on rat CA1 and CA3 pyramidal neurons. *Neurosci. Res.* **2003**, *45*, 459–472. [CrossRef]
7. Ying, S.W.; Abbas, S.Y.; Harrison, N.L.; Goldstein, P.A. Propofol block of I(h) contributes to the suppression of neuronal excitability and rhythmic burst firing in thalamocortical neurons. *Eur. J. Neurosci.* **2006**, *23*, 465–480. [CrossRef]
8. Krajčová, A.; Waldauf, P.; Anděl, M.; Duška, F. Propofol infusion syndrome: A structured review of experimental studies and 153 published case reports. *Crit. Care* **2015**, *19*, 398. [CrossRef]
9. Kevin, L.G.; Novalija, E.; Stowe, D.F. Reactive oxygen species as mediators of cardiac injury and protection: The relevance to anesthesia practice. *Anesth. Analg.* **2005**, *101*, 1275–1287. [CrossRef]
10. Robinson, R.B.; Siegelbaum, S.A. Hyperpolarization-activated cation currents: From molecules to physiological function. *Annu. Rev. Physiol.* **2003**, *65*, 453–480. [CrossRef]
11. Milanesi, R.; Baruscotti, M.; Gnecchi-Ruscone, T.; DiFrancesco, D. Familial sinus bradycardia associated with a mutation in the cardiac pacemaker channel. *N. Engl. J. Med.* **2006**, *354*, 151–157. [CrossRef] [PubMed]
12. Tosaki, A. ArrhythmoGenoPharmacoTherapy. *Front. Pharmacol.* **2020**, *11*, 616. [CrossRef] [PubMed]
13. Lewis, A.S.; Chetkovich, D.M. HCN channels in behavior and neurological disease: Too hyper or not active enough? *Mol. Cell. Neurosci.* **2011**, *46*, 357–367. [CrossRef]
14. Zhou, C.; Liang, P.; Liu, J.; Ke, B.; Wang, X.; Li, F.; Li, T.; Bayliss, D.A.; Chen, X. HCN1 Channels Contribute to the Effects of Amnesia and Hypnosis but not Immobility of Volatile Anesthetics. *Anesth. Analg.* **2015**, *121*, 661–666. [CrossRef] [PubMed]

15. Kojima, A.; Ito, Y.; Kitagawa, H.; Matsuura, H. Ionic mechanisms underlying the negative chronotropic action of propofol on sinoatrial node automaticity in guinea pig heart. *Br. J. Pharmacol.* **2015**, *172*, 799–814. [CrossRef]
16. Kojima, A.; Ito, Y.; Kitagawa, H.; Matsuura, H.; Nosaka, S. Direct negative chronotropic action of desflurane on sinoatrial node pacemaker activity in the guinea pig heart. *Anesthesiology* **2014**, *120*, 1400–1413. [CrossRef]
17. Lee, C.H.; MacKinnon, R. Structures of the Human HCN1 Hyperpolarization-Activated Channel. *Cell* **2017**, *168*, 111–120.e111. [CrossRef]
18. VanSchouwen, B.; Akimoto, M.; Sayadi, M.; Fogolari, F.; Melacini, G. Role of Dynamics in the Autoinhibition and Activation of the Hyperpolarization-activated Cyclic Nucleotide-modulated (HCN) Ion Channels. *J. Biol. Chem.* **2015**, *290*, 17642–17654. [CrossRef]
19. Wainger, B.J.; DeGennaro, M.; Santoro, B.; Siegelbaum, S.A.; Tibbs, G.R. Molecular mechanism of cAMP modulation of HCN pacemaker channels. *Nature* **2001**, *411*, 805–810. [CrossRef]
20. Xu, X.; Vysotskaya, Z.V.; Liu, Q.; Zhou, L. Structural basis for the cAMP-dependent gating in the human HCN4 channel. *J. Biol. Chem.* **2010**, *285*, 37082–37091. [CrossRef]
21. Dai, G.; Aman, T.K.; DiMaio, F.; Zagotta, W.N. Electromechanical coupling mechanism for activation and inactivation of an HCN channel. *Nat. Commun.* **2021**, *12*, 2802. [CrossRef] [PubMed]
22. Flynn, G.E.; Zagotta, W.N. Insights into the molecular mechanism for hyperpolarization-dependent activation of HCN channels. *Proc. Natl. Acad. Sci. USA* **2018**, *115*, E8086–E8095. [CrossRef] [PubMed]
23. James, Z.M.; Borst, A.J.; Haitin, Y.; Frenz, B.; DiMaio, F.; Zagotta, W.N.; Veesler, D. CryoEM structure of a prokaryotic cyclic nucleotide-gated ion channel. *Proc. Natl. Acad. Sci. USA* **2017**, *114*, 4430–4435. [CrossRef] [PubMed]
24. Kojima, A.; Fukushima, Y.; Ito, Y.; Ding, W.G.; Ueda, R.; Seto, T.; Kitagawa, H.; Matsuura, H. Interactions of Propofol With Human Voltage-gated Kv1.5 Channel Determined by Docking Simulation and Mutagenesis Analyses. *J. Cardiovasc. Pharmacol.* **2018**, *71*, 10–18. [CrossRef] [PubMed]
25. Yuasa, M.; Kojima, A.; Mi, X.; Ding, W.G.; Omatsu-Kanbe, M.; Kitagawa, H.; Matsuura, H. Characterization and functional role of rapid- and slow-activating delayed rectifier K(+) currents in atrioventricular node cells of guinea pigs. *Pflug. Arch. Eur. J. Physiol.* **2021**, *473*, 1885–1898. [CrossRef]
26. Horrigan, F.T.; Aldrich, R.W. Coupling between voltage sensor activation, Ca2+ binding and channel opening in large conductance (BK) potassium channels. *J. Gen. Physiol.* **2002**, *120*, 267–305. [CrossRef]
27. Tibbs, G.R.; Rowley, T.J.; Sanford, R.L.; Herold, K.F.; Proekt, A.; Hemmings, H.C., Jr.; Andersen, O.S.; Goldstein, P.A.; Flood, P.D. HCN1 channels as targets for anesthetic and nonanesthetic propofol analogs in the amelioration of mechanical and thermal hyperalgesia in a mouse model of neuropathic pain. *J. Pharmacol. Exp. Ther.* **2013**, *345*, 363–373. [CrossRef]
28. Müller-Wirtz, L.M.; Maurer, F.; Brausch, T.; Kiefer, D.; Floss, M.; Doneit, J.; Volk, T.; Sessler, D.I.; Fink, T.; Lehr, T.; et al. Exhaled Propofol Concentrations Correlate With Plasma and Brain Tissue Concentrations in Rats. *Anesth. Analg.* **2021**, *132*, 110–118. [CrossRef]
29. Stieber, J.; Thomer, A.; Much, B.; Schneider, A.; Biel, M.; Hofmann, F. Molecular basis for the different activation kinetics of the pacemaker channels HCN2 and HCN4. *J. Biol. Chem.* **2003**, *278*, 33672–33680. [CrossRef]
30. Cacheaux, L.P.; Topf, N.; Tibbs, G.R.; Schaefer, U.R.; Levi, R.; Harrison, N.L.; Abbott, G.W.; Goldstein, P.A. Impairment of hyperpolarization-activated, cyclic nucleotide-gated channel function by the intravenous general anesthetic propofol. *J. Pharmacol. Exp. Ther.* **2005**, *315*, 517–525. [CrossRef]
31. Ludwig, A.; Zong, X.; Stieber, J.; Hullin, R.; Hofmann, F.; Biel, M. Two pacemaker channels from human heart with profoundly different activation kinetics. *EMBO J.* **1999**, *18*, 2323–2329. [CrossRef] [PubMed]
32. Ishihara, M.; Kojima, A.; Ding, W.G.; Kitagawa, H.; Matsuura, H. Dexmedetomidine Exerts a Negative Chronotropic Action on Sinoatrial Node Cells Through the Activation of Imidazoline Receptors. *J. Cardiovasc. Pharmacol.* **2021**, *78*, 826–838. [CrossRef] [PubMed]
33. Novella Romanelli, M.; Sartiani, L.; Masi, A.; Mannaioni, G.; Manetti, D.; Mugelli, A.; Cerbai, E. HCN Channels Modulators: The Need for Selectivity. *Curr. Top. Med. Chem.* **2016**, *16*, 1764–1791. [CrossRef] [PubMed]
34. Ramírez, D.; Zúñiga, R.; Concha, G.; Zúñiga, L. HCN Channels: New Therapeutic Targets for Pain Treatment. *Molecules* **2018**, *23*, 2094. [CrossRef] [PubMed]
35. Lyashchenko, A.K.; Redd, K.J.; Yang, J.; Tibbs, G.R. Propofol inhibits HCN1 pacemaker channels by selective association with the closed states of the membrane embedded channel core. *J. Physiol.* **2007**, *583*, 37–56. [CrossRef]
36. Craven, K.B.; Zagotta, W.N. CNG and HCN channels: Two peas, one pod. *Annu. Rev. Physiol.* **2006**, *68*, 375–401. [CrossRef]
37. Lolicato, M.; Bucchi, A.; Arrigoni, C.; Zucca, S.; Nardini, M.; Schroeder, I.; Simmons, K.; Aquila, M.; DiFrancesco, D.; Bolognesi, M.; et al. Cyclic dinucleotides bind the C-linker of HCN4 to control channel cAMP responsiveness. *Nat. Chem. Biol.* **2014**, *10*, 457–462. [CrossRef] [PubMed]
38. Weißgraeber, S.; Saponaro, A.; Thiel, G.; Hamacher, K. A reduced mechanical model for cAMP-modulated gating in HCN channels. *Sci. Rep.* **2017**, *7*, 40168. [CrossRef] [PubMed]
39. Saponaro, A.; Cantini, F.; Porro, A.; Bucchi, A.; DiFrancesco, D.; Maione, V.; Donadoni, C.; Introini, B.; Mesirca, P.; Mangoni, M.E.; et al. A synthetic peptide that prevents cAMP regulation in mammalian hyperpolarization-activated cyclic nucleotide-gated (HCN) channels. *eLife* **2018**, *7*, e35753. [CrossRef]

40. Chen, X.; Sirois, J.E.; Lei, Q.; Talley, E.M.; Lynch, C., 3rd; Bayliss, D.A. HCN subunit-specific and cAMP-modulated effects of anesthetics on neuronal pacemaker currents. *J. Neurosci. Off. J. Soc. Neurosci.* **2005**, *25*, 5803–5814. [CrossRef]
41. Wang, Z.J.; Blanco, I.; Hayoz, S.; Brelidze, T.I. The HCN domain is required for HCN channel cell-surface expression and couples voltage- and cAMP-dependent gating mechanisms. *J. Biol. Chem.* **2020**, *295*, 8164–8173. [CrossRef] [PubMed]
42. Kusumoto, F.M.; Schoenfeld, M.H.; Barrett, C.; Edgerton, J.R.; Ellenbogen, K.A.; Gold, M.R.; Goldschlager, N.F.; Hamilton, R.M.; Joglar, J.A.; Kim, R.J.; et al. 2018 ACC/AHA/HRS Guideline on the Evaluation and Management of Patients With Bradycardia and Cardiac Conduction Delay: A Report of the American College of Cardiology/American Heart Association Task Force on Clinical Practice Guidelines and the Heart Rhythm Society. *Circulation* **2019**, *140*, e382–e482. [CrossRef] [PubMed]
43. Baruscotti, M.; Bucchi, A.; Viscomi, C.; Mandelli, G.; Consalez, G.; Gnecchi-Rusconi, T.; Montano, N.; Casali, K.R.; Micheloni, S.; Barbuti, A.; et al. Deep bradycardia and heart block caused by inducible cardiac-specific knockout of the pacemaker channel gene Hcn4. *Proc. Natl. Acad. Sci. USA* **2011**, *108*, 1705–1710. [CrossRef] [PubMed]
44. Herrmann, S.; Stieber, J.; Stöckl, G.; Hofmann, F.; Ludwig, A. HCN4 provides a 'depolarization reserve' and is not required for heart rate acceleration in mice. *EMBO J.* **2007**, *26*, 4423–4432. [CrossRef]
45. Hoesl, E.; Stieber, J.; Herrmann, S.; Feil, S.; Tybl, E.; Hofmann, F.; Feil, R.; Ludwig, A. Tamoxifen-inducible gene deletion in the cardiac conduction system. *J. Mol. Cell. Cardiol.* **2008**, *45*, 62–69. [CrossRef]
46. Kozasa, Y.; Nakashima, N.; Ito, M.; Ishikawa, T.; Kimoto, H.; Ushijima, K.; Makita, N.; Takano, M. HCN4 pacemaker channels attenuate the parasympathetic response and stabilize the spontaneous firing of the sinoatrial node. *J. Physiol.* **2018**, *596*, 809–825. [CrossRef]
47. Mesirca, P.; Bidaud, I.; Briec, F.; Evain, S.; Torrente, A.G.; Le Quang, K.; Leoni, A.L.; Baudot, M.; Marger, L.; Chung You Chong, A.; et al. G protein-gated IKACh channels as therapeutic targets for treatment of sick sinus syndrome and heart block. *Proc. Natl. Acad. Sci. USA* **2016**, *113*, E932–E941. [CrossRef]
48. Horiguchi, T.; Nishikawa, T. Propofol-nitrous oxide anesthesia enhances the heart rate response to intravenous isoproterenol infusion. *Anesth. Analg.* **2003**, *96*, 132–135. [CrossRef]
49. Eschenhagen, T.; Mende, U.; Diederich, M.; Hertle, B.; Memmesheimer, C.; Pohl, A.; Schmitz, W.; Scholz, H.; Steinfath, M.; Böhm, M.; et al. Chronic treatment with carbachol sensitizes the myocardium to cAMP-induced arrhythmia. *Circulation* **1996**, *93*, 763–771. [CrossRef]
50. Ukai, M.; Ogawa, K. Cyclic nucleotides concentrations in the canine heart with regional ischemia. The role of cyclic AMP in ventricular fibrillation and the effect of dibutyryl cyclic AMP. *Jpn. Circ. J.* **1984**, *48*, 247–252. [CrossRef]
51. Mangoni, M.E.; Couette, B.; Bourinet, E.; Platzer, J.; Reimer, D.; Striessnig, J.; Nargeot, J. Functional role of L-type Cav1.3 Ca2+ channels in cardiac pacemaker activity. *Proc. Natl. Acad. Sci. USA* **2003**, *100*, 5543–5548. [CrossRef]
52. Toyoda, F.; Ding, W.G.; Matsuura, H. Responses of the sustained inward current to autonomic agonists in guinea-pig sino-atrial node pacemaker cells. *Br. J. Pharmacol.* **2005**, *144*, 660–668. [CrossRef] [PubMed]
53. Toyoda, F.; Ding, W.G.; Matsuura, H. Heterogeneous functional expression of the sustained inward Na(+) current in guinea pig sinoatrial node cells. *Pflug. Arch. Eur. J. Physiol.* **2018**, *470*, 481–490. [CrossRef] [PubMed]
54. Toyoda, F.; Mesirca, P.; Dubel, S.; Ding, W.G.; Striessnig, J.; Mangoni, M.E.; Matsuura, H. Ca(V)1.3 L-type Ca(2+) channel contributes to the heartbeat by generating a dihydropyridine-sensitive persistent Na(+) current. *Sci. Rep.* **2017**, *7*, 7869. [CrossRef] [PubMed]
55. Ebert, T.J.; Muzi, M.; Berens, R.; Goff, D.; Kampine, J.P. Sympathetic responses to induction of anesthesia in humans with propofol or etomidate. *Anesthesiology* **1992**, *76*, 725–733. [CrossRef] [PubMed]

biomolecules

MDPI

Article

The Effect of a Synthetic Estrogen, Ethinylestradiol, on the hERG Block by E-4031

Fumiya Tamura [1,†], Shintaro Sugimoto [2,†], Mana Sugimoto [2], Kazuho Sakamoto [2], Masahiko Yamaguchi [2], Takeshi Suzuki [3], Keiichi Fukuda [1], Masaki Ieda [4] and Junko Kurokawa [2,*]

1 Department of Cardiology, Keio University School of Medicine, 35 Shinanomachi, Shinjuku-ku, Tokyo 160-8582, Japan; fumiya.tamura@keio.jp (F.T.); kfukuda@a2.keio.jp (K.F.)
2 Department of Bio-Informational Pharmacology, School of Pharmaceutical Sciences, University of Shizuoka, 52-1 Yada, Suruga-ku, Shizuoka-shi, Shizuoka 422-8526, Japan; s21808@u-shizuoka-ken.ac.jp (S.S.); m17062@u-shizuoka-ken.ac.jp (M.S.); kazuho@u-shizuoka-ken.ac.jp (K.S.); masahiko-y@u-shizuoka-ken.ac.jp (M.Y.)
3 Division of Basic Biological Sciences, Faculty of Pharmacy, Keio University, 1-5-30 Shibakoen, Minato-ku, Tokyo 105-8512, Japan; suzuki-tk@pha.keio.ac.jp
4 Department of Cardiology, Faculty of Medicine, University of Tsukuba, 1-1-1 Tennoudai, Tsukuba City, Ibaraki 305-8575, Japan; mieda@md.tsukuba.ac.jp
* Correspondence: junkokuro@u-shizuoka-ken.ac.jp
† These authors contributed equally to this work.

Citation: Tamura, F.; Sugimoto, S.; Sugimoto, M.; Sakamoto, K.; Yamaguchi, M.; Suzuki, T.; Fukuda, K.; Ieda, M.; Kurokawa, J. The Effect of a Synthetic Estrogen, Ethinylestradiol, on the hERG Block by E-4031. *Biomolecules* **2021**, *11*, 1385. https://doi.org/10.3390/biom11091385

Academic Editors: Yosuke Okamoto and Kyoichi Ono

Received: 30 July 2021
Accepted: 18 September 2021
Published: 20 September 2021

Abstract: Inhibition of K^+-conductance through the human ether-a-go-go related gene (hERG) channel leads to QT prolongation and is associated with cardiac arrhythmias. We previously reported that physiological concentrations of some estrogens partially suppress the hERG channel currents by interacting with the S6 residue F656 and increase the sensitivity of hERG blockade by E-4031. Although these studies suggested that clinically used synthetic estrogens with similar structures have the marked potential to alter hERG functions, the hERG interactions with synthetic estrogens have not been assessed. We therefore examined whether ethinylestradiol (EE2), a synthetic estrogen used in oral contraceptives, affects hERG function and blockade by drugs. Supratherapeutic concentrations of EE2 did not alter amplitudes or kinetics of the hERG currents elicited by train pulses at 20 mV (0.1 Hz). On the other hand, EE2 at therapeutic concentrations reduced the degree of hERG current suppression by E-4031. The administration of EE2 followed by E-4031 blockade reversed the current suppression, suggesting that the interaction of EE2 and E-4031 alters hERG at the drug-binding site. The effects of EE2 on hERG blockade raised the possibility that other estrogens, including synthetic estrogens, can alter hERG blockade by drugs that cause QT prolongation and ventricular arrhythmias.

Keywords: cardiac potassium channel; hERG blocker; synthetic estrogen; QT intervals; drug interaction

1. Introduction

The rapid component of the delayed rectifier potassium current (I_{Kr}) plays an essential role in cardiac repolarization [1]. The pore-forming subunit of the human I_{Kr} channel is encoded by the human ether-a-go-go-related gene (hERG; KCNH2) [2]. A decrease in I_{Kr} as a result of drug blockade is a major cause of acquired (drug-induced) long QT syndrome (LQTS), which is associated with electrocardiographical QT_C prolongation and lethal ventricular arrhythmias, presenting as torsades de pointes (TdP) [3]. Thus, the hERG channel has become a primary anti-target in drug development. However, drug-induced QT_C prolongation is a complicated phenomenon that is related not just to unintended hERG blockade, but also to multi-channel blockade, drug–drug interactions, and a variety of patient factors, including sex [4–7].

Women are more prone to developing TdP in response to QT-prolonging drugs than men [4–7], and the mechanism may be related to baseline QT_C intervals, which are approximately 20 ms longer in women than in men [7]. The mechanism of sex differences in baseline QT_C intervals involves the shortening of QT_C intervals, mainly by endogenous testosterone and progesterone [8–11]. Although the effects of estrogen on QT intervals may not be as dominant as those of testosterone or progesterone, studies of menopausal hormone therapy (MHT) in the form of estrogen-alone therapy (ET) and estrogen plus progesterone therapy (EPT) suggested a counterbalancing effect of exogenous estrogen and progesterone on QT intervals [12]. Specifically, ET lengthens the QT interval, whereas EPT has no effect. In our animal studies [13,14], estrogen at a physiological concentration lengthened QT intervals by suppressing I_{Kr} in a receptor-independent manner [13]. Our previous patch-clamp analysis with hERG channel-expressing cells revealed that some estrogens, estradiol and estrone sulfate, interact with the hERG channel, and alter the effects of a selective hERG blocker, E-4031 [13,15]. This study is consistent with the clinical QT prolongation by estrogens used for ET. Recently, oral contraceptives were reported to increase the risk of TdP based on the administration of *d,l*-sotalol to healthy female volunteers [16]. However, to date, no studies have assessed whether oral contraceptive ingredients affect the hERG channel function. Thus, we investigated the effects of ethinylestradiol (EE2), which is used in almost all modern formulations of combined oral contraceptive pills, on the hERG channel currents and hERG blockade by E-4031 in stable hERG-expressing HEK293 cells.

2. Materials and Methods

2.1. Materials

Stock solutions of EE2 (Tokyo Chemical Industry Co. Ltd., Tokyo, Japan) at 10 mM (in ethanol) and E-4031 (Eisai Co. Ltd., Tokyo, Japan) at 10 mM (in H_2O) were diluted to final concentrations in the external solutions. The final concentration of solvent (ethanol) was confirmed to have no effects on hERG currents [13]. All other materials were of reagent grade quality and obtained from standard sources.

2.2. Cell Culture

Human embryonic kidney (HEK) 293 cells stably expressing hERG [17] were cultured in phenol red-free D-MEM supplemented with 10% charcoal-treated FBS and 200 μg/mL of geneticin, G418. All cells were kept in an incubator at 37 °C with 5% CO_2, and plated on culture dishes the day before electrophysiological experiments.

2.3. Electrophysiology

Methods were described in detail previously [13]. In brief, hERG channel currents were recorded at room temperature (22 ± 2 °C) using the perforated patch-clamp technique with an Axopatch 200B amplifier (Molecular Devices, San Jose, CA, USA). The control bath solution contained 132 mM NaCl, 4.8 mM KCl, 1.2 mM $MgCl_2$, 2 mM $CaCl_2$, 5 mM glucose, and 10 mM HEPES, pH 7.4. Pipettes (2–4 MΩ resistance) were filled with a solution containing 110 mM K-aspartate, 5 mM ATP-K_2, 1 mM $CaCl_2$, 1 mM $MgCl_2$, 11 mM EGTA, and 5 mM HEPES, pH 7.3. To achieve patch perforation (series resistance: 10–20 MΩ), amphotericin B (0.3 mg/mL) (Nacalai Tesque, Inc., Kyoto, Japan) was added to the pipette solution. Signals were low-pass filtered at 5 kHz, sampled at 2 kHz, and compensated for cell capacitance (10–40 pF), but not for series resistance.

To investigate the effects of EE2 on hERG current amplitudes, the peak deactivating tail current was recorded at the repolarizing steps to −40 mV subsequent to 2-s depolarizing test pulses to 20 mV from a holding potential of −80 mV, as described previously [13]. The hERG channel tail-current amplitude was monitored at 0.1 Hz. In order to examine the hERG activation curves, the data of the normalized tail current amplitudes of I_{hERG} were fitted to the Boltzmann equation: $I/I_{max} = G/G_{max} = \{1 + \exp[-(V_m - V_{0.5})/k] - 1\}$, where

G/G_{max} is normalized chord conductance at V_m to the maximum chord conductance, $V_{0.5}$ is the potential where the conductance is half-maximally activated, and k is the slope factor.

2.4. Data Analysis

All values are presented as the mean ±S.E. pCLAMP 10.7 software (Molecular Devices, San Jose, CA, USA) was used to both acquire and analyze data for the patch-clamp experiments. Graphical and statistical analyses were carried out using OriginPro 2021 software (OriginLab Corporation, Northampton, MA, USA). Significant differences for multiple comparisons in Figures 1C and 2 were assessed using one-way ANOVA followed by Tukey's post hoc test. Statistical analysis was carried out using SPSS software Ver27.0 (IBM, Armonk, NY, USA). $p < 0.05$ was considered to be significant.

3. Results

To assess the effects of ethinylestradiol (EE2, Figure 1A) on the hERG current, patch clamp analysis in stable hERG-HEK cells was performed as described in the Methods. The hERG channel function was measured as the peak tail current at −40 mV following a 2-s voltage step from −80 mV to the test voltages (V_t). As shown in Figure 1B,C, exposure of EE2 (0.1–10 nM) had no effect on hERG channel current traces elicited by a test pulse at 20 mV. Under the same experimental condition, exposure to 3 nM 17-β-estradiol (E2, Figure 1A,C) for 4 min significantly blocked hERG tail currents by 21.6 ± 3.2% ($p < 0.00001$ ANOVA with repeated measures, vs. EE2), confirming the acute and partial hERG block by E2 [13]. The effects of EE2 on current–voltage (I–V) relationships for peak outward hERG currents and peak tail currents were investigated by applying EE2 at 1 nM and 10 nM, respectively (Figure 1D–F). For each cell, the respective tail peak amplitudes at each step were normalized to the amplitudes of maximum tail current elicited by a strong V_t to the plateau level of channel activation, allowing us to estimate the macroscopic channel availability at the end of the preceding V_t. As shown in Table 1, which summarizes the data analysis, the 5-min application of EE2 did not affect the maximum hERG current density or voltage dependence of the activation. These analyses for voltage-dependent activation of the hERG channel revealed that EE2 at clinical dosages has no direct effects on the hERG channel activity.

Table 1. Effects of EE2 on voltage-dependence of hERG activation. Before and after a cumulative 5-min exposure from 1 nM to 10 nM EE2, I-V relationships were obtained by the same voltage protocol described in Figure 1D–G. Eighteen cells were used. No significant difference was detected (ANOVA with repeated measures).

	Control (Before)	EE2 at 1 nM	EE2 at 10 nM
(n = 18)			
$V_{0.5}$ (mV)	3.5 ± 1.5	4.4 ± 1.4	5.4 ± 1.3
Slope factor (k)	10.3 ± 0.3	10.2 ± 0.3	10.5 ± 0.3
I_{max} (pA/pF)	46.5 ± 3.5	47.3 ± 3.9	47.7 ± 3.9

We next evaluated the effects of EE2 on the sensitivity of a hERG blocker, E-4031, to hERG currents in HEK293 cells. In the presence or absence of hormones, plots of tail amplitudes were normalized relative to the values just before the application of E-4031. Experiments in the presence or absence of hormones were performed on the same day to avoid the effects of unexpected batch-to-batch variation between cell cultures. As shown in Figure 2, the presence of EE2 in the external solution significantly reduced the fractional inhibition of hERG currents induced by E-4031 at 300 nM.

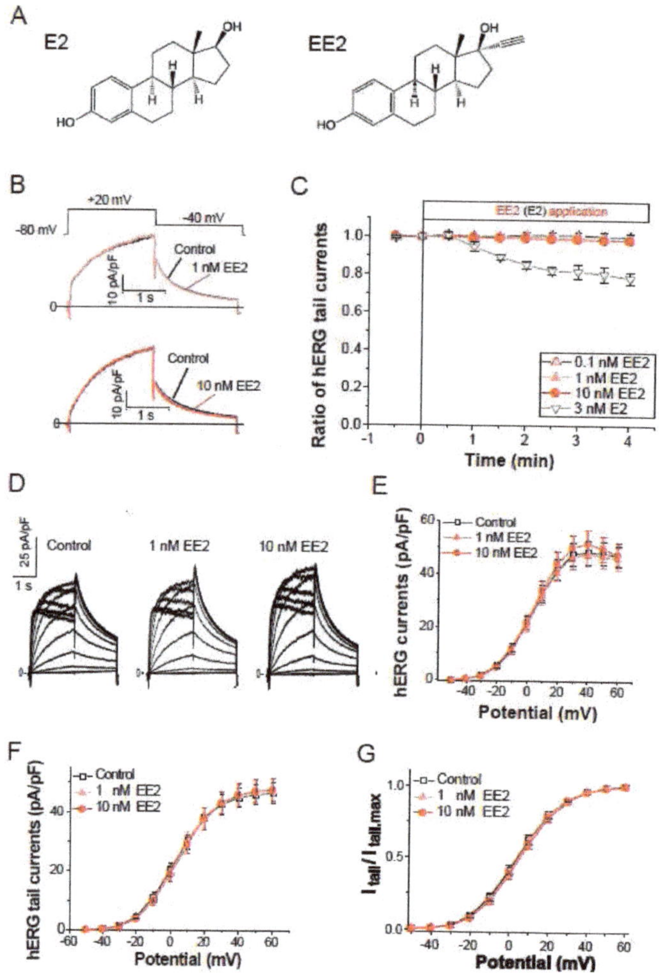

Figure 1. Effects of the external application of ethinylestradiol, EE2, on hERG channel currents. Membrane currents were recorded from HEK293 cells stably expressing hERG. The hERG channels were sequentially activated at 0.1 Hz by 2-s test pulses from a holding potential (V_h) at −80 mV. (**A**), Chemical structures of E2 (left) and EE2 (right). (**B**), Representative traces before and after the application of EE2 (upper; 1 nM, lower; 10 nM) for 4 min. (**C**), Time courses of the effects of EE2 at 0.1 nM (open triangles, $n = 5$), 1 nM (closed triangles, $n = 8$), and 10 nM ($n = 7$), and E2 at 3 nM (closed triangles, $n = 5$). After stabilizing the tail amplitudes for 1 min, currents were recorded in the presence of each concentration of EE2 in the bath solution. Plots (means ± S.E.M.) are shown as ratios of the peak tail amplitudes just before the application of estrogens (control at time zero). (**D–G**), No effect on current-voltage relationships of the hERG channel elicited by a series of 2-s test pulses from −50 to 60 mV (10-mV increments, 0.1 Hz). (**D**), Representative superimposed traces at step pulses (2-s test pulses, −40 mV return, $V_h = -80$ mV) from a single cell before (left) and after a 5-min cumulative application of EE2 from 1 nM (middle) to 10 nM (right). (**E**), Current–voltage relationship at the end of test pulses. (**F**), Current–voltage relationship of tail peak currents recorded at −40 mV. (**G**), Normalized peaks of tail currents were plotted as a function of the hERG activation (Boltzmann fitting). Eighteen cells were used. Data are summarized in Table 1.

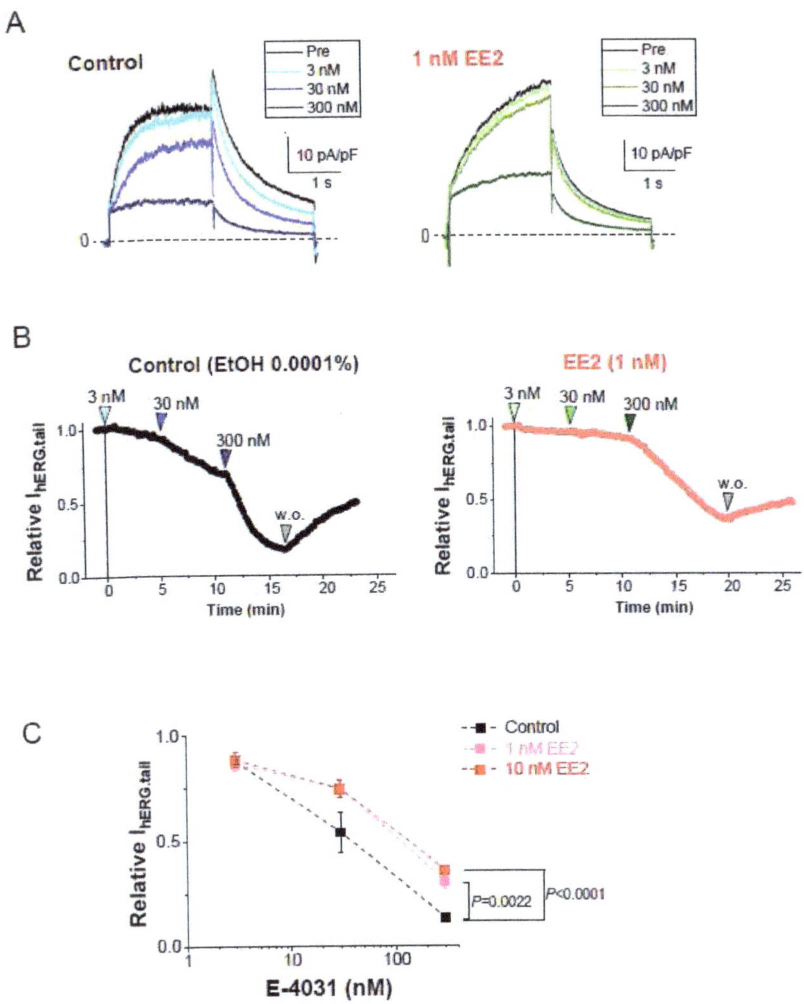

Figure 2. Blockade of hERG currents by E-4031 with or without EE2. hERG currents were recorded from HEK293 cells using the scheme as that described in Figure 1D. The hERG channels were sequentially activated at 0.1 Hz by 20-mV test pulses for 2 s from a holding potential (V_h) at −80 mV, whereas tail currents were recorded with a repolarizing step to −40 mV. EE2 at 1 nM was administered 5 min prior to the cumulative application of E-4031. (**A**). Representative traces in the presence of EtOH (left) and EE2 (right) are shown by superimposing the traces before (control) and after the addition of E-4031 (3, 30 and 300 nM). Scale, 10 pA/pF, 1 s. (**B**). Time course of hERG inhibition by cumulative application of E-4031 in the presence of 0.0001% EtOH (left) and 1 nM EE2 (right). After the current amplitudes were stabilized for 1 min, E-4031 was added in the presence of EtOH or EE2. The timing of E-4031 applications (3, 30, and 300 nM) indicated by the arrowheads above the plot. (**C**). Concentration-dependent inhibition of E-4031 was plotted as relative values of the tail amplitudes compared to the right before the application of E-4031. Colored lines indicate the presence of EE2. Control (no EE2); $n = 8$, EE2 at 1 nM; $n = 8$, EE2 at 10 nM; $n = 8$.

To assess whether later-administered EE2 can alter the hERG blockade by E-4031, EE2 at 1 nM was added after the hERG blockade was stabilized by E-4031 at 30 nM (Figure 3). A time course of the peak tail hERG currents at −40 mV is depicted in Figure 3A. To investigate the effects of EE2 on the hERG channel blockade by E-4031, the peak tail hERG amplitudes were compared under each administrative condition (Figures 3 and 4). E-4031 at 30 nM near IC_{50} [13] inhibited the peak tail currents of the hERG channel (Figure 3A, blue plots), and the following addition of EE2 at 1 nM restored the tail amplitudes in a time-dependent manner (Figure 3A, green plots). To examine the voltage dependence of the effects, I-V relationships of the peak tail current amplitudes before and after the administration of E-4031 and addition of EE2 were averaged (Figure 4B) and plotted versus the preceding test pulses from −40 mV to 60 mV. E-4031 reduced the peak hERG tail amplitudes measured after voltage steps to 20–60 mV, and the addition of EE2 at 1 nM partially recovered the reduced current amplitudes measured after voltage steps to 40–60 mV (Figure 4A,B). After normalization by the maximal peak amplitude recorded during the preceding voltage step at 60 mV in the control (Figure 4C), EE2-induced recovery from inhibition was observed in the tail amplitudes measured after voltage steps to −10 mV and 0 mV, where the hERG channel played a significant role in repolarization of the cardiac action potential. The tail current amplitude normalized to the maximum tail current amplitude at each I-V challenge was used to fit the activation curves shown in Figure 4D. No significant changes in midpoint voltage (V_{50}) or slope factor (k) for activation were found by the application of drugs, whereas the decreased Imax by E-4031 significantly recovered after the addition of EE2 (Table 2), confirming the results shown in Figure 4.

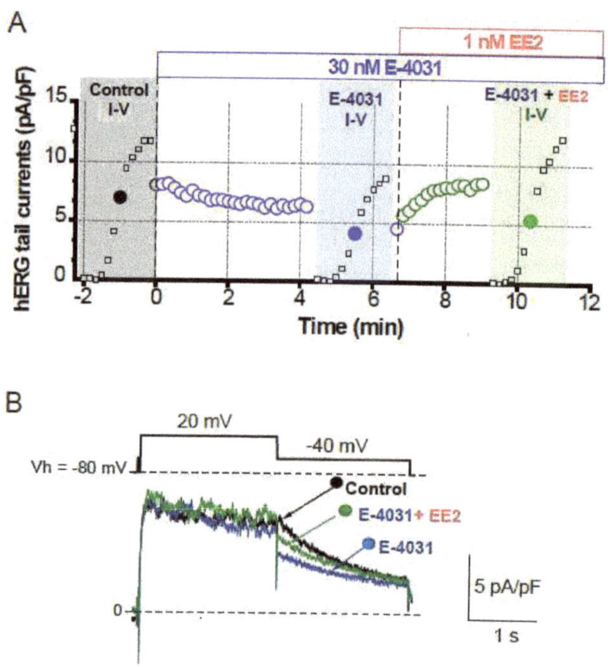

Figure 3. Effects of EE2 on blockade of hERG currents by E-4031. HERG currents were recorded from HEK293 cells stably expressing hERG as described in Figure 1. (**A**). Time course of hERG tail current blockade by E-4031 at 30 nM and the following recovery by the addition of EE2 at 1 nM. Plots from a representative experiment are normalized by the tail amplitude before E-4031 application (time 0). The current-voltage (I–V) relationships were tested before drug application (control, black), after E-4031 block (blue) and after the addition of EE2 (E-4031 + EE2, green). (**B**). Representative traces elicited by 20-mV test pulse (closed circles in I–V relationships) are shown by superimposing the traces before (control, black) and after the application of E-4031 only (blue) and E-4031 plus EE2 (green). Scale, 5 pA/pF, 1 s.

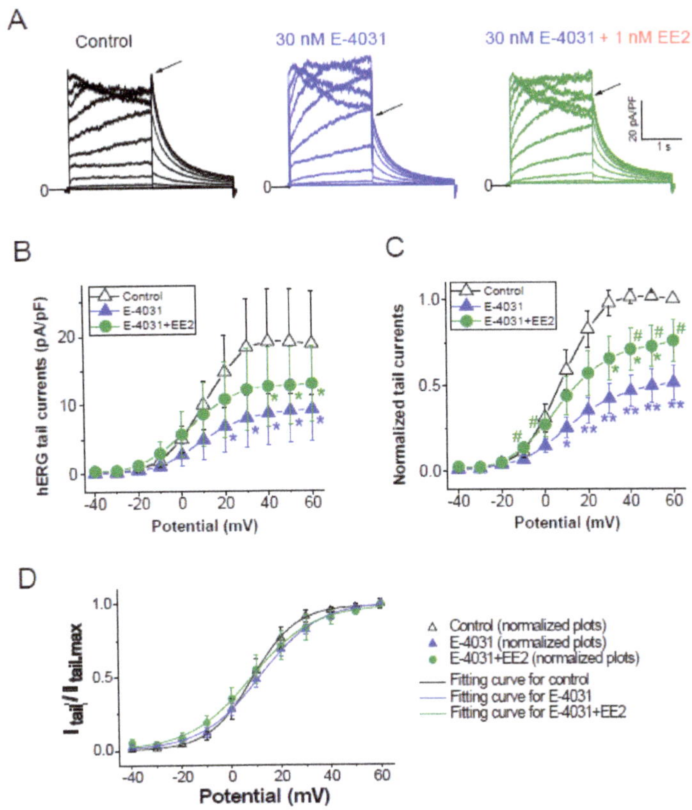

Figure 4. Effects of EE2 on I–V relationships of hERG currents by E-4031. I–V relationships of the hERG channel were activated by the same voltage protocol described in Figure 1D, and compared before drug application (control, black), after 30 nM E-4031 block (blue), and after the addition of EE2 (30 nM E-4031 + 1 nM EE2, green). (**A**), Representative traces elicited by step pulses (−40 mV to 60 mV). Tail peaks are indicated by arrows. Scale, 20 pA/pF, 1 s. (**B**), Tail I-V curves before (control), after the application of E-4031, and after the addition of EE2 (E-4031 + EE2). $p < 0.05$ ANOVA with repeated measures. * $p < 0.05$ vs. control. (**C**), Respective tail peak amplitudes were normalized to the tail amplitude preceding 60 mV before drug applications (control). $p < 0.05$ ANOVA with repeated measures. * $p < 0.05$, ** $p < 0.01$, vs. control, # $p < 0.05$ vs. E-4031. (**D**), Comparison of channel availability curves obtained from tail I-Vs. Smooth lines are Boltzmann fits, as described in the Methods, that generated V0.5 of activation in Control (black line), in E-4031 (blue line), and in E-4031/EE2 (green line); ns, ANOVA. Seven experiments were performed. Significance was evaluated using Student's t-tests after repeated measures one-way ANOVA.

Table 2. Effects of EE2 at 1 nM on the change in voltage dependence of the hERG activation by E-4031 at 30 nM. The experimental condition and results are also shown in Figure 4. Seven experiments were performed.

	Control	E-4031	E-4031 ± EE2
($n = 7$)			
$V_{0.5}$ (mV)	9.6 ± 3.2	11.1 ± 2.2	8.7 ± 5.3
Slope factor (k)	8.1 ± 0.4	11.4 ± 0.9	10.9 ± 1.3
I_{max} (pA/pF)	20.0 ± 7.7	9.2 ± 4.6 *	13.3 ± 5.7 #

$p < 0.05$ ANOVA with repeated measures. * $p < 0.05$ Control vs. E-4031, # $p < 0.05$ E-4031 vs. E-4031 + EE2.

4. Discussion

Synthetic estrogens, such as oral contraceptives, have medical applications worldwide [18]. Ethinylestradiol (EE2) is an estrogen medication used widely in oral contraceptives in combination with progestins [18]. In this study, EE2 reduced the hERG block by E-4031 without exhibiting detectable effects on the hERG channel function. As the hERG blockade by selective I_{Kr} inhibitors like E-4031 is associated with QT prolongation and an increased risk of arrhythmia, interaction with EE2 may reduce this risk. Because the risk of drug-induced QT prolongation is associated with physiological conditions such as heart rate [19] and genetic substrates such as LQT1 mutations [20], the clinical relevance of the degree of recovery of partial hERG block by EE2 may be case-by-case. The concentrations of EE2 used in this study (1–10 nM) are near or greater than peak concentrations in common clinical doses (1–5 nM) [21]. Thus, oral contraceptives containing EE2 may alter the risk of arrhythmia induction by a hERG inhibitor, E-4031.

We previously reported that 17-β-estradiol (E2), the most bio-active estrogen, promotes hERG inhibition and rodent QT interval prolongation caused by E-4031 [13,14]. As shown in Figure 1A, E2 alone partially inhibited the hERG channel described initially by Kurokawa et al. [13], which is different from the interaction with EE2 described in the present study. The partial inhibition of hERG channels by physiological serum concentrations of E2 [13] or estrone sulfate [15], but not by its precursors dihydrotestosterone or progesterone, was abolished when the aromatic side chain F656 located in the inner cavity of the hERG channel was mutated to threonine or methionine [13], suggesting that the aromatic ring in the chemical structure of E2 is the site of interaction. The interaction of E2 with hERG inhibition via the aromatic ring has also been demonstrated in vivo, as the QT-prolonging effects of E-4031 were increased by E2 in aromatase knockout mice lacking endogenous estrogen [14]. The effects of E2 on the risk of QT prolongation by hERG inhibition [14] suggest the general involvement of aromatic estrogens with or without independent inhibition of hERG [15], but other interactions between estrogens and hERG blockers have not been confirmed and require further investigation.

On the molecular level, we discuss the present results based on our previous MD simulations of the interaction between E2 and dofetilide [22]. The MD simulation using an hERG open state model identified a binding site for E2 which is defined by the aromatic F656 side-chain, supplemented by a number of hydrophobic residues including L650 and A653. Moreover, Y652 is also transiently involved at this site. The E2 intracellular cavity binding site is adjacent to that of dofetilide, suggesting a likely interaction. Because the chemical structure that binds it to the common binding site of hERG is almost identical for dofetilide and E-4031, a similar interaction between E2 and E-4031could be speculated. It is apparent that 17α-ethynylation of estradiol in EE2 alters the hydrophobic environment around side chains of Y652 and F656, implying that the effects of E2 and EE2 are so different. In the present study, we found that EE2 reduced the effects of E-4031 on hERG channels, as opposed to enhancing them, suggesting that estrogens can alter the action of hERG inhibitors in both directions. Whether EE2 was administered prior to (Figure 2) or after (Figures 3 and 4) E-4031, EE2 similarly attenuated hERG inhibition. In addition, alanine substitution of Y652, a typical drug-binding site of the hERG channel, did not affect the action of E2 [13]. This suggests that the site of action of estrogens, including F656 of the hERG channel, is distinct from the typical binding site of hERG inhibitors, which includes both F656 and Y652. This is a reasonable hypothesis considering that hERG potassium channels exhibit numerous interactions with diverse chemical scaffolds [23]. Further studies on the binding site of estrogens in the hERG channel are expected in the future.

The clinical significance of our cell-based assay is that we confirmed that synthetic estrogens used in medicine can influence the action of hERG blockers. There is a worldwide debate concerning whether oral contraceptives increase the risk of arrhythmias caused by QT-prolonging drugs, and there are reports that some oral contraceptives increase the risk of TdP caused by *d,l*-sotalol [16]. In summary, this study suggests that the accuracy of the assessment of cardiotoxicity due to arrhythmogenesis caused by hERG inhibition can be improved by considering the interaction with estrogens.

Author Contributions: Conceptualization, J.K.; methodology, J.K.; software, J.K. and K.S.; validation, J.K., F.T., S.S., and K.S.; formal analysis, K.S.; investigation, F.T. and S.S.; resources, J.K., K.S., and M.Y.; writing—original draft preparation, F.T., S.S., and J.K.; writing—review and editing, J.K. and F.T.; visualization, J.K., F.T., S.S. and M.S.; supervision, J.K., T.S., K.F., and M.I.; funding acquisition, J.K and K.S. All authors have read and agreed to the published version of the manuscript.

Funding: This study was supported by JSPS KAKENHI 17K19499, 19H03380 and 20K21487 to J.K., 18K06897 to K.S., the Research on Regulatory Harmonization and Evaluation of Pharmaceuticals, Medical Devices, Regenerative and Cellular Therapy Products, Gene Therapy Products, and Cosmetics (18mk0104117h0703, 20mk0104117h0703, 21mk0101189h0301) from the Japan Agency for Medical Research and Development (AMED). This study was supported by the Adaptable and Seamless Technology Transfer Program through Target-driven R&D (A-STEP) from the Japan Science and Technology Agency (JST, VP30118068096), collaborative Joint Research for Exploratory Research Center on Life and Living Systems (ExCELLS), and a grant from the University of Shizuoka.

Institutional Review Board Statement: Not applicable.

Informed Consent Statement: Not applicable.

Data Availability Statement: Not applicable.

Conflicts of Interest: The authors declare no conflict of interest.

References

1. Nerbonne, J.M.; Kass, R.S. Molecular Physiology of Cardiac Repolarization. *Physiol. Rev.* **2005**, *85*, 1205–1253. [CrossRef]
2. Sanguinetti, M.C.; Jiang, C.; Curran, M.E.; Keating, M.T. A mechanistic link between an inherited and an acquired cardiac arrhythmia: HERG encodes the IKr potassium channel. *Cell* **1995**, *81*, 299–307. [CrossRef]
3. Clancy, C.E.; Kurokawa, J.; Tateyama, M.; Wehrens, X.H.; Kass, R.S. K+ channel structure-activity relationships and mechanisms of drug-induced QT prolongation. *Annu. Rev. Pharmacol. Toxicol.* **2003**, *43*, 441–461. [CrossRef] [PubMed]
4. Drici, M.D.; Arrighi, I.; Chouabe, C.; Mann, J.R.; Lazdunski, M.; Romey, G.; Barhanin, J. Involvement of IsK-associated K+ channel in heart rate control of repolarization in a murine engineered model of Jervell and Lange-Nielsen syndrome. *Circ. Res.* **1998**, *83*, 95–102. [CrossRef] [PubMed]
5. Lehmann, M.H.; Hardy, S.; Archibald, D.; Quart, B.; MacNeil, D.J. Sex Difference in Risk of Torsade de Pointes with *d,l*- Sotalol. *Circulation* **1996**, *94*, 2535–2541. [CrossRef] [PubMed]
6. Makkar, R.R.; Fromm, B.S.; Steinman, R.T.; Meissner, M.D.; Lehmann, M.H. Female Gender as a Risk Factor for Torsades de Pointes Associated with Cardiovascular Drugs. *JAMA* **1993**, *270*, 2590–2597. [CrossRef] [PubMed]
7. Kurokawa, J.; Kodama, M.; Furukawa, T.; Clancy, C.E. Sex and Gender Aspects in Antiarrhythmic Therapy. *Handb. Exp. Pharmacol.* **2012**, *214*, 237–263. [CrossRef]
8. Bai, C.-X.; Kurokawa, J.; Tamagawa, M.; Nakaya, H.; Furukawa, T. Nontranscriptional Regulation of Cardiac Repolarization Currents by Testosterone. *Circulation* **2005**, *112*, 1701–1710. [CrossRef]
9. Nakamura, H.; Kurokawa, J.; Bai, C.X.; Asada, K.; Xu, J.; Oren, R.V.; Zhu, Z.I.; Clancy, C.E.; Isobe, M.; Furukawa, T. Progesterone regulates cardiac repolarization through a nongenomic pathway: An in vitro patch-clamp and computational modeling study. *Circulation* **2007**, *116*, 2913–2922. [CrossRef]
10. Nakagawa, M.; Ooie, T.; Takahashi, N.; Taniguchi, Y.; Anan, F.; Yonemochi, H.; Saikawa, T. Influence of Menstrual Cycle on QT Interval Dynamics. *Pacing Clin. Electrophysiol.* **2006**, *29*, 607–613. [CrossRef]
11. Rodriguez, I.; Kilborn, M.J.; Liu, X.-K.; Pezzullo, J.C.; Woosley, R.L. Drug-Induced QT Prolongation in Women During the Menstrual Cycle. *JAMA* **2001**, *285*, 1322–1326. [CrossRef]
12. Kadish, A.H.; Greenland, P.; Limacher, M.C.; Frishman, W.H.; Daugherty, S.A.; Schwartz, J.B. Estrogen and Progestin Use and the QT Interval in Postmenopausal Women. *Ann. Noninvasive Electrocardiol.* **2004**, *9*, 366–374. [CrossRef] [PubMed]
13. Kurokawa, J.; Tamagawa, M.; Harada, N.; Honda, S.; Nakaya, H.; Furukawa, T. Acute effects of oestrogen on the guinea pig and human IKr channels and drug-induced prolongation of cardiac repolarization. *J. Physiol.* **2008**, *586*, 2961–2973. [CrossRef] [PubMed]
14. Kurokawa, J.; Sasano, T.; Kodama, M.; Li, M.; Ebana, Y.; Harada, N.; Honda, S.-I.; Nakaya, H.; Furukawa, T. Aromatase knockout mice reveal an impact of estrogen on drug-induced alternation of murine electrocardiography parameters. *J. Toxicol. Sci.* **2015**, *40*, 339–348. [CrossRef] [PubMed]
15. Kakusaka, S.; Asayama, M.; Kaihara, A.; Sasano, T.; Suzuki, T.; Kurokawa, J.; Furukawa, T. A Receptor-Independent Effect of Estrone Sulfate on the hERG Channel. *J. Pharmacol. Sci.* **2009**, *109*, 152–156. [CrossRef]

16. Salem, J.-E.; Dureau, P.; Bachelot, A.; Germain, M.; Voiriot, P.; Lebourgeois, B.; Trégouët, D.-A.; Hulot, J.-S.; Funck-Brentano, C. Association of Oral Contraceptives with Drug-Induced QT Interval Prolongation in Healthy Nonmenopausal Women. *JAMA Cardiol.* **2018**, *3*, 877–882. [CrossRef]
17. Wu, L.-M.; Orikabe, M.; Hirano, Y.; Kawano, S.; Hiraoka, M. Effects of Na+ Channel Blocker, Pilsicainide, on HERG Current Expressed in HEK-293 Cells. *J. Cardiovasc. Pharmacol.* **2003**, *42*, 410–418. [CrossRef]
18. Maier, W.E.; Herman, J.R. Pharmacology and Toxicology of Ethinyl Estradiol and Norethindrone Acetate in Experimental Animals. *Regul. Toxicol. Pharmacol.* **2001**, *34*, 53–61. [CrossRef]
19. Ishii, K.; Nagai, M.; Takahashi, M.; Endoh, M. Dissociation of E-4031 from the HERG channel caused by mutations of an amino acid results in greater block at high stimulation frequency. *Cardiovasc. Res.* **2003**, *57*, 651–659. [CrossRef]
20. Napolitano, C.; Schwartz, P.J.; Brown, A.M.; Ronchetti, E.; Bianchi, L.; Pinnavaia, A.; Acquaro, G.; Priori, S.G. Evidence for a Cardiac Ion Channel Mutation Underlying Drug-Induced QT Prolongation and Life-Threatening Arrhythmias. *J. Cardiovasc. Electrophysiol.* **2000**, *11*, 691–696. [CrossRef]
21. Van den Heuvel, M.W.; van Bragt, A.J.; Alnabawy, A.K.; Kaptein, M.C. Comparison of ethinylestradiol pharmacokinetics in three hormonal contraceptive formulations: The vaginal ring, the transdermal patch and an oral contraceptive. *Contraception* **2005**, *72*, 168–174. [CrossRef] [PubMed]
22. Yang, P.-C.; Perissinotti, L.L.; Lopez-Redondo, F.; Wang, Y.; Demarco, K.R.; Jeng, M.-T.; Vorobyov, I.; Harvey, R.D.; Kurokawa, J.; Noskov, S.Y.; et al. A multiscale computational modelling approach predicts mechanisms of female sex risk in the setting of arousal-induced arrhythmias. *J. Physiol.* **2017**, *595*, 4695–4723. [CrossRef] [PubMed]
23. Yu, H.-B.; Zou, B.-Y.; Wang, X.-L.; Xiao-Liang, W. Investigation of miscellaneous hERG inhibition in large diverse compound collection using automated patch-clamp assay. *Acta Pharmacol. Sin.* **2016**, *37*, 111–123. [CrossRef] [PubMed]

biomolecules

MDPI

Article

Comparative Study of Transcriptome in the Hearts Isolated from Mice, Rats, and Humans

Daigo Okada [1], Yosuke Okamoto [2,*], Toshiro Io [3], Miho Oka [3], Daiki Kobayashi [2], Suzuka Ito [2], Ryo Yamada [1], Kuniaki Ishii [4] and Kyoichi Ono [2]

1 Center for Genomic Medicine, Graduate School of Medicine, Kyoto University, Shogoinkawahara-cho, Kyoto 606-8507, Japan; dokada@genome.med.kyoto-u.ac.jp (D.O.); yamada.ryo.5u@kyoto-u.jp (R.Y.)
2 Department of Cell Physiology, Akita Graduate School of Medicine, Hondo, Akita 010-8543, Japan; dkobayashi@med.akita-u.ac.jp (D.K.); s4020512@s.akita-u.ac.jp (S.I.); onok@med.akita-u.ac.jp (K.O.)
3 Research Department, Ono Pharmaceutical Co., Ltd., Kyutaromachi, Osaka 618-8585, Japan; io@ono.co.jp (T.I.); m.oka@ono.co.jp (M.O.)
4 Department of Pharmacology, Faculty of medicine, Yamagata University, Iida-Nishi, Yamagata 990-9585, Japan; kuishii@med.id.yamagata-u.ac.jp
* Correspondence: okamoto@med.akita-u.ac.jp

Abstract: The heart is a significant organ in mammalian life, and the heartbeat mechanism has been an essential focus of science. However, few studies have focused on species differences. Accordingly, challenges remain in studying genes that have universal functions across species and genes that determine species differences. Here, we analyzed transcriptome data in mouse, rat, and human atria, ventricles, and sinoatrial nodes (SA) obtained from different platforms and compared them by calculating specificity measure (SPM) values in consideration of species differences. Among the three heart regions, the species differences in SA were the greatest, and we searched for genes that determined the essential characteristics of SA, which was *SHOX2* in our criteria. The SPM value of *SHOX2* was prominently high across species. Similarly, by calculating SPM values, we identified 3 atrial-specific, 11 ventricular-specific, and 17 SA-specific markers. Ontology analysis identified 70 cardiac region- and species-specific ontologies. These results suggest that reanalyzing existing data by calculating SPM values may identify novel tissue-specific genes and species-dependent gene expression. This study identified the importance of *SHOX2* as an SA-specific transcription factor, a novel cardiac regional marker, and species-dependent ontologies.

Keywords: transcriptome; heart; SHOX2

Citation: Okada, D.; Okamoto, Y.; Io, T.; Oka, M.; Kobayashi, D.; Ito, S.; Yamada, R.; Ishii, K.; Ono, K. Comparative Study of Transcriptome in the Hearts Isolated from Mice, Rats, and Humans. *Biomolecules* **2022**, *12*, 859. https://doi.org/10.3390/biom12060859

Academic Editor: Pietro Scicchitano

Received: 31 March 2022
Accepted: 17 June 2022
Published: 20 June 2022

1. Introduction

The heart is the principal organ of mammalian lives, which functions first in an early developmental stage, and never stops for the rest of life. Thus, the molecular processes underlying heart function have been a significant focus of science for a long time. From the 1950s to the 1970s, actin-myosin sliding filament theory revealed the contraction mechanisms in response to intracellular Ca^{2+} ($[Ca^{2+}]_i$) elevation accompanied by the membrane depolarization via voltage-gated ion channels [1–3]. From the 1980s to the 1990s, molecular cloning strategies identified the gene sequences and primary structures of the ion channels [4–7]. The hyperpolarization-activated current (I_h) in the sinoatrial node (SA) was recorded in 1976 [8], which become known as the I_f current [9] via the HCN4 channel [10]. It is thought that $[Ca^{2+}]_i$ initiates the SA beating [11] as well as the ventricular contraction. These cardiac activities are regulated by the autonomic nervous system. One is the sympathetic control with the adrenergic stimulation. Intracellular signaling under the β-adrenaline receptor reinforces the ion currents and $[Ca^{2+}]_i$ movements, increasing the contraction's capability and accelerating the heart rate. The other is the parasympathetic control with the cholinergic signaling, which oppositely suppresses the sympathetic effects [12,13]. As such, muscle filament, ion channels, Ca^{2+} handling, and cell signaling networks are regarded as fundamental factors in cardiac physiology.

However, while the essential functions of the heart are believed to be well established, biophysical cardiac parameters at the molecular and cellular levels are not consistent throughout the species. For example, a Na^+-Ca^{2+} exchanger (NCX) is a device that converts $[Ca^{2+}]_i$ elevation into membrane potential, which is essential for driving SA automaticity in rabbits and mice, as mentioned above. Meanwhile, the guinea-pig SA beats independently of changes in $[Ca^{2+}]_i$ [14] and is not suppressed by an NCX inhibitor [15]. The L-type Ca^{2+} current (LCC) responses to intracellular cGMP differ in guinea-pigs, rats, and humans. cGMP magnifies the β-adrenergic LCC enhancement in guinea-pigs [16] while it exhibits the opposite effect in rats [17]. Further study using human myocytes uncovered that cGMP itself increased LCC without other effectors. Rivet-Bastide et al. proposed that different actions of cGMP on LCC were mainly attributable to the difference in the phosphodiesterase isotype [17]. In the case of potassium current, the 4-aminopyridine sensitive transient outward current observed in mouse, rat, and human cardiomyocytes [18,19] is unrecognizable in guinea-pigs [20]. By contrast, an isoprenaline-stimulable Cl^- current of the heart is prominent in guinea-pigs rather than in other mammals [21,22]. A comparative study of I_h was recently demonstrated in rat, guinea pig, and rabbit pulmonary vein cardiomyocytes [23,24]. Surprisingly, I_h in rats is neither I_f nor I_{KH} [25], but ClC-2 current [26]. In this careful and refined consideration of the data, species differences in molecular function within the heart have generally been observed. Therefore, a comprehensive comparison of species differences in molecular information within the heart is a critical challenge for future cardiac research. Thus, to compare species differences in gene expressions in a high-throughput manner, we acquired unpublished transcriptome data. We acquired these data from academic institutions or a pharmaceutical company that had launched experiments and remained unpublished because they failed to achieve their original purposes.

In the current study, we present an analysis of the transcriptomes in three different animal species—mice, rats, and humans—separately for atria, ventricles, and SA. The bioinformatical evaluation across the species disclosed the essential genes of the heart and the crucial difference among the species.

2. Materials and Methods

2.1. RNA Extractions from Experimental Animals

In order to extract endogenous RNAs from different parts of cardiac tissues, 16 C57/BL6 male mice (5–10 weeks) and 12 Wistar male rats (8–11 weeks) were sacrificed. Three mice and two rats were required to prepare an adequate RNA sample for the sinoatrial node (SA). In the surgical operation, the animals were deeply anesthetized by 320–450 mg/kg pentobarbital sodium with 100 UI heparin. After checking suppression of the nociceptive reflex, the chest cavity was opened, and the heart and lungs were excised in a block and then perfused sequentially with ice-cold, Ca^{2+} free, and heparinized external solution until most of the blood was washed out from the heart and lungs. The composition of the external solution (mM) was: NaCl 136.9, KCl 5.4, NaH_2PO_4 0.33, HEPES 5.0, MgCl 0.5 and glucose 5.5 (pH 7.4 with NaOH). The heart-lung block was pinned to the tissue bath. Soft tissue containing the vagus nerve and adipose tissue was trimmed off under a stereomicroscope, then cardiopulmonary tissues were isolated one by one. The left atrium (LA) adjacent to the pulmonary vein (PV), a mass of the left ventricle (LV), and the free-wall of the right ventricle (RV) were isolated. Each LV mass was dissected into three pieces as samples. Because the SAN isolation procedure takes approximately 20 min, SA was isolated separately. The SA region was delimited by the borders of the crista terminalis, the interatrial septum, the superior vena cava, and the right atrium (RA) (Supplementary Figure S1). Our SA-cutting technique was confirmed by quantitative PCR for mice and by the microarray data itself for rats (Supplementary Figure S2). Because the SAN is a tiny cardiac area, three mice and two rats were needed to extract the required amount of RNA for the microarray. All cardiopulmonary tissues were isolated within 30 min after the heart was removed from the body. During tissue isolation, the external solution was perfused at a rate of 10 mL/min on ice. In addition to PV, LV, RV, SA, and RA samples, pulmonary arteries were added to the samples for the rat microarray. All tissue samples were fresh-frozen in liquid nitrogen and stored at −80 °C for later RNA extraction. RNAs were extracted using QIAGEN RNeasy mini columns (QIAGEN, Venlo, The Netherland).

2.2. *RNA Extractions from Heart Donations*

Human heart samples were purchased from AnaBios Corporation (San Diego, CA, USA), which provides heart organs through the organ procurement organization of the USA in compliance with the Health Insurance Portability and Accountability Act. All personal information of donors is protected. In the current study, three normal hearts were separated into LA, RA, LV, RV, and SAN. These samples were frozen in RNAlater (Thermo Fisher Science, Waltham, MA, USA) at −20 °C. RNAs were extracted in our laboratory using QIAGEN RNeasy mini columns (QIAGEN).

2.3. *Mouse Gene Expression Microarrays (Illumina)*

Five hundred ng total RNA was labeled according to the manufacturer's instructions using the Illumina TotalPrepTM RNA amplification kit (Illumina, San Diego, CA, USA), and 750 ng biotinylated RNA per sample hybridized overnight to Illumina Mouse Ref-8 v2 BeadChips. Following post-hybridization rinses, arrays were incubated with streptavidin-conjugated Cy3, and scanned at a resolution of 0.53 μm using an Illumina iScan scanner. Hybridization intensity data were extracted using Illumina BeadStudio GenomeStudio software, ver. 2011.1 (Illumina).

A total of 18 microarray data samples were used for the mouse data, consisting of 4, 2, 4, 4, 4 samples of LA, LV, RA, RV, and SA. Signal values were transformed by log2(intensity + 1) and standardized using the normalizeBetweenArrays function in the limma package to align the medians of all arrays [27]. The probe information was converted to gene information corresponding to unique Ensemble Gene IDs. This conversion was done via Entrez Gene ID using DAVID web tool [27]. When multiple probes corresponded to a single gene, the maximum signal intensity value was adopted.

2.4. *Rat Gene Expression Microarrays (Agilent)*

RNAs from rat samples were amplified as complementary RNA (cRNA) and labeled with Cy3 using the Low-Input QuickAmp Labeling kit (Agilent Technologies, Santa Clara, CA, USA). Cy3-labeled cRNA was fragmented and hybridized to SurePrint G3 Rat GE microarray 8 × 60K v2 (Agilent Technologies). SurePrint Array was scanned by the Agilent DNA Microarray Scanner (G2505C).

A total of 23 microarray data samples were used for the rat data, with 4, 4, 4, 4, 4, 4, 3, 4 samples of LA, LV, PA, PV, RV, and SA. All background-subtracted data were detrended and normalized by 90 percentile normalization. Probes containing missing values were removed, and probe information was converted to gene information corresponding to unique Ensemble Gene IDs. This conversion was done via the Ensemble Transcriptome ID. Here, when multiple probes corresponded to a single gene, the maximum signal intensity value was adopted. The gene expression signal values thus obtained were converted by log2(intensity + 1) and used as gene expression intensity in subsequent analyses.

2.5. *RNA-Seq on the Donated Hearts*

The high-throughput sequence was demonstrated with NovaSeq 6000 (Illumina). First, extracted RNAs were purified by poly(A) capture. Resultant mRNAs were then fragmented and reverse-transcribed into single-stranded complementary DNAs (cDNAs). Subsequently, cDNAs were double-stranded by a DNA polymerase. During the polymerase reactions, deoxy UTP (dUTP) was mixed in nucleotide materials. Both ends of double-stranded DNA (ds DNA) were ligated to a 13 bp adapter sequence. Next, the ds DNAs were subjected to PCR amplification for the multi-sized DNA library preparation. NovaSeq

Control software v1.4.0 analyzed the sequencing runs and tag sequences classified each read in the raw sequencing data. A total of 21 RNA-seq data were used for human data, three samples each of LA, left atrial appendage (LAA), LV, pulmonary vein (PV), RA, RV, and SA. Fastp software (version 0.12.4, see reference [28]) was used for reading quality control and adapter removal [28]. In the raw data, the number of the generated pair-end reads per sample ranges between 49,720,696 to 368,429,512 reads (92,267,671 reads per sample on average). Further, we confirmed that Q30 > 90% is satisfied in all samples. Reads were aligned using STAR software (version 2.7.0a, see [29]). The fasta and gtf files of GRCh38/release105 obtained from the Ensemble database were used as reference genome and gene annotation information. The count values of each gene were quantified from the alignment results by featureCounts software (see [30]). We calculated the Transcripts Per Kilobase Million (TPM) values based on the calculated gene lengths and gene counts [31,32]. These values were converted by log2(intensity + 1) and used as gene expression intensity in subsequent analyses. As the quality check of this data, we compared our data with the public RNA-seq data in human RA, RV, LA, and LV samples (GSE112339 [33]). We confirmed the high correlation (Spearman's correlation coefficient > 0.9) between the mean expression value of each region of our and previous data (Supplementary Figure S3).

2.6. The Evaluation of Similarities of Expression Intensity among Species or Regions

Hierarchical clustering was applied to the gene expression matrices of humans, mice, and rats to evaluate the similarity of regions. It was performed using the R package TCC's clusterSample function with default setting [34]. The dendextend package was used to draw the results [35]. A comparison of transcriptome data among species was performed using gene ortholog relationships. Human–mouse, human–rat, and mouse–rat ortholog conversion tables were obtained from Ensemble Biomart. Using this table, scatter plots of gene expression profiles between species were created with one species as the x-axis and another species as the y-axis. We calculated Spearman's correlation coefficient of expression values. For each gene of the x-axis species, an orthologous counterpart gene of the y-axis species was mapped. If more than one gene of the y-axis species corresponded to one gene of the x-axis species, the average value was applied.

2.7. Identification of Heart Region-Specific Expressed Genes

Heart region-specific genes were identified in each species. Three heart regions were targeted: LA, SA, and V. Here, LV and RV were grouped together as V. Genes whose expression values exceeded a certain threshold in at least one sample were used for subsequent analysis. These threshold values were defined as >1, >4, and >6 for humans, rats, and mice, respectively. They are determined based on the distribution of expression levels. For human RNA-seq data, one was added to all elements of the gene expression matrix before subsequent analysis to avoid zero values affecting the results. For each gene, differences in expression levels among regions were tested by analysis of variance (ANOVA), and p-values were calculated. The p-values were adjusted by the Benjamini–Hochberg (BH) method considering multiple testing problems. In addition to the statistical tests, the region specificity of LA, V, and SA was calculated as the specificity measure (SPM) value for each gene [36]. SPM is obtained by dividing the average expression value of a region by the absolute values of the average expression vector of three regions. SPM values range from zero to one; the larger the value, the more specific the region. A gene was defined as a specifically expressed gene at a region if SPM was above a certain threshold and BH adjusted p-value < 0.05. This threshold was defined as the 95th percentile point of the null SPM values calculated from one label permutation per gene. The same procedure as in the evaluation of similarities of expression intensity was used to calculate the correspondence of SPM values in human–mouse and human–rat. We conducted the clustering analysis for nine species-region pairs for the SPM values obtained for the human genes with orthologs in both rats and mice. The distance metric used in the clustering was the 1—Spearman's correlation coefficient for SPM values, and the hclust function in R was used with default settings. The z-score normalized SPM was calculated by the standard normalization of SPM values for each species-region pair.

2.8. Comparison of Region-Specific Expressed Genes among Three Species

We conducted the biological interpretations of the region-specific gene sets (LA-specific, V-specific, and SA-specific) and compared them among species. First, we focused on commonly LA-specific, V-specific, or SA-specific gene sets in all three species. These are three gene sets: common LA-specific, common V-specific, and common SA-specific. This was defined as a region-specific gene in humans whose orthologous gene is also identically region-specific in both mice and rats. In addition, we focused on the species-selective region-specific gene set, which is a region-specific gene set for only one species. These are nine gene sets: human-selective LA-specific, human-selective V-specific, and human-selective SA-specific, mouse-selective LA-specific, mouse-selective V-specific, and mouse-selective SA-specific, rat-selective LA-specific, rat-selective V-specific, and rat-selective SA-specific. We focused on the transcription factor genes defined as the genes with Gene Ontology term GO:0003700 (DNA-binding transcription factor activity). Enrichment analysis for gene ontology was performed on the gene sets using the clusterProfiler package in R to identify the Gene Ontology (BH adjusted p-value < 0.05) associated with each gene set [37]. The background gene set of this enrichment analysis is all genes included in the region-specific gene analysis for the other two species. We used the R package (The R Project for Statistical Computing, https://www.r-project.org/) org.Hs.eg.db (version 3.13.0), org.Rn.eg.db (version 3.13.0), org.Mm.eg.db (version 3.13.0) as the databases for Gene Ontology and gene annotation.

2.9. Reverse Transcription-Polymerase Chain Reaction (RT-PCR)

Quantitative RT-PCR was performed using SYBR Green (BioRad, Hercules, CA, USA or Roche, Basel, Switzerland) on an ABI PRISM 799HT Sequence Detection System (Applied Biosystems, Waltham, MA, USA). PCR efficiency was evaluated by using a standard curve of four serial dilution points. Data were analyzed using Applied Biosystems software, and mRNA was normalized to the housekeeping genes, H1 or cardiac β-tubulin. All reactions were carried out in quadruplicate, and each average was in use. The PCR result measures the expression of the ventricle with arbitrary unit 1 and shows how many times more than the expression of the ventricle. Statistical significances of RT-PCR results were determined by an ordinary one-way analysis of variance (ANOVA) with Tukey's multiple comparisons test, using GraphPad Prism ver. 9.0.0 (GraphPad Software Inc., San Diego, CA, USA).

3. Results

3.1. Classification of Cardiac Regions Based on Gene Expression Patterns

Microarray data of mice using Illumina's Bead chip clustered gene expression patterns according to anatomic segments of the heart. The ventricles were further divided into the left ventricle (LV) and right ventricle (RV), and the atria into the right atrium (RA) and left atrium (LA) (Figure 1A). The rat microarray data using Agilent's SurePrint included pulmonary artery (PA) and pulmonary vein (PV) samples and no RA sample compared to the mouse microarray sample. PA, being extracardiac tissue, is naturally clustered outside of the cardiac zone. Although clusters of ventricles were present, LV and RV were not cleanly separated. LA, SA, and PV were roughly clustered (Figure 1B).

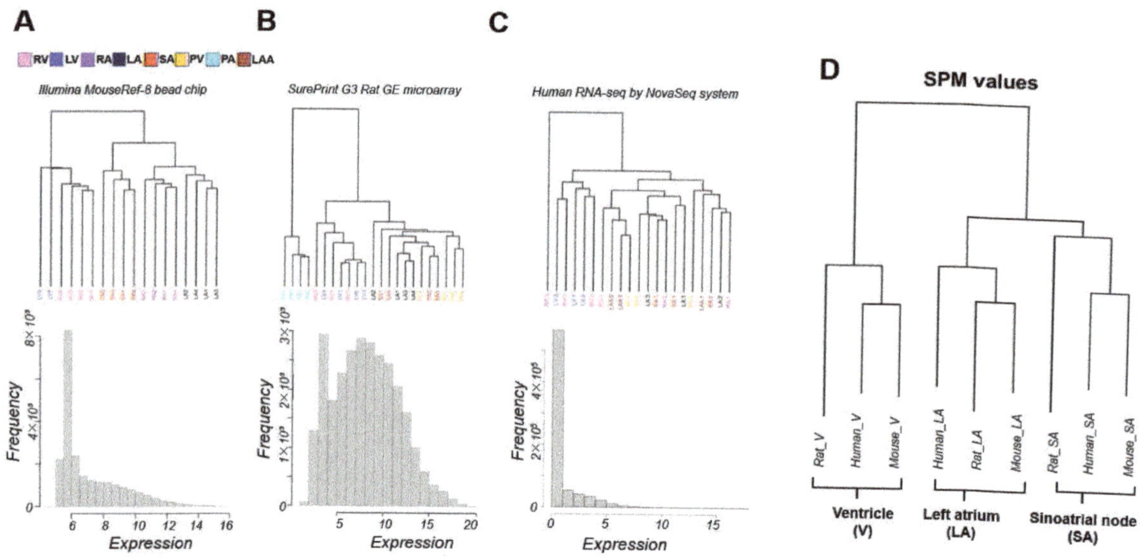

Figure 1. The clustering for the transcriptome samples and the distributions of expression intensity of gene expression intensities for mouse (**A**) and rat (**B**) array data (Illumina bead chip or SurePrint GE microarray) and (**C**) human RNA-seq data. (**D**) Interspecies comparisons of transcriptome data were performed using human gene ortholog relationships. For each gene, differences in expression levels between sites were tested by analysis of variance (ANOVA), and p values were calculated. For each gene, the site-specificity of the three sites was calculated as the specificity measure (SPM) value. Most of the analysis was performed on a single platform by SPM values. The results of hierarchical clustering of SPM values for site-specificity. Human genes with orthologs in mice and rats were targeted. The analysis was limited to the left atrium (LA), ventricles (V) (right ventricle (RV), and left ventricle (LV) data were combined), and sinoatrial nodes (SA), which are cardiac regions that are common in the transcriptome data. LAA; left atrial appendage, RA; right atrium, PA; pulmonary artery, PV; pulmonary vein. Each cardiopulmonary region is color-coded, with a corresponding panel in (**A**).

On the other hand, unlike experimental animals, for which we controlled for strain and age in months, the human RNA-seq data had sizable individual variability, and it was difficult to achieve statistically superior clustering with only three samples for each cardiac region. The expression pattern was roughly divided into ventricular and non-ventricular, and ventricles were not separated into LV and RV (Figure 1C). We challenged academic discovery by cross-platform analysis of these unpublished transcriptome data, which were acquired for several different research projects. Samples were limited to LA, V, and SA; RA was not present in the rat microarray; LV and RV were computationally indistinguishable from the data in our possession. The specificity measure (SPM) values (see Section 2) were then calculated for each cardiac region (LA, V, SA) for a unified analysis of transcriptome data with different medians and units for expression levels. Cross-platform clustering of the SPM values of orthologous genes among mice, rats, and humans clustered the data by cardiac region rather than by platform (Figure 1D). Hierarchical clustering reflects the relative distance of relationships within the samples. In Figure 1A–C, the transcriptome sample in humans and rats did not separate well because the differences between regions were not very large compared to the differences between individuals. Our data processing method using SPM averages out the inter-individual differences, removes some of the platform differences and reflects the differences among regions in Figure 1D.

3.2. Interspecies Correlation of the Transcriptome

Figure 2 displays scatter plots visualizing the relationship of gene expressions for each animal pair. The upper three panels show the LA, the middle three panels show the V, and the lower three panels show the SA from left to right the gene expression relationships in mouse–rat, human–mouse, and human–rat. The transcriptome data demonstrated a moderate correlation with Spearman's correlation coefficient of 0.53–0.63 among the species. The correlation between mice and rats, both of which are rodents, was relatively high, being more than 0.6 in each cardiac region. Although rodents and humans are different order animals, the correlation coefficients between LA and V transcriptomes were 0.56–0.58. It is noteworthy that the SA transcriptome showed a correlation coefficient of less than 0.55 between rodents and humans and had the largest species difference in the heart region analyzed in this study (Figure 2 bottom, Supplementary Table S1).

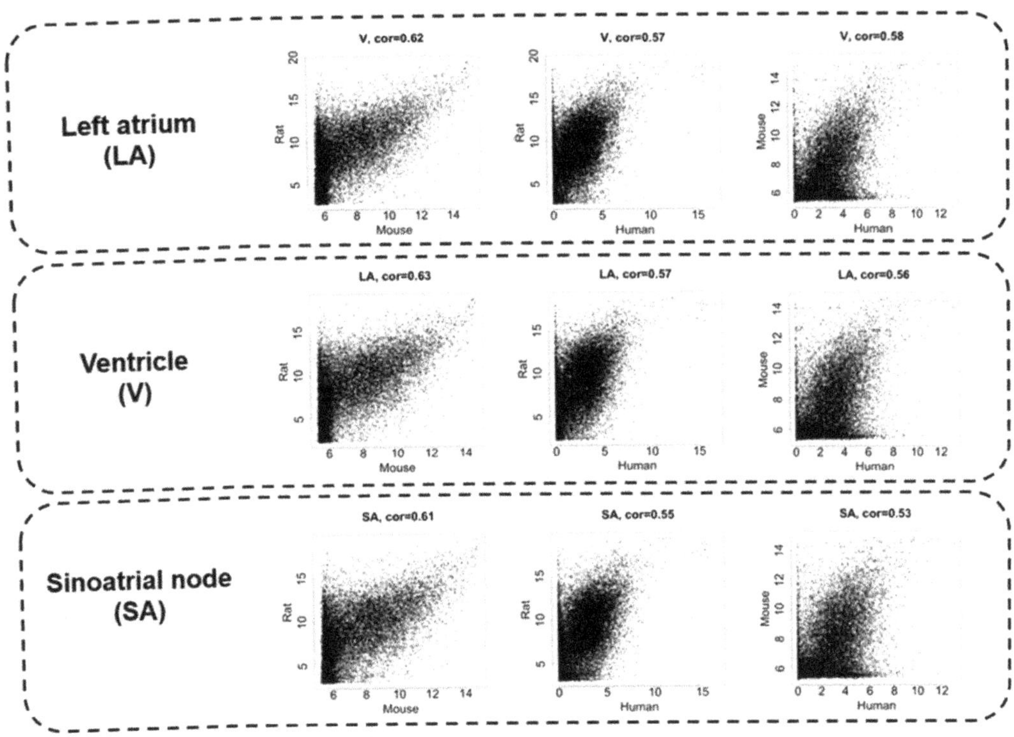

Figure 2. The expression intensity correlations between species are compared in the left atrium (LA), ventricle (V), and sinus node (SA), in order from top to bottom. The leftmost panels show the correlation between mice and rats, the middle panels show the correlation between rats and humans, and the rightmost panels show the correlation between humans and mice. Spearman's correlation coefficient is displayed at the top of each graph.

3.3. Identification of Cardiac Region-Specific Transcription Factors

By calculating the SPM value, we can identify the cardiac region-specific genes in each platform (i.e., each animal species). 762 LA-specific genes were identified in mice, 220 in rats, and 29 in humans (Figure 3A). 748 V-specific genes were identified in mice, 355 in rats, and 21 in humans (Figure 3B). SA-specific genes were identified as 1220 in mice, 150 in rats, and 170 in humans (Figure 3C). The area-proportional Venn diagram in Figure 3 displays up to five transcription factors in each space, representing the specific genes of each animal species in each cardiac region. Some specific genes overlap between species; others do not.

Three common LA-specific genes were identified across the species, *MYBPHL*, *IGFBPL1*, and *KCNJ3*; no transcription factors were among them (Figure 3A). There were 99 LA-specific genes common to mice and rats, and among these, nine transcription factors that dominantly regulate LA gene expression were found, including *PITX2*, *FOXC1*, *BCN1*, *SOX5*, *TBX5*, *FOXC2*, *ETV1*, *NR2F2*, and *ZBTB7C*. Figure 3A highlights three transcription factors (*PITX2*, *FOXC1*, and *BCN1*) that commonly rank in the top five SPM values in mouse and rat LA. 11 V-specific genes were identified throughout the species, of which *IRX5*, *IRX2*, and *HEY2* were transcription factors, and were considered to characterize the cardiac ventricle functionally or anatomically [38–40]. There were 81 V-specific genes common to mice and rats, of which three were transcription factors in *IRX4*, *ZSCAN20*, and *EBF3*. 17 SA-specific genes were common throughout the species, of which *SHOX2* and *HOXA2* were transcription factors. Transcription factors strongly expressed in SA have been studied for nearly 20 years as genes controlling cardiac automaticity. Not only *SHOX2* [41,42] and *HOXA2* [43] but also *TBX18* [44–47], *ISL1* [48,49], *TBX3* [46,50,51], and *TBX5* [52] have been reported. In our analysis, there were 81 SA-specific genes common to mice and humans, including seven transcription factors such as *SHOX2* and *HOXA2*, and *TBX18* and *ISL1* (Supplementary File S1). There were 80 SA-specific genes common to mice and rats, including six transcription factors (Supplementary File S1). There were 1075 mouse SA-specific genes not classified as human or rat SA-specific genes, 39 of which were transcription factors, and *TBX3* was among these 39 genes. *TBX5* was one of the atrium-specific genes common to mice and rats, as described above. Figure 4 plots the SPM values of these six transcription factors of interest in SA. Each axis of the 3D or 2D plot corresponds to the animal species from which the data originated, with the SPM values increasing from left to right or from bottom to top. Therefore, *SHOX2* in the upper right corner of the 3D plot in Figure 1A is a transcription factor with high SA specificity across species. Upon interpretation of the data, it is important to note that the selection of specific genes is based on an artificial threshold. Two continuous values, the *p*-value of ANOVA and the SPM value, are referenced to determine the threshold value. Not only *SHOX2* and *HOXA2*, which were selected as specific transcription factors by our criteria, but also *ISL1* satisfied the criteria for specific genes in terms of SPM value. The reason why *ISL1* cannot be concluded as an SA-specific transcription factor was due to its relatively low expression in rat SA, which failed to reach the threshold for the adjusted *p*-value (see Section 4). To confirm SA-specific transcription factors, quantitative PCR was performed using mouse samples. The relative PCR results in Figure 5 indicate how many times more than the ventricular expression level as an arbitrary unit. An analysis of variance accompanied by Tukey's multiple comparisons test revealed that *Etv1* was predominantly expressed in LA and RA ($p < 0.0001$), while *Tbx3*, *Shox2*, and *Isl1* were predominantly expressed in SA ($p < 0.0001$). *Tbx5* was predominantly expressed in SA ($p = 0.0002$ vs. LA, $p = 0.0032$ vs. RA, $p < 0.0001$ vs. LV, RV). No significant differences in *Hoxa2* expression were observed.

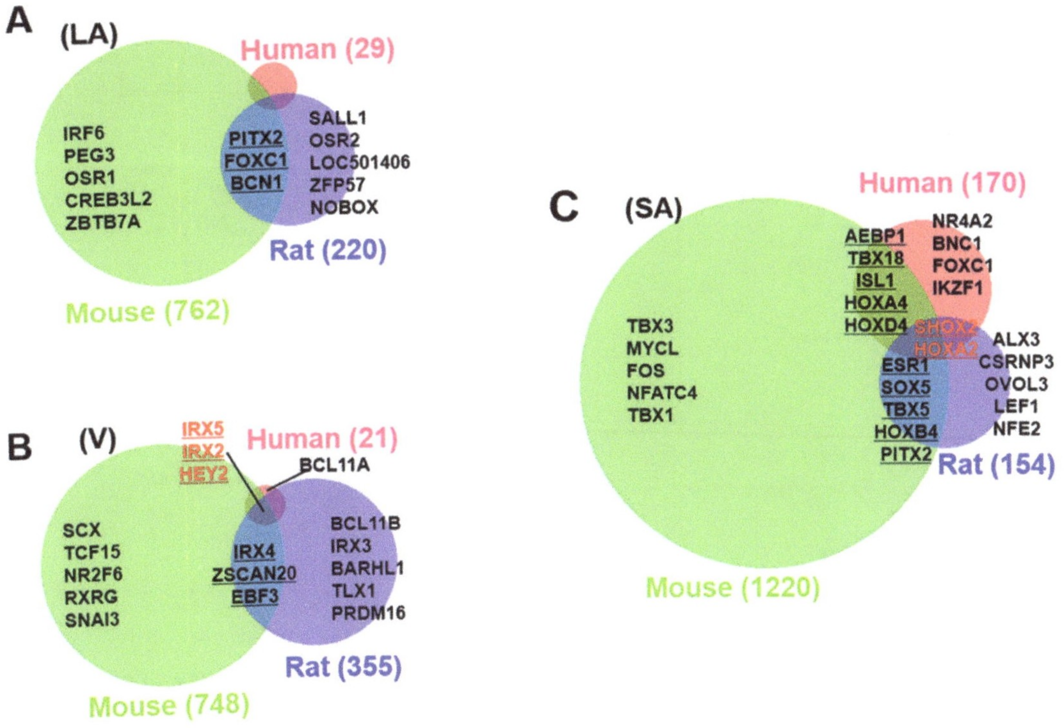

Figure 3. Cardiac region-specific genes. The number of left atrial (LA, (**A**)), ventricular (V, (**B**)), and sinoatrial node (SA, (**C**)) specific genes selected based on SPM values are indicated by numbers in parentheses beside the area-proportional Venn diagram. Animal species are color-coded as human (pink), mouse (green), and rat (blue). Transcription factors are shown as representatives of specific genes, with a maximum of five top-ranked transcription factors in each space. The rank is determined by the magnitude of the SPM value. Specific transcription factors that are conserved across more than two species are underlined; specific transcription factors that are conserved among the three species are shown in red.

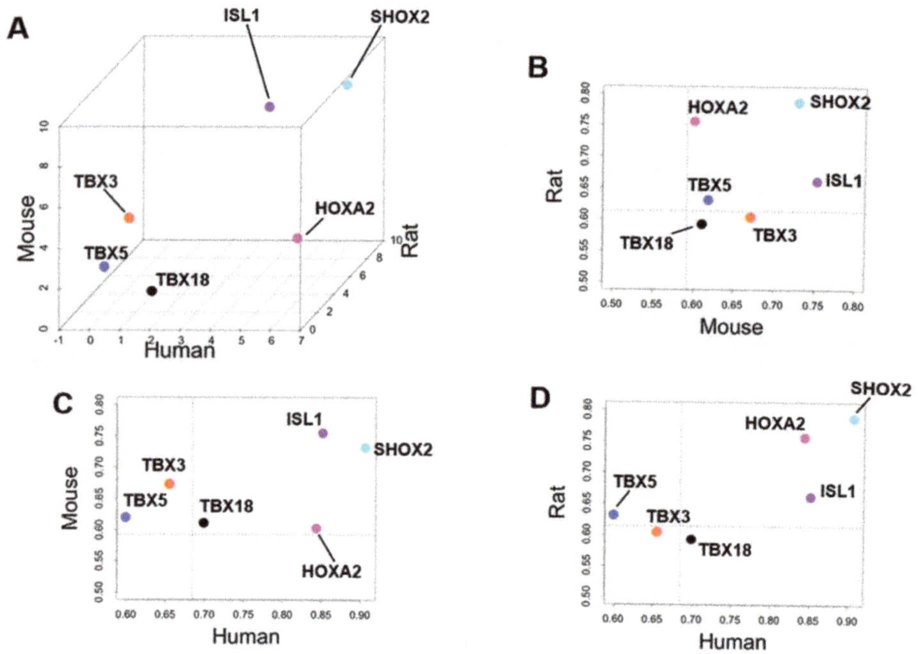

Figure 4. Specificity of six selected transcription factors with respect to sinoatrial nodes across three animal species. (**A**) Plot of z-score normalized specificity measure (SPM) values for three animal species: human on the *x*-axis, rat on the *y*-axis, and mouse on the *z*-axis. Scatter plots showing the relationship between (**B**) mouse–rat, (**C**) human–mouse, and (**D**) human–rat raw SPM values. The dotted lines indicate the threshold SPM values for specificity in this study. *SHOX2*, *HOXA2*, and *ISL1* exceed the threshold values in all species. *SHOX2*, *HOXA2*, and *ISL1* exceed the threshold values in all species.

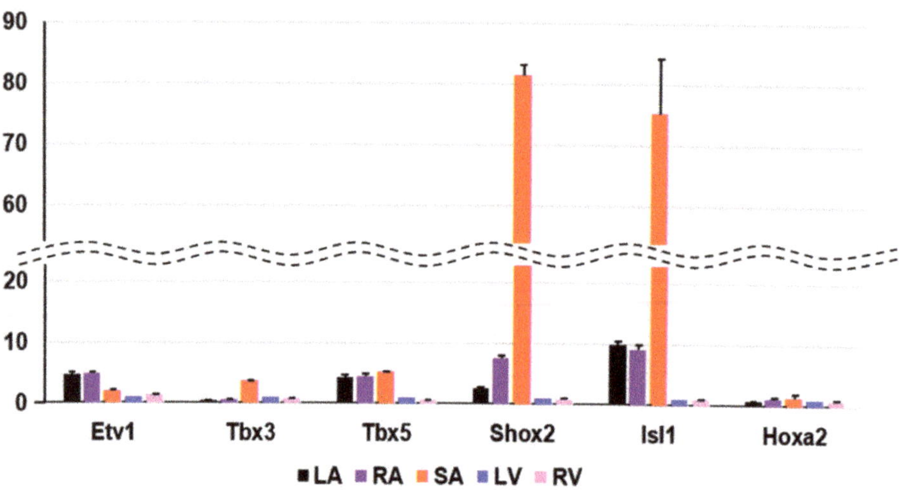

Figure 5. Quantitative PCR validation of selected transcripts. The PCR result measures the expression of the ventricle with arbitrary unit 1, and shows how many times more than the expression of the ventricle.

In our transcriptome analysis, *Etv1* is atrium-specific, *Tbx3*, *Shox2* and *Isl1* are SA-specific genes in mice. With respect to *Hoxa2*, our specificity measure (SPM) criteria could not be confirmed. Each heart region is color-coded at the bottom. LA; left atrium, RA; right atrium, SA; sinoatrial node, LV; left ventricle, RV; right ventricle.

3.4. Heart Region-Specific Genes across Species, and Ontology Analysis of Heart Region and Species-Specific Genes

By calculating SPM values, we identified three atrial-specific, 11 ventricular-specific, and 17 SA-specific markers displayed in Figure 6A–C, respectively. These include *MYBPHL*, a known atrial marker ([53], Figure 6A), and *MYL3*, a known ventricular marker ([54], Figure 6B), which may have found novel cardiac region-specific markers. Unfortunately, *HCN4*, the most well-known marker of SA, was not on the mouse microarray chip and was therefore missing from our analysis; among the SA-specific genes, *GAP-43* is a marker of innervation in SA ([55,56]), suggesting similar innervation across species in SA. Ontology analysis revealed 70 ontologies that were cardiac region-specific and even species-specific (Supplementary File S2). Three examples of these are displayed in Figure 7A–C. Among human atrial-specific genes, *MYOT* and *KCNA5* both have the function of binding to the actin skeleton (Figure 7A), and although *KCNA5* is often considered an atrial marker, it is not an atrial-specific gene in mice or rats in our analysis. In rat ventricles, P2Y receptors (*P2ry10*, *P2yr13*, *P2yr2*), Toll-like receptors (*Tlr2*), G protein-coupled receptors (*Gpr171*, *Gpr65*, *Gpr176*), and IL18 receptors seem to be selectively expressed (Figure 7B). Although *HCN4*, an essential marker of SA, was dropped from the analysis as previously mentioned above, *HCN2* was a mouse-selective SA-specific gene (Figure 7C).

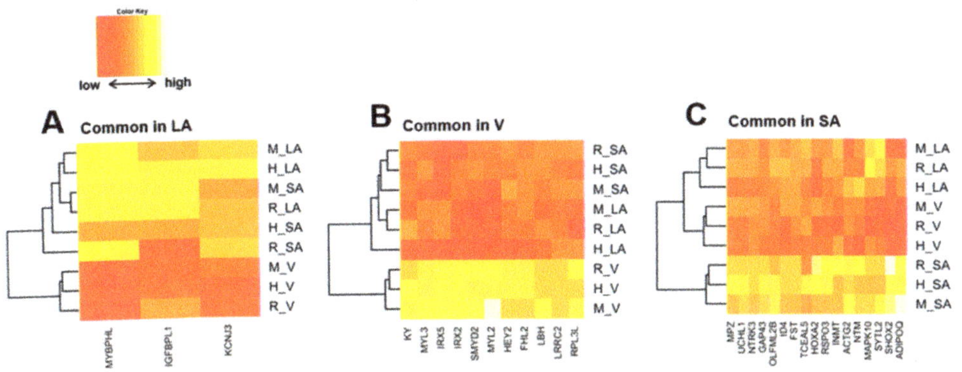

Figure 6. (**A**) is a heat map of z-score normalized SPM values in left atrial (LA)-specific genes that are common across species. Similarly, (**B**,**C**) show ventricle-specific gene groups and sinoatrial node (SA)-specific gene groups, respectively.

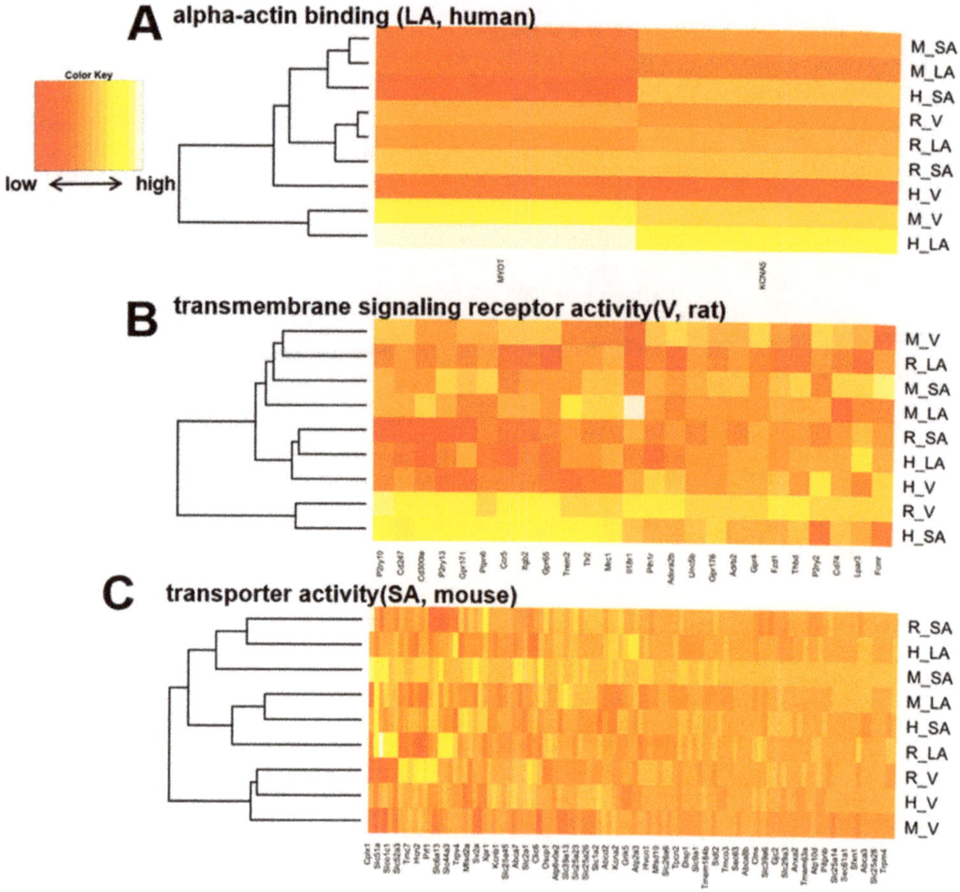

Figure 7. (**A**), A heat map of the human-selective LA-specific genes with the representative enriched gene ontology. Similarly, (**B**,**C**) show representative rat ventricle-specific ontologies and mouse SA-specific ontologies, respectively.

4. Discussions

4.1. Major Findings

This is the first study to analyze the transcriptomes of three separate cardiac regions in three different animal species. SPM values are a powerful normalization method, enabling region-specific gene analysis on cross-platform (Figure 1). The transcriptome, a comprehensive set of gene expression data, showed moderate correlation across species, but among the cardiac regions, SA showed the largest species differences (Figure 2, Supplementary Table S1). Therefore, we attempted to identify SA-specific transcription factors as a model for identifying special genes by cross-platform analysis using SPM values. Although transcription factors that characterize the SA function have been intensively investigated for a couple of decades, these studies technically had to take into account the differences among animal species. For examples, *TBX18* [44–47], *TBX3* [46,50,51], and *TBX5* [52,57,58] have been repeatedly reported as important transcription factors characterizing SA. *Tbx18* is identified by in situ hybridizations from the superior vena cava to 75% of the SA head, and *Tbx5* is expressed throughout the SA at 14.5 days of mouse embryonic development. Gene deletion of *Tbx18* results in abnormal SA morphology, while knockout of *Tbx3* results in no abnormal morphology but *Cx40* and *Cx43* gene, unrequired for SA function, is expressed in mice [46]. In 2013, Kapoor et al. selected *Tbx18* among five transcription factors, including *Shox2*, *Tbx3*, and *Tbx5*, as a transcription factor with a potent ability to form cardiac rhythm in screening using cultured cardiac cells [47]. They reported that the transfection of *Tbx18* can convert quiescent cells into pacemaker cells, and the *Tbx18* transfection strategy is becoming the basis of bioengineering to create bio-pacemakers [45]. *Tbx3* is believed to be important for the functional maturation of SA, as ectopic arrhythmias occur in conditionally knockout mice [51], and pacemaker currents are observed when *Tbx3* is overexpressed in stem cells [50].

TBX5 is a transcription factor that regulates the electrical conduction system, rather than an SA-specific transcription factor. In human disease [57] and mouse models of *TBX5* deficiency [52,58], the PR interval on electrocardiogram is prolonged.

On the other hand, these three T-box transcription factors did not correspond to SA-specific genes in our criteria (Figure 3C and Supplementary File S1). *Tbx18* did not have a high enough SPM value in rat SAN, and *TBX3* did not have a high enough SPM value in both rat, and human SA. *TBX5* probably underlies the entire supraventricular area of the heart rather than the SA-specific (Figures 3C and 5).

Interestingly, these T-box genes (*TBX18*, *TBX3*, and *TBX5*) exceed the criteria in mouse SA. In other words, our analysis reproduces the results of previous studies while at the same time suggesting that the previous studies have relied on mouse data. In our cross-species analysis, the only SA-specific transcription factors that are common across species are *SHOX2* and *HOXA2*, and the importance of *SHOX2* exceeds that of other transcription factors, especially given its high SPM value (Figures 3C and 4, and Supplementary File S1). The SPM value is indicated as a measure of the degree of regional specificity. With this value, new candidate markers for atria, ventricles, and SA are shown in Figure 6. It is expected to be utilized in future studies. *KCNA5*, previously considered an atrial-specific ion channel [59], is an example. This gene is supposed to be a human-only atrial marker (Figure 6A). It is thought that the difficulty in developing atrial antiarrhythmic drugs targeting *KCNA5* is due to the fact that different K^+ channels from *KCNA5* are the main source of outward current in experimental animals such as rats and mice [18].

4.2. *Priority of SHOX2 over HOXA2 and ISL1*

SHOX2 is highly expressed in SA compared to other cardiac regions in every species, and both the ANOVA and SPM values strongly suggest its SA-specificity. *Shox2* was first reported as a transcription factor for SA differentiation 15 years ago. In situ hybridization identified *Shox2* as being expressed locally in the SA region, and *Shox2* knockout mice express *Nkx2.5*, a transcription factor required for differentiation of the working cardiac muscle, rather than SA, and promote expression of *Cx40* and *Cx43*, which interfere with SA function [41]. As a result, regular heartbeat is impaired in *Shox2* knockout mice in vivo [42]. In humans, mutations that disrupt *SHOX2* function also affect heart rhythm and are inherited as familial atrial fibrillation [60]. Our analysis reinforces that *Shox2* is a critically important gene for the healthy heartbeat. On the other hand, the calculation results concerning *HOXA2* and *ISL1* are not assured of reproducibility in the present analysis alone. *HOXA2* exhibits high SPM values in rat and human SA but only marginally above the threshold in mouse SA (Figure 4). In addition, significantly high expression in SA of this gene could not be confirmed by RT-PCR. It is not easy to define *HOXA2* as a common SA-specific transcription factor from our study. The number of samples and selected cardiac regions varies from experiment to experiment, and the threshold of the SPM value can change in future studies. *Hoxa2* has been previously mentioned once in other SA transcriptome analyses [43] and has rarely been studied as a gene involved in the cardiac beating. The function and locational information of *HOXA2* in the heart may be a research topic for future studies. *ISL1* had a high SPM value in SA in all species (Figure 4). In 2013, *Isl1* was identified as a binding target of *Shox2* [61]. Two years later, using a combination of genetically engineered mice and high-throughput analysis, two teams reported that *Isl1* is a genetic marker for SA and is involved in SA function [48,49].

Isl1 may play a role in the maintenance of SA function comparable to *Shox2*. However, this well-known gene was not qualified as an SA-specific transcription factor in the current study because, as shown in Supplementary File S3, *ISL1* is not significantly higher in the rat expression intensity data. Furthermore, it is interesting that *HOXA2* and *ISL1* are most highly expressed in the pulmonary artery (PA), which is an extracardiac tissue (Supplementary File S3). Since the pulmonary artery is part of the secondary heart field in early development, *HOXA2* and *ISL1* may be genes whose expression was suppressed as the heart field differentiated [62,63]. In contrast, *SHOX2* was specifically and strongly expressed in the regions responsible for cardiac automaticity, including SA and pulmonary veins [64,65].

4.3. Validation of Our Analysis from Other Transcriptome Analysis

We checked the previous bulk-RNA-seq data set (GSE112339), which measured the transcriptome of human LA, LV, RV, and RA [33]. We compared the mean expression values of each region between our study and this previous study (Supplementary Figure S3). Spearman's correlation coefficients are all greater than 0.9 (0.91 in LA, 0.90 in LV, 0.94 in RV, 0.93 in RA), indicating that although the sample size of our data is relatively small, it is sufficiently representative of previous data to be considered reliable. Recently, single-cell genomics analysis has become more prevalent in genomics analysis. Single-cell RNA-seq (scRNA-seq) data can provide information on cellular heterogeneity and quantify the cellular subset [66]. For example, the scRNA-seq study for the human heart reported V-common transcription factors *IRX5 and HEY2* as the transcription factors enriched in ventricular cardiomyocytes [67] as the same as our results. From a future perspective, it would be interesting to expand our computational workflow to single cell RNA-seq data and perform interspecies comparisons of heart single cell expression profiles. Such analysis may lead to further findings on the mechanisms behind the region-specific genes and heart biology interspecies differences.

4.4. Limitations

The present analysis is methodologically indirect since the raw data were normalized in multi-steps. The selection criteria for statistically significant specific genes were arbitrarily determined with consulting distributions of expression levels. For example, even though *TBX3* has recently been reported to be an important transcription factor for SA function in humans as well as mice [43], our analysis did not detect this gene as an SA-specific transcription factor. As such, our artificial threshold settings have limitations, and even if we fail to detect a gene as specific, the analytic results will not be universally reproduced. Despite this, the PCR data and the literature seem to support the certainty of our analysis to some extent, and the priority of *SHOX2* as SA-specific transcription factor may be established. Our strategy of using SPM values to identify unique genes is beneficial as long as we are aware of the limitations. From our view, the major limitation in the current study is that the number of the region-specific genes were small in humans. The cause could be the more considerable genetic or environmental diversity compared to laboratory animals, in addition to the limited sample size of human data, which leads to low detection power in human region-specific gene analysis. The region-specific genes with adjusted p-value < 0.05 were 43% in mice, and 36% in rats, compared to 6% in humans. In the case of human data analysis, the variation due to genetic or environmental effects is more significant than that of laboratory animals. Therefore, the detection power in the human data set would be smaller than laboratory animals, and the analysis of human samples requires a larger sample size than the analysis of laboratory animal samples to obtain the same detection power. Our analysis possibly did not contain enough region-specific human genes. In addition, because we limited our analysis to genes that are orthologous among the three species, we ignored genes expressed in mice and rats that are not orthologous. In other words, genes that are genuinely selective for mice and rats have been omitted.

Furthermore, microarray platforms often do not carry determinant genes, and the absence of *HCN4* in the mouse array in our analysis damages the reliability of the SA data.

4.5. Future Perspectives

We identified species-dependent or cross-species information in the present study by comprehensively analyzing species differences in three cardiac regions. Such methods for refining information on genes of interest are not only species difference-based analyses but have also been used in previous analyses based on differences among animal strains. For example, Swindell et al. conducted the comparative transcriptomes between calorie-restricted and control mice using several strains of mice and reported the genes that respond in a strain-specific manner [68]. Another genomics study has reported the vital conservation of RNA editing sites among mouse strains, including wild-derived strains [69]. Such strain comparisons for the heart region-specific genes are a candidate application of the computational approach used in this study as a future perspective because it is already known that there are differences in heart size and function between strains of mice or rats. In a study comparing eight strains of mice, three strains had significantly reduced cardiac function after ex vivo ischemia, while two of these strains had significantly preserved cardiac function during in vivo acute hypoxia [70]. Twenty-three strains and a four-way crossbred strain of rats were examined for heart weight, and the strain differences in heart weight were significant, with the degree of genetic determination estimated to be 65% to 75% [71]. These differences between strains may be attributable to differences in genetic background. Considering that our analysis resulted in greater power to detect specific genes in laboratory animal samples than in human samples, it may be meaningful to identify common genes and strain-dependent genes based on strain differences.

5. Conclusions

In this study, we reanalyzed unpublished transcriptome data from three different platforms and employed the method of calculating the SPM values to examine species differences in transcriptomes from different cardiac regions. The analytic results were sufficiently robust to include findings that challenge existing knowledge. *TBX18*, *TBX3*, and *ISL1*, which were previously thought to be sinus node (SA)-specific transcription factors, are not classified as SA-specific transcription factors in humans or rats, and *SHOX2* and *HOX2A* are common SA marker candidates across species. In particular, the predominance of *SHOX2* as an SA-specific transcription factor was definitive. It was strongly likely that this transcription factor was essential for inducing biological pacemakers from stem cells. Thus, the identification of specific genes using SPM values can reveal tissue-specific and animal species-dependent genes. Novel cardiac region-specific genes may have been discovered in our study. It, thus, was suggested that what had been previously thought to be atrial and SA markers may be restricted to animal species such as humans and mice.

Supplementary Materials: The following supporting information can be downloaded at: https://www.mdpi.com/article/10.3390/biom12060859/s1, Figure S1: Preparation of mouse sinoatrial node; Figure S2: Validation of SA node isolation technique; Figure S3: Validation of our human RNA-seq data by another study; Table S1: Correlation coefficient of transcriptomes among species; File S1: Sinoatrial node-specific transcription factors among species; File S2: Cardiac region-specific and species dependent ontologies. File S3: Rat microarray raw intensity data for six transcription factors. Reference [72] is cited in the Supplementary Materials.

Author Contributions: D.O. analyzed transcriptome data and wrote the manuscript. Y.O. wrote the manuscript and designed and performed experiments and funding. D.K. performed experiments. S.I. performed experiments. T.I. designed experiment and funding. M.O. processed transcriptome data. R.Y., K.I., K.O. and Y.O. supervised the study. All authors have read and agreed to the published version of the manuscript.

Funding: This research was funded by JSPS KAKENHI (Grant-in-Aid for Young Scientists (B)), grant numbers 17H15554 and 20K16116 to Y.O., JSPS KAKENHI Grant-in-Aid for Research Activity Start-up (Grant number 21K21316) to D.O., and by collaboration with Ono Pharmaceutical Cooperation, Ltd. (Japan).

Institutional Review Board Statement: The protocols used in this study were approved by the Animal Care and Use Committee of the National Institutes of Health (protocol #034LCS2016), the Animal Ethics Committees of the Akita University School of Medicine (a-1-2953), the Ethics Review Committees of the Yamagata University Faculty of Medicine (#110), the Ethics Committees of Kyoto University Graduate School and Faculty of Medicine (R3118) and Akita University Certified Clinical Research Revies Board (#1982).

Informed Consent Statement: Informed consent was obtained appropriately from all subjects involved in the study by AnaBios Corporation along to the Health Insurance Portability and Accountability Act.

Data Availability Statement: The analytic results in this study are available on a request basis. All transcriptome data that we analyzed in the current study are uploaded into the GEO database. GSE200326, GSE203369, and GSE203367 are accession numbers for mouse Illumina array, rat Agilent SurePrint array, and human RNA-seq of heart transcriptome, respectively.

Acknowledgments: We thank Kirill V. Tarasov, Yelena S. Tarasova, Bruce D. Ziman, and Edward G. Lakatta (Laboratory of Cardiovascular Science, National Institute on Aging, National Institutes of Health) for providing mouse microarray data and performing experiments, Elin Lehrmann (Laboratory of Genetics and Genomics, National Institute on Aging, National Institutes of Health) for performing microarray experiments, curating and depositing the mouse microarray data into the GEO database, and Yongqing Zhang (Computational Genomics Unit, National Institute on Aging, National Institutes of Health) for performing the initial bioinformatic analysis for the mouse experiments. This research was also supported in part by the Intramural research Program of the NIH, National Institute on Aging.

Conflicts of Interest: The authors declare no conflict of interest.

References

1. Huxley, A.F.; Niedergerke, R. Structural changes in muscle during contraction: Interference microscopy of living muscle fibres. *Nature* **1954**, *173*, 971–973. [CrossRef] [PubMed]
2. Huxley, H.; Hanson, J. Changes in the Cross-Striations of Muscle during Contraction and Stretch and their Structural Interpretation. *AAPG Mem.* **2011**, *42*, 61–102. [CrossRef] [PubMed]
3. Ebashi, S.; Endo, M.; Ohtsuki, I. Control of muscle contraction. *Q. Rev. Biophys.* **1969**, *2*, 351–384. [CrossRef] [PubMed]
4. Noda, M.; Shimizu, S.; Tanabe, T.; Takai, T.; Kayano, T.; Ikeda, T.; Takahashi, H.; Nakayama, H.; Kanaoka, Y.; Minamino, N.; et al. Primary structure of Electrophorus electricus sodium channel deduced from cDNA sequence. *Nature* **1984**, *312*, 121–127. [CrossRef]
5. Tanabe, T.; Takeshima, H.; Hikami, A.; Flockerzi, V.; Takahashi, H.; Kangawa, K.; Kojima, M.; Matsuo, H.; Hirose, T.; Numa, S. Primary structure of the receptor for calcium channel blockers from skeletal muscle. *Jpn. J. Pharmacol.* **1988**, *46*, 33. [CrossRef]
6. Login, S.; Takeshima, H.; Nishimura, S.; Matsumoto, T.; Ishida, H.; Kangawa, K.; Minamino, N.; Matsuo, H.; Hanaoka, M.; Hirose, T.; et al. Primary structure and expression from complementary DNA of skeletal muscle ryanodine receptor Hiroshi. *Nature* **1989**, *339*, 439–445.
7. Wang, Q.; Curran, M.E.; Splawski, I.; Burn, T.C.; Millholland, J.M.; VanRaay, T.J.; Shen, J.; Timothy, K.W.; Vincent, G.M.; de Jager, T.; et al. Positional cloning of a novel potassium channel gene: KVLQT1 mutations cause cardiac arrhythmias. *Nat. Genet.* **1996**, *12*, 17–23. [CrossRef]
8. Noma, A.; Irisawa, H. Membrane currents in the rabbit sinoatrial node cell as studied by the double microelectrode method. *Pflügers Arch. Eur. J. Physiol.* **1976**, *364*, 45–52. [CrossRef]
9. Brown, B.Y.H.; Difrancesco, D. Voltege-clamp investigation of membrane current underlying pace-maker activity in rabbit sino-atrial node. *J. Physiol.* **1980**, *308*, 331–351. [CrossRef]
10. Stieber, J.; Herrmann, S.; Feil, S.; Löster, J.; Feil, R.; Biel, M.; Hofmann, F.; Ludwig, A. The hyperpolarization-activated channel HCN4 is required for the generation of pacemaker action potentials in the embryonic heart. *Proc. Natl. Acad. Sci. USA* **2003**, *100*, 15235–15240. [CrossRef]
11. Vinogradova, T.M.; Lyashkov, A.E.; Zhu, W.; Ruknudin, A.M.; Sirenko, S.; Yang, D.; Deo, S.; Barlow, M.; Johnson, S.; Caffrey, J.L.; et al. High basal protein kinase A-dependent phosphorylation drives rhythmic internal Ca^{2+} store oscillations and spontaneous beating of cardiac pacemaker cells. *Circ. Res.* **2006**, *98*, 505–514. [CrossRef] [PubMed]
12. Gordan, R.; Gwathmey, J.K.; Xie, L.-H. Autonomic and endocrine control of cardiovascular function. *World J. Cardiol.* **2015**, *7*, 204. [CrossRef] [PubMed]
13. Lewis, M.E.; Al-Khalidi, A.H.; Bonser, R.S.; Clutton-Brock, T.; Morton, D.; Paterson, D.; Townend, J.N.; Coote, J.H. Vagus nerve stimulation decreases left ventricular contractility in vivo in the human and pig heart. *J. Physiol.* **2001**, *534*, 547–552. [CrossRef]

14. Himeno, Y.; Toyoda, F.; Satoh, H.; Amano, A.; Cha, C.Y.; Matsuura, H.; Noma, A. Minor contribution of cytosolic Ca2+ transients to the pacemaker rhythm in guinea pig sinoatrial node cells. *Am. J. Physiol.—Heart Circ. Physiol.* **2011**, *300*, 251–261. [CrossRef] [PubMed]
15. Namekata, I.; Jitsukata, K.; Fukuda, A.; Odaka, R.; Hamaguchi, S.; Tanaka, H. Intracellular Ca2+-Mediated Mechanisms for the Pacemaker Depolarization of the Mouse and Guinea Pig Sinus Node Tissue. *Biomolecules* **2022**, *12*, 377. [CrossRef]
16. Ono, K.; Trautweint, W. Potentiation by cyclic GMP of beta-adrenergic effect on Ca2+ current in guinea-pig ventricular cells. *J. Physiol.* **1991**, *443*, 387–404. [CrossRef]
17. Mary, P.O.I.S.; Lohmannt, S.M.; Waltert, U.; Fischmeister, R. Ca2+ current is regulated by cyclic GMP-dependent protein kinase in mammalian cardiac myocytes. *Proc. Natl. Acad. Sci. USA* **1991**, *88*, 1197–1201. [CrossRef]
18. London, B.; Wang, D.W.; Hill, J.A.; Bennett, P.B. The transient outward current in mice lacking the potassium channel gene Kv1.4. *J. Physiol.* **1998**, *509*, 171–182. [CrossRef]
19. Wettwer, E.; Amos, G.; Gath, J.; Zerkowski, H.R.; Reidemeister, J.C.; Ravens, U. Transient outward current in human and rat ventricular myocytes. *Cardiovasc. Res.* **1993**, *27*, 1662–1669. [CrossRef]
20. Li, G.R.; Yang, B.; Sun, H.; Baumgarten, C.M. Existence of a transient outward K+ current in guinea pig cardiac myocytes. *Am. J. Physiol.—Heart Circ. Physiol.* **2000**, *279*, 130–138. [CrossRef]
21. Matsuoka, S.; Ehara, T.; Noma, A. Chloride-sensitive nature of the adrenaline-induced current in guinea-pig cardiac myocytes. *J. Physiol.* **1990**, *425*, 579–598. [CrossRef] [PubMed]
22. Takano, M.; Noma, A. Distribution of the isoprenaline-induced chloride current in rabbit heart. *Pflügers Arch. Eur. J. Physiol.* **1992**, *420*, 223–226. [CrossRef] [PubMed]
23. Okamoto, Y.; Kawamura, K.; Nakamura, Y.; Ono, K. Pathological impact of hyperpolarization-activated chloride current peculiar to rat pulmonary vein cardiomyocytes. *J. Mol. Cell. Cardiol.* **2014**, *66*, 53–62. [CrossRef] [PubMed]
24. Takagi, D.; Okamoto, Y.; Ohba, T.; Yamamoto, H.; Ono, K. Comparative study of hyperpolarization-activated currents in pulmonary vein cardiomyocytes isolated from rat, guinea pig, and rabbit. *J. Physiol. Sci.* **2020**, *70*, 6. [CrossRef]
25. Ehrlich, J.R.; Cha, T.-J.; Zhang, L.; Chartier, D.; Villeneuve, L.; Hébert, T.E.; Nattel, S. Characterization of a hyperpolarization-activated time-dependent potassium current in canine cardiomyocytes from pulmonary vein myocardial sleeves and left atrium. *J. Physiol.* **2004**, *557*, 583–597. [CrossRef]
26. Okamoto, Y.; Nagasawa, Y.; Obara, Y.; Ishii, K.; Takagi, D.; Ono, K. Molecular identification of HSPA8 as an accessory protein of a hyperpolarization-activated chloride channel from rat pulmonary vein cardiomyocytes. *J. Biol. Chem.* **2019**, *294*, 16049–16061. [CrossRef]
27. Ritchie, M.E.; Phipson, B.; Wu, D.; Hu, Y.; Law, C.W.; Shi, W.; Smyth, G.K. Limma powers differential expression analyses for RNA-sequencing and microarray studies. *Nucleic Acids Res.* **2015**, *43*, e47. [CrossRef]
28. Chen, S.; Zhou, Y.; Chen, Y.; Gu, J. Fastp: An ultra-fast all-in-one FASTQ preprocessor. *Bioinformatics* **2018**, *34*, i884–i890. [CrossRef]
29. Dobin, A.; Davis, C.A.; Schlesinger, F.; Drenkow, J.; Zaleski, C.; Jha, S.; Batut, P.; Chaisson, M.; Gingeras, T.R. STAR: Ultrafast universal RNA-seq aligner. *Bioinformatics* **2013**, *29*, 15–21. [CrossRef]
30. Liao, Y.; Smyth, G.K.; Shi, W. FeatureCounts: An efficient general purpose program for assigning sequence reads to genomic features. *Bioinformatics* **2014**, *30*, 923–930. [CrossRef]
31. Li, B.; Ruotti, V.; Stewart, R.M.; Thomson, J.A.; Dewey, C.N. RNA-Seq gene expression estimation with read mapping uncertainty. *Bioinformatics* **2009**, *26*, 493–500. [CrossRef] [PubMed]
32. Wagner, G.P.; Kin, K.; Lynch, V.J. Measurement of mRNA abundance using RNA-seq data: RPKM measure is inconsistent among samples. *Theory Biosci.* **2012**, *131*, 281–285. [CrossRef] [PubMed]
33. Johnson, E.K.; Matkovich, S.J.; Nerbonne, J.M. Regional Differences in mRNA and lncRNA Expression Profiles in Non-Failing Human Atria and Ventricles. *Sci. Rep.* **2018**, *8*, 13919. [CrossRef] [PubMed]
34. Sun, J.; Nishiyama, T.; Shimizu, K.; Kadota, K. TCC: An R package for comparing tag count data with robust normalization strategies. *BMC Bioinform.* **2013**, *14*, 219. [CrossRef] [PubMed]
35. Galili, T. dendextend: An R package for visualizing, adjusting and comparing trees of hierarchical clustering. *Bioinformatics* **2015**, *31*, 3718–3720. [CrossRef] [PubMed]
36. Pan, J.B.; Hu, S.C.; Wang, H.; Zou, Q.; Ji, Z.L. PaGeFinder: Quantitative identification of spatiotemporal pattern genes. *Bioinformatics* **2012**, *28*, 1544–1545. [CrossRef] [PubMed]
37. Wu, T.; Hu, E.; Xu, S.; Chen, M.; Guo, P.; Dai, Z.; Feng, T.; Zhou, L.; Tang, W.; Zhan, L.; et al. clusterProfiler 4.0: A universal enrichment tool for interpreting omics data. *Innovation* **2021**, *2*, 100141. [CrossRef]
38. Costantini, D.L.; Arruda, E.P.; Agarwal, P.; Kim, K.H.; Zhu, Y.; Zhu, W.; Lebel, M.; Cheng, C.W.; Park, C.Y.; Pierce, S.A.; et al. The homeodomain transcription factor Irx5 establishes the mouse cardiac ventricular repolarization gradient. *Cell* **2005**, *123*, 347–358. [CrossRef]
39. Kim, K.H.; Rosen, A.; Bruneau, B.G.; Hui, C.C.; Backx, P.H. Iroquois homeodomain transcription factors in heart development and function. *Circ. Res.* **2012**, *110*, 1513–1524. [CrossRef]
40. Gibb, N.; Lazic, S.; Yuan, X.; Deshwar, A.R.; Leslie, M.; Wilson, M.D.; Scott, I.C. Hey2 regulates the size of the cardiac progenitor pool during vertebrate heart development. *Development* **2018**, *145*, dev167510. [CrossRef]
41. Blaschke, R.J.; Hahurij, N.D.; Kuijper, S.; Just, S.; Wisse, L.J.; Deissler, K.; Maxelon, T.; Anastassiadis, K.; Spitzer, J.; Hardt, S.E.; et al. Targeted Mutation Reveals Essential Functions of the Homeodomain Transcription Factor Shox2 in Sinoatrial and Pacemaking Development. *Circulation* **2007**, *115*, 1830–1838. [CrossRef] [PubMed]

42. Espinoza-Lewis, R.A.; Yu, L.; He, F.; Liu, H.; Tang, R.; Shi, J.; Sun, X.; Martin, J.F.; Wang, D.; Yang, J.; et al. Shox2 is essential for the differentiation of cardiac pacemaker cells by repressing Nkx2-5. *Dev. Biol.* **2009**, *327*, 376–385. [CrossRef] [PubMed]
43. van Eif, V.W.W.; Stefanovic, S.; van Duijvenboden, K.; Bakker, M.; Wakker, V.; de Gier-de Vries, C.; Zaffran, S.; Verkerk, A.O.; Boukens, B.J.; Christoffels, V.M. Transcriptome analysis of mouse and human sinoatrial node cells reveals a conserved genetic program. *Development* **2019**, *146*, dev173161. [CrossRef] [PubMed]
44. Hu, Y.; Li, N.; Liu, L.; Zhang, H.; Xue, X.; Shao, X.; Zhang, Y.; Lang, X. Genetically Modified Porcine Mesenchymal Stem Cells by Lentiviral Tbx18 Create a Biological Pacemaker. *Stem Cells Int.* **2019**, *2019*, 3621314. [CrossRef] [PubMed]
45. Grijalva, S.I.; Gu, J.M.; Li, J.; Fernandez, N.; Fan, J.; Sung, J.H.; Lee, S.Y.; Herndon, C.; Buckley, E.M.; Park, S.J.; et al. Engineered Cardiac Pacemaker Nodes Created by TBX18 Gene Transfer Overcome Source–Sink Mismatch. *Adv. Sci.* **2019**, *6*, 1901099. [CrossRef] [PubMed]
46. Wiese, C.; Grieskamp, T.; Airik, R.; Mommersteeg, M.T.M.; Gardiwal, A.; de Gier-de Vries, C.; Schuster-Gossler, K.; Moorman, A.F.M.; Kispert, A.; Christoffels, V.M. Formation of the Sinus Node Head and Differentiation of Sinus Node Myocardium Are Independently Regulated by Tbx18 and Tbx3. *Circ. Res.* **2009**, *104*, 388–397. [CrossRef]
47. Kapoor, N.; Liang, W.; Marbán, E.; Cho, H.C. Direct conversion of quiescent cardiomyocytes to pacemaker cells by expression of Tbx18. *Nat. Biotechnol.* **2013**, *31*, 54–62. [CrossRef]
48. Liang, X.; Zhang, Q.; Cattaneo, P.; Zhuang, S.; Gong, X.; Spann, N.J.; Jiang, C.; Cao, X.; Zhao, X.; Zhang, X.; et al. Transcription factor ISL1 is essential for pacemaker development and function. *J. Clin. Investig.* **2015**, *125*, 3256–3268. [CrossRef]
49. Vedantham, V.; Galang, G.; Evangelista, M.; Deo, R.C.; Srivastava, D. RNA sequencing of mouse sinoatrial node reveals an upstream regulatory role for Islet-1 in cardiac pacemaker cells. *Circ. Res.* **2015**, *116*, 797–803. [CrossRef]
50. Zhao, H.; Wang, F.; Zhang, W.; Yang, M.; Tang, Y.; Wang, X.; Zhao, Q.; Huang, C. Overexpression of TBX3 in human induced pluripotent stem cells (hiPSCs) increases their differentiation into cardiac pacemaker-like cells. *Biomed. Pharmacother.* **2020**, *130*, 110612. [CrossRef]
51. Hoogaars, W.M.H.; Engel, A.; Brons, J.F.; Verkerk, A.O.; de Lange, F.J.; Wong, L.Y.E.; Bakker, M.L.; Clout, D.E.; Wakker, V.; Barnett, P.; et al. Tbx3 controls the sinoatrial node gene program and imposes pacemaker function on the atria. *Genes Dev.* **2007**, *21*, 1098–1112. [CrossRef] [PubMed]
52. Arnolds, D.E.; Liu, F.; Fahrenbach, J.P.; Kim, G.H.; Schillinger, K.J.; Smemo, S.; McNally, E.M.; Nobrega, M.A.; Patel, V.V.; Moskowitz, I.P. TBX5 drives Scn5a expression to regulate cardiac conduction system function. *J. Clin. Investig.* **2012**, *122*, 2509–2518. [CrossRef] [PubMed]
53. Barefield, D.Y.; Puckelwartz, M.J.; Kim, E.Y.; Wilsbacher, L.D.; Vo, A.H.; Waters, E.A.; Earley, J.U.; Hadhazy, M.; Dellefave-Castillo, L.; Pesce, L.L.; et al. Experimental Modeling Supports a Role for MyBP-HL as a Novel Myofilament Component in Arrhythmia and Dilated Cardiomyopathy. *Circulation* **2017**, *136*, 1477–1491. [CrossRef]
54. Harris, S.P.; Lyons, R.G.; Bezold, K.L. In the thick of it: HCM-causing mutations in myosin binding proteins of the thick filament. *Circ. Res.* **2011**, *108*, 751–764. [CrossRef] [PubMed]
55. Yang, Y.; Yuan, Y.; Wong, J.; Fishbein, M.C.; Chen, P.S.; Everett, T.H. Recording Intrinsic Nerve Activity at the Sinoatrial Node in Normal Dogs with High-Density Mapping. *Circ. Arrhythm. Electrophysiol.* **2021**, *14*, e008610. [CrossRef] [PubMed]
56. Brahmajothi, M.V.; Morales, M.J.; Campbell, D.L.; Steenbergen, C.; Strauss, H.C. Expression and distribution of voltage-gated ion channels in ferret sinoatrial node. *Physiol. Genom.* **2010**, *42 A*, 131–140. [CrossRef]
57. Bruneau, B.G.; Nemer, G.; Schmitt, J.P.; Charron, F.; Robitaille, L.; Caron, S.; Conner, D.A.; Gessler, M.; Nemer, M.; Seidman, C.E.; et al. A murine model of Holt-Oram syndrome defines roles of the T-Box transcription factor Tbx5 in cardiogenesis and disease. *Cell* **2001**, *106*, 709–721. [CrossRef]
58. Moskowitz, I.P.G.; Pizard, A.; Patel, V.V.; Bruneau, B.G.; Kim, J.B.; Kupershmidt, S.; Roden, D.; Berul, C.I.; Seidman, C.E.; Seidman, J.G. The T-Box transcription factor Tbx5 is required for the patterning and maturation of the murine cardiac conduction system. *Development* **2004**, *131*, 4107–4116. [CrossRef]
59. Olson, T.M.; Alekseev, A.E.; Liu, X.K.; Park, S.; Zingman, L.V.; Bienengraeber, M.; Sattiraju, S.; Ballew, J.D.; Jahangir, A.; Terzic, A. Kv1.5 channelopathy due to KCNA5 loss-of-function mutation causes human atrial fibrillation. *Hum. Mol. Genet.* **2006**, *15*, 2185–2191. [CrossRef]
60. Li, N.; Wang, Z.S.; Wang, X.H.; Xu, Y.J.; Qiao, Q.; Li, X.M.; Di, R.M.; Guo, X.J.; Li, R.G.; Zhang, M.; et al. A SHOX2 loss-of-function mutation underlying familial atrial fibrillation. *Int. J. Med. Sci.* **2018**, *15*, 1564–1572. [CrossRef]
61. Hoffmann, S.; Berger, I.M.; Glaser, A.; Bacon, C.; Li, L.; Gretz, N.; Steinbeisser, H.; Rottbauer, W.; Just, S.; Rappold, G. Islet1 is a direct transcriptional target of the homeodomain transcription factor Shox2 and rescues the Shox2-mediated bradycardia. *Basic Res. Cardiol.* **2013**, *108*, 339. [CrossRef] [PubMed]
62. Herriges, J.C.; Yi, L.; Hines, E.A.; Harvey, J.F.; Xu, G.; Gray, P.A.; Ma, Q.; Sun, X. Genome-scale study of transcription factor expression in the branching mouse lung. *Dev. Dyn.* **2012**, *241*, 1432–1453. [CrossRef] [PubMed]
63. Gao, R.; Liang, X.; Cheedipudi, S.; Cordero, J.; Jiang, X.; Zhang, Q.; Caputo, L.; Günther, S.; Kuenne, C.; Ren, Y.; et al. Pioneering function of Isl1 in the epigenetic control of cardiomyocyte cell fate. *Cell Res.* **2019**, *29*, 486–501. [CrossRef]
64. Doisne, N.; Maupoil, V.; Cosnay, P.; Findlay, I. Catecholaminergic automatic activity in the rat pulmonary vein: Electrophysiological differences between cardiac muscle in the left atrium and pulmonary vein. *Am. J. Physiol. Heart Circ. Physiol.* **2009**, *297*, H102–H108. [CrossRef] [PubMed]

65. Okamoto, Y.; Takano, M.; Ohba, T.; Ono, K. Arrhythmogenic coupling between the Na+–Ca2+ exchanger and inositol 1,4,5-triphosphate receptor in rat pulmonary vein cardiomyocytes. *J. Mol. Cell. Cardiol.* **2012**, *52*, 988–997. [CrossRef]
66. Okada, D.; Zheng, C.; Cheng, J.H.; Yamada, R. Cell population-based framework of genetic epidemiology in the single-cell omics era. *BioEssays* **2022**, *44*, 2100118. [CrossRef]
67. Litviňuková, M.; Talavera-López, C.; Maatz, H.; Reichart, D.; Worth, C.L.; Lindberg, E.L.; Kanda, M.; Polanski, K.; Heinig, M.; Lee, M.; et al. Cells of the adult human heart. *Nature* **2020**, *588*, 466–472. [CrossRef]
68. Swindell, W.R.; List, E.O.; Berryman, D.E.; Kopchick, J.J. Transcriptional profiling identifies strain-specific effects of caloric restriction and opposite responses in human and mouse white adipose tissue. *Aging* **2018**, *10*, 701–746. [CrossRef]
69. Danecek, P.; Nellåker, C.; McIntyre, R.E.; Buendia-Buendia, J.E.; Bumpstead, S.; Ponting, C.P.; Flint, J.; Durbin, R.; Keane, T.M.; Adams, D.J. High levels of RNA-editing site conservation amongst 15 laboratory mouse strains. *Genome Biol.* **2012**, *13*, r26. [CrossRef]
70. Barnabei, M.S.; Palpant, N.J.; Metzger, J.M. Influence of genetic background on ex vivo and in vivo cardiac function in several commonly used inbred mouse strains. *Physiol. Genom.* **2010**, *42 A*, 103–113. [CrossRef]
71. Tanase, H.; Yamori, Y.; Hansen, C.T.; Lovenberg, W. Heart Size in Inbred Strains of Rats. *Hypertension* **1982**, *4*, 864–873. [CrossRef] [PubMed]
72. Aziz, Q.; Nobles, M.; Tinker, A. Acute Isolation of Cells from Murine Sino-atrial Node. *Bio Protoc.* **2020**, *10*, e3477. [CrossRef]

biomolecules

MDPI

Article

Intracellular Ca^{2+}-Mediated Mechanisms for the Pacemaker Depolarization of the Mouse and Guinea Pig Sinus Node Tissue

Iyuki Namekata *, Kento Jitsukata, Ayumi Fukuda, Ryosuke Odaka, Shogo Hamaguchi and Hikaru Tanaka

Department of Pharmacology, Faculty of Pharmaceutical Sciences, Toho University, 2-2-1 Miyama Funabashi, Chiba 274-8510, Japan; kento.jitsukata@gmail.com (K.J.); 1015188f@st.toho-u.jp (A.F.); 3021002o@st.toho-u.jp (R.O.); shogo.hamaguchi@phar.toho-u.ac.jp (S.H.); htanaka@phar.toho-u.ac.jp (H.T.)
* Correspondence: iyuki@phar.toho-u.ac.jp; Tel.: +81-474722088

Abstract: Intracellular Ca^{2+}-mediated mechanisms for pacemaker depolarization were studied in sinus node tissue preparations from mice and guinea pigs. Microelectrode recordings revealed that the sinus node of the mouse, which had a higher beating rate, had a steeper slope of the pacemaker depolarization than that of the guinea pig. BAPTA and ryanodine, agents that interfere with intracellular Ca^{2+}, significantly decreased the slope of the pacemaker depolarization in both species. In contrast, SEA0400, a specific inhibitor of the Na^+-Ca^{2+} exchanger (NCX), as well as change to low Na^+ extracellular solution, significantly decreased the slope in the mouse, but not in the guinea pig. Niflumic acid, a blocker of the Ca^{2+} activated Cl^- channel, decreased the slope in both species. Confocal microscopy revealed the presence of spontaneous Ca^{2+} oscillations during the interval between Ca^{2+} transients; such phenomenon was more pronounced in the mouse than in the guinea pig. Thus, although intracellular Ca^{2+}-mediated mechanisms were involved in the pacemaker depolarization of the sinus node in both species, the NCX current was involved in the mouse but not in the guinea pig.

Keywords: cardiac pacemaking; Na^+-Ca^{2+} exchanger; pacemaker depolarization; sinus node

Citation: Namekata, I.; Jitsukata, K.; Fukuda, A.; Odaka, R.; Hamaguchi, S.; Tanaka, H. Intracellular Ca^{2+}-Mediated Mechanisms for the Pacemaker Depolarization of the Mouse and Guinea Pig Sinus Node Tissue. *Biomolecules* **2022**, *12*, 377. https://doi.org/10.3390/biom12030377

Academic Editors: Yosuke Okamoto, Kyoichi Ono and Samuel Dudley

Received: 30 November 2021
Accepted: 11 February 2022
Published: 28 February 2022

1. Introduction

The contraction of the myocardium is driven by the action potential originating in the sinus node, the cardiac pacemaker. The action potential of the sinus node has a characteristic pacemaker depolarization phase (phase 4 depolarization) in the membrane voltage range of −60–−40 mV which leads the membrane potential to the threshold level for the rapid action potential upstroke (phase 0 depolarization). The pacemaker depolarization is formed by multiple inward membrane currents including the hyperpolarization-activated current (I_f), L-type and T-type Ca^{2+} currents, and the sustained inward current (I_{st}) [1–5]. The combination of membrane currents involved in the pacemaker depolarization appears to be different among animal species. For example, the I_f current manifests in sinus node cells of most species including mice, guinea pigs, and rabbits, while it was masked by the inwardly rectifying K^+ current (I_{K1}) in rats and monkeys [6]. The current density and function of the T-type Ca^{2+} current was reported to be larger in the sinus nodes from smaller animals [3,7].

It has been postulated that the pacemaker depolarization is also influenced by intracellular Ca^{2+}. The Ca^{2+} released from the sarcoplasmic reticulum (SR) causes the Na^+-Ca^{2+} exchanger (NCX) to generate a depolarizing current and accelerate the pacemaker depolarization. This mechanism was referred to as the Ca^{2+} clock and was considered to function in coordination with the ion channels of the cell membrane (membrane clock) to form the pacemaker depolarization [8,9]. However, the related experimental results appear to vary among researchers, and definitive conclusions have not been reached for the involvement of intracellular Ca^{2+}-mediated mechanisms or the precise pacemaking mechanisms in each animal species. For example, in the guinea pig sinus node, some researchers emphasize the involvement of the Ca^{2+}-clock while others reported negative results [4,9,10].

One of the possible reasons for this discrepancy is the use of isolated sinus node cells. The sinus node consists of cardiomyocytes with different electrophysiological properties [5,11], and these cells work as a functional syncytium to form the pacemaker depolarization. Thus, in a strict sense, none of the cells isolated from the sinus node region exactly represent the sinus node pacemaker. Thus, it is of value to obtain information on the function of the sinus node as a whole. In the present study, we intended to clarify the involvement of intracellular Ca^{2+}-mediated mechanisms in the pacemaker depolarization of the mouse and guinea pig sinus node using tissue preparations. We performed standard microelectrode measurements of action potentials with sinus node tissue preparations and applied selective pharmacological agents.

2. Materials and Methods

Standard glass microelectrode experiments were performed with isolated sinus node tissue from the mouse and the guinea pig as described previously [2,3]. Microelectrode penetrations into the sinus node region were made from the epicardial surface. The extracellular solution was of the following composition (mM concentration): NaCl 118.4, KCl 4.7, $CaCl_2$ 2.5, $MgSO_4$ 1.2, KH_2PO_4 1.2, $NaHCO_3$ 24.9, and glucose 11.1 (pH 7.4), and the solution was gassed with 95% O_2-5% CO_2 and maintained at 36 $\pm$ 0.5 °C. The low Na^+ extracellular solution was prepared with the equimolar substitution of NaCl with LiCl so that the final Na^+ concentration was 70 mM. Change to low Na^+ solution and return to normal solution were performed with a flow-switching device which enabled change of the extracellular solution within approximately 1 s. To chelate intracellular Ca^{2+} with O,O′-Bis (2-aminophenyl) ethyleneglycol-N,N,N′,N′-tetraacetic acid (BAPTA), its cell-permeable tetraacetoxymethyl ester (BAPTA-AM) was applied to the preparations at a final concentration of 300 μM.

The action potential parameters measured were firing rate, cycle length, maximum diastolic potential, threshold potential, the slope of the pacemaker depolarization (slope), maximum rate of rise of the action potential upstroke (maximum rate of rise; $(dV/dt)_{max}$), peak potential, and duration at 50% repolarization (APD_{50}). To obtain the slope value, the middle portion of the pacemaker depolarization phase was fitted by a straight line; the curved regions close to the maximum diastolic potential and threshold potential were not included in the fitting. The spontaneous firing of the sinus node tissue preparations was well maintained; the change in firing rate at 10 and 30 min after the addition of vehicle was less than 1% and 2% of the basal value, respectively.

For the analysis of intracellular Ca^{2+} movements, the sinus node tissue preparations were incubated with a Ca^{2+} indicator (Cal-590 AM) at 10 μM for 1 h at 37 °C in the following composition: NaCl 143, KCl 4.7, $MgCl_2$ 1.0, NaH_2PO_4 0.33, glucose 5.5, and HEPES 5 (mM). The recording chamber, the bottom of which was a coverslip, was placed on the stage of an inverted microscope. Preparations were placed at the bottom of the recording chamber epicardial surface down. The normal extracellular solution (NaCl 143, KCl 4.7, $MgCl_2$ 1.0, $CaCl_2$ 1.8, NaH_2PO_4 0.33, glucose 5.5, and HEPES 5 (mM) maintained at 32 °C) was continuously perfused. Confocal microscopic analyses of the Ca^{2+} oscillations in the myocardial layer were performed with a rapid scanning confocal microscope A1R (Nikon, Tokyo, Japan). The emission 570–620 nm on excitation at 561 nm was detected for Cal-590 fluorescence, and the scanning was performed at a speed of frame/16.9–33.8 ms. The fluorescent intensity of Cal-590 at each time point was normalized against the basal intensity.

BAPTA-AM (Tokyo Chemical Industry, Tokyo, Japan), ryanodine (Wako Pure Chemical Industries, Osaka, Japan), SEA0400 (synthesized in our faculty), niflumic acid (Sigma-Aldrich, St. Louis, MO, USA), and Cal-590 AM (Cosmo Bio, Tokyo, Japan) were dissolved in dimethyl sulfoxide (DMSO). All data were expressed as the mean $\pm$ standard error of the mean (S.E.M). Data were analyzed by the paired *t*-test. A p value less than 0.05 was considered statistically significant.

3. Results

Microelectrode recordings showed that the sinus node of the mouse, which had a higher firing rate, had a steeper slope of the pacemaker depolarization and a shorter action potential duration than that of the guinea pig. There was no difference in the maximum diastolic potential and threshold potential between the mouse and guinea pig; the maximum diastolic potential was about −60 mV, and the threshold potential was about −40 mV in both species.

In the mouse sinus node, chelation of intracellular Ca^{2+} with BAPTA (300 μM) induced significant decreases in the firing rate and the slope of the pacemaker depolarization (Figure 1 and Table 1). BAPTA had no significant effect on other parameters in the mouse. In the guinea pig sinus node, BAPTA induced significant decreases in the firing rate and the slope of the pacemaker depolarization (Figure 1 and Table 1). BAPTA shifted the maximum diastolic potential and the threshold potential towards depolarized potentials in the guinea pig.

In the mouse and guinea pig sinus node, ryanodine (0.1 μM) significantly decreased the firing rate and the slope of the pacemaker depolarization (Figure 1 and Table 1). Ryanodine had no significant effect on other parameters in both species.

In the mouse sinus node, SEA0400 (1 and 10 μM) decreased the firing rate and the slope of the pacemaker depolarization (Figure 2 and Table 2). SEA0400 shifted the maximum diastolic potential and the threshold potential towards depolarized potentials and decreased the maximum rate of rise and the peak potential in the mouse. Further, when the effect of SEA0400 on the beating rate of the mouse right atria was examined after the beating rate was decreased with carbachol, SEA0400 significantly decreased the beating rate; the beating rate in the presence of 0.3 μM carbachol before and after the application of SEA0400 (1 μM) was 234.1 ± 10.7 and 163.8 ± 29.1 (n = 6), respectively.

In the guinea pig sinus node, SEA0400 had no effect on the action potential waveform (Figure 2 and Table 2). Further, when the effect of SEA0400 on the beating rate of the guinea pig right atria was examined after the beating rate was increased with noradrenaline, SEA0400 had no effect on the beating rate; the beating rate in the presence of 3 μM noradrenaline before and after the application of SEA0400 (1 μM) was 328.4 ± 8.9 and 324.6 ± 8.9 (n = 5), respectively.

To support the results obtained with SEA0400, we intended to reduce the inward NCX current with low Na^+ extracellular solution. In the mouse sinus node, changing the extracellular solution to low Na^+ solution significantly decreased the firing rate and the slope of the pacemaker depolarization (Figure 3 and Table 3). The low Na^+ solution shifted the maximum diastolic potential and the threshold potential towards depolarized potentials and decreased the maximum rate of rise and the peak potential. In the guinea pig, changing the extracellular solution to low Na^+ solution had no significant effect on the firing rate and the slope of the pacemaker depolarization. The low Na^+ solution did not affect any of the action potential parameters (Figure 3 and Table 3).

In the mouse sinus node, niflumic acid (30 μM), a blocker of the Ca^{2+}-activated Cl^- channel, significantly decreased the firing rate and the slope of the pacemaker depolarization (Figure 4 and Table 4). Niflumic acid had no significant effect on other parameters in the mouse. In the guinea pig sinus node, niflumic acid significantly decreased the firing rate and the slope of the pacemaker depolarization (Figure 4 and Table 4). The action potential duration was shortened by niflumic acid in the guinea pig.

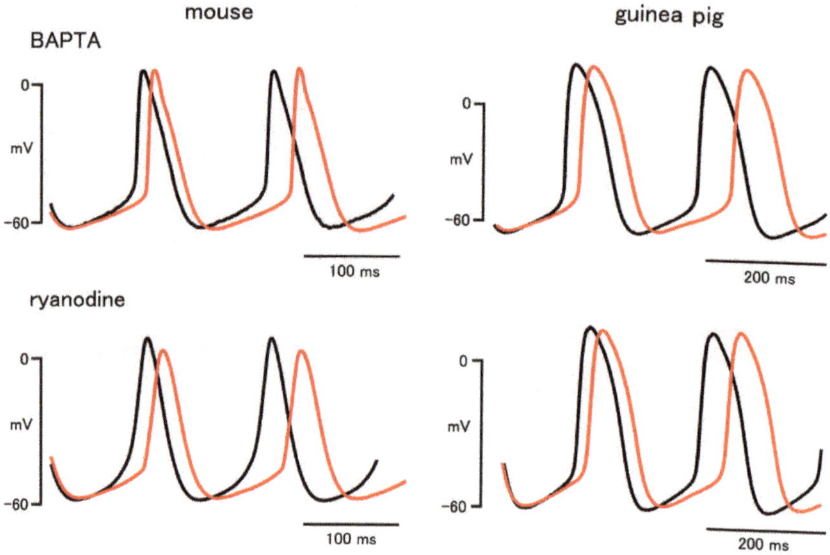

Figure 1. Effects of BAPTA and ryanodine on the sinus node action potential of the mouse and guinea pig. Typical traces before (black lines) and after (red lines) application of 300 μM BAPTA (upper) and 0.1 μM ryanodine (lower).

Table 1. Action potential parameters before and after application of BAPTA and ryanodine.

	BAPTA		Ryanodine	
	Mouse	**Guinea Pig**	**Mouse**	**Guinea Pig**
Firing rate (bpm)	423.7 ± 10.1 396.0 ± 6.4 *	241.4 ± 16.1 213.1 ± 12.2 *	453.6 ± 10.0 414.3 ± 13.9 *	248.7 ± 11.7 237.4 ± 12.6 *
Cycle length (ms)	141.9 ± 3.4 151.7 ± 2.4 *	254.2 ± 17.0 286.7 ± 17.2 *	132.5 ± 2.9 145.5 ± 4.7 *	241.3 ± 13.4 256.7 ± 13.6 *
Maximum diastolic potential (mV)	−60.6 ± 1.4 −60.0 ± 1.7	−62.9 ± 1.2 −58.9 ± 2.1 *	−58.8 ± 1.2 −55.9 ± 1.7	−63.2 ± 1.3 −60.9 ± 1.7
Slope of pacemaker depolarization (mV/s)	192.0 ± 26.2 130.0 ± 17.7 *	147.4 ± 14.7 104.8 ± 9.8 *	223.5 ± 25.8 110.0 ± 24.5 *	138.2 ± 10.4 119.2 ± 9.1 *
Threshold potential (mV)	−50.4 ± 1.6 −51.6 ± 1.5	−49.0 ± 1.6 −46.4 ± 2.1 *	−46.9 ± 1.9 −45.8 ± 3.0	−52.6 ± 1.9 −49.1 ± 2.3
Maximum rate of rise (V/s)	10.3 ± 2.6 9.0 ± 2.2	9.0 ± 2.2 8.1 ± 1.9	8.7 ± 2.7 7.8 ± 2.2	18.7 ± 9.4 15.6 ± 7.8
Peak potential (mV)	0.9 ± 2.5 −0.9 ± 3.2	18.9 ± 1.1 17.7 ± 1.5	4.6 ± 1.2 4.2 ± 1.0	18.3 ± 1.7 17.8 ± 1.6
Duration at 50% repolarization (ms)	30.3 ± 1.7 30.2 ± 1.4	81.8 ± 2.0 85.6 ± 6.3	28.3 ± 0.9 28.9 ± 1.4	75.4 ± 5.9 77.4 ± 6.4

Upper and lower values in each row indicate the parameters obtained before and after (10 min for BAPTA and 5 min for ryanodine) the application of agents, respectively. Values indicate the mean ± S.E.M. from 5–6 preparations. Asterisks indicate significant difference from corresponding values before the application of agents as evaluated by the paired *t*-test ($p < 0.05$).

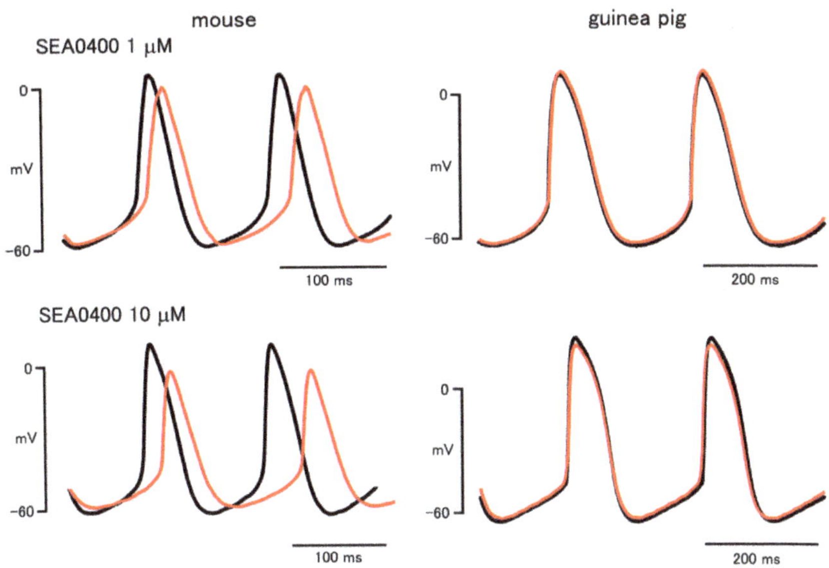

Figure 2. Effect of SEA0400 on the sinus node action potential of the mouse and guinea pig. Typical traces before (black lines) and after (red lines) application of 1 μM (upper) and 10 μM (lower).

Table 2. Action potential parameters before and after application of SEA0400.

	SEA0400 (1 μM)		SEA0400 (10 μM)	
	Mouse	**Guinea Pig**	**Mouse**	**Guinea Pig**
Firing rate (bpm)	449.7 ± 24.9 396.8 ± 29.0 *	239.0 ± 9.2 240.1 ± 9.2	428.2 ± 27.2 363.8 ± 17.7 *	239.4 ± 8.9 237.4 ± 11.6
Cycle length (ms)	135.2 ± 8.1 155.0 ± 12.2 *	253.0 ± 10.1 251.8 ± 9.7	142.1 ± 8.0 166.4 ± 7.6 *	252.0 ± 9.0 255.1 ± 12.3
Maximum diastolic potential (mV)	−55.8 ± 1.8 −51.7 ± 1.9 *	−62.6 ± 2.5 −62.1 ± 2.4	−59.2 ± 1.0 −55.9 ± 1.3 *	−61.7 ± 2.1 −59.1 ± 3.0
Slope of pacemaker depolarization (mV/s)	217.0 ± 22.2 173.2 ± 22.9 *	148.3 ± 16.2 150.8 ± 13.8	215.4 ± 26.6 161.0 ± 20.9 *	172.5 ± 6.0 177.0 ± 16.3
Threshold potential (mV)	−46.0 ± 1.2 −41.9 ± 1.4 *	−49.8 ± 2.7 −48.7 ± 2.5	−48.6 ± 1.2 −45.5 ± 1.3 *	−48.3 ± 2.5 −43.9 ± 3.1
Maximum rate of rise (V/s)	6.7 ± 0.6 5.6 ± 0.7 *	7.0 ± 1.5 7.7 ± 2.2	7.9 ± 1.6 4.5 ± 0.8 *	5.8 ± 1.5 4.8 ± 1.1
Peak potential (mV)	4.8 ± 0.7 −0.5 ± 1.3 *	12.0 ± 2.1 12.8 ± 1.8	4.7 ± 1.6 −11.2 ± 3.0 *	11.9 ± 1.6 12.0 ± 1.6
Duration at 50% repolarization (ms)	33.5 ± 2.1 35.5 ± 1.5	80.4 ± 5.4 79.8 ± 4.4	31.3 ± 2.2 36.2 ± 2.2	85.4 ± 6.2 83.6 ± 5.3

Upper and lower values in each row indicate the parameters obtained before and after (10 min) the application of agents, respectively. Values indicate the mean ± S.E.M. from 5–6 preparations. Asterisks indicate significant difference from corresponding values before the application of agents as evaluated by the paired *t*-test ($p < 0.05$).

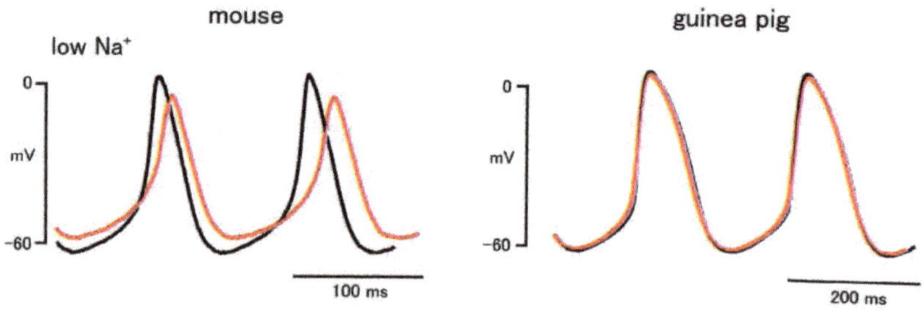

Figure 3. Effect of low Na^+ solution on the sinus node action potential of the mouse and guinea pig. Typical traces before (black lines) and after (red lines) rapid change of the extracellular solution to low Na^+ solution.

Table 3. Action potential parameters before and after the rapid change of the extracellular solution to low Na^+ solution.

	Low Na^+	
	Mouse	**Guinea Pig**
Firing rate (bpm)	498.4 ± 17.1 455.5 ± 20.5 *	232.7 ± 9.5 232.7 ± 9.9
Cycle length (ms)	121.1 ± 4.0 133.1 ± 5.9 *	260.2 ± 12.0 260.5 ± 12.6
Maximum diastolic potential (mV)	−60.4 ± 1.1 −56.9 ± 0.9 *	−63.2 ± 0.9 −64.0 ± 1.6
Slope of pacemaker depolarization (mV/s)	251.4 ± 25.3 206.7 ± 30.7 *	166.3 ± 17.4 187.3 ± 20.4
Threshold potential (mV)	−49.8 ± 1.1 −47.6 ± 1.0 *	−48.2 ± 1.1 −48.1 ± 1.8
Maximum rate of rise (V/s)	8.6 ± 1.1 5.6 ± 0.6 *	4.0 ± 0.7 3.6 ± 0.6 *
Peak potential (mV)	1.9 ± 1.3 −5.1 ± 2.4 *	11.0 ± 1.2 8.3 ± 2.4
Duration at 50% repolarization (ms)	28.7 ± 1.3 32.2 ± 1.3 *	89.3 ± 3.5 88.8 ± 4.2

Upper and lower values in each row indicate the parameters obtained before and after (1 min) the application of low Na^+ solution. Values indicate the mean ± S.E.M. from 6 preparations. Asterisks indicate significant difference from corresponding values before the application of agent as evaluated by the paired *t*-test ($p < 0.05$).

To analyze intracellular Ca^{2+} movements, the sinus node tissue preparations loaded with the Ca^{2+} sensitive fluoroprobe, Cal-590, were observed with a rapid scanning confocal microscope. In both species, Ca^{2+} transients, synchronized elevations of Ca^{2+} fluorescence throughout the myocardial tissue, were observed at a constant frequency. In mouse sinus node, spontaneous Ca^{2+} oscillations, non-synchronized local rises in Ca^{2+} fluorescence, were observed in roughly one third of the cells (Figure 5). Such Ca^{2+} oscillations were less frequently observed in the guinea pig (Figure 6).

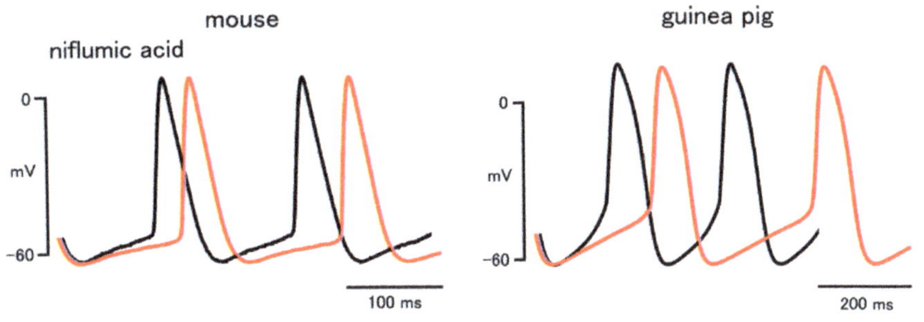

Figure 4. Effect of niflumic acid on the sinus node action potential of the mouse and guinea pig. Typical traces before (black lines) and after (red lines) application of 30 μM niflumic acid.

Table 4. Action potential parameters before and after application of niflumic acid.

	Niflumic Acid	
	Mouse	**Guinea Pig**
Firing rate (bpm)	444.5 ± 18.1 418.4 ± 21.0 *	236.6 ± 14.1 205.0 ± 12.5 *
Cycle length (ms)	136.1 ± 5.5 145.3 ± 7.6 *	258.4 ± 15.9 298.0 ± 17.7 *
Maximum diastolic potential (mV)	−61.0 ± 1.1 −60.6 ± 1.8	−62.7 ± 3.1 −61.8 ± 2.8
Slope of pacemaker depolarization (mV/s)	221.5 ± 40.9 173.8 ± 35.7 *	152.2 ± 20.2 104.8 ± 12.4 *
Threshold potential (mV)	−50.5 ± 1.7 −51.7 ± 2.5	−48.8 ± 3.2 −49.0 ± 3.5
Maximum rate of rise (V/s)	14.5 ± 6.5 14.7 ± 6.8	8.4 ± 2.1 6.9 ± 1.3
Peak potential (mV)	5.8 ± 0.8 5.2 ± 1.6	19.4 ± 4.0 17.8 ± 3.4
Duration at 50% repolarization (ms)	31.3 ± 1.2 32.1 ± 1.0	78.5 ± 2.6 74.3 ± 3.2 *

Upper and lower values in each row indicate the parameters obtained before and after (10 min) the application of niflumic acid. Values indicate the mean ± S.E.M. from 6 preparations. Asterisks indicate significant difference from corresponding values before the application of agent as evaluated by the paired *t*-test ($p < 0.05$).

In the mouse sinus node, SEA0400 (1 μM) markedly reduced the firing frequency of spontaneous Ca^{2+} transients (Figure 5); the frequency before and 10 min after the addition of 1 μM SEA0400 was 185.8 ± 24.1 and 110.3 ± 23.2 bpm ($n = 5$; $p < 0.05$), respectively. SEA0400 caused an increase in basal Ca^{2+} concentration; the basal Ca^{2+} fluorescence at 10 min after the addition of 1 μM SEA0400 was 110.3 ± 8.7% ($n = 5$) of the value before application. SEA0400 had no apparent effect on the local Ca^{2+} oscillations.

In the guinea pig sinus node, SEA0400 had no effect on the firing frequency of spontaneous Ca^{2+} transients (Figure 6); the frequency before and 10 min after the addition of 1 μM SEA0400 was 157.2 ± 5.9 and 158.2 ± 8.5 bpm ($n = 5$), respectively. SEA0400 caused an increase in basal Ca^{2+} concentration; the basal Ca^{2+} fluorescence at 10 min after the addition of 1 μM SEA0400 was 112.0 ± 4.5% ($n = 5$) of the value before application.

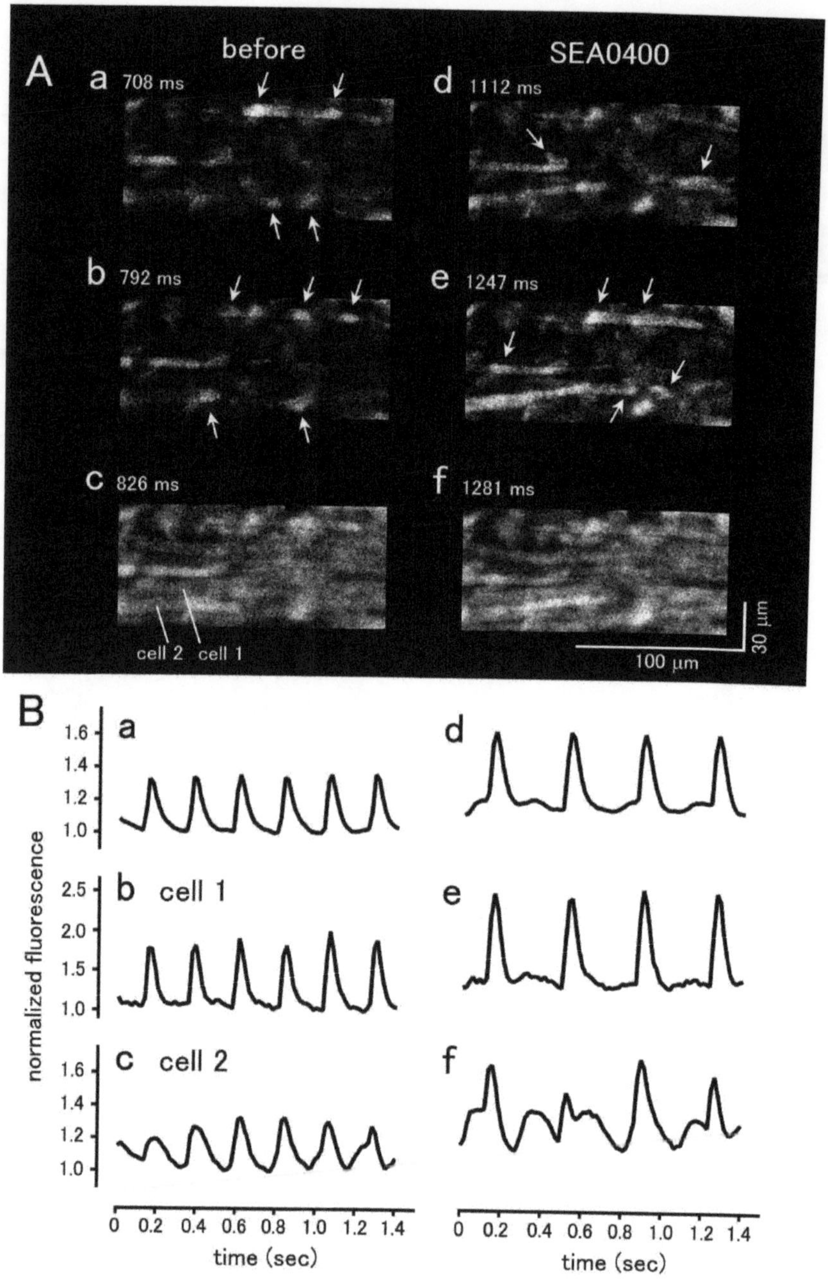

Figure 5. Effect of SEA0400 on spontaneously occurring Ca^{2+} transients in the mouse sinus node tissue. (**A**) Typical x-y images of the myocardium loaded with Cal-590 before (**a–c**) and 10 min after the addition of 1 µM SEA0400 (**d–f**). Panels c and f are the images at the peak of the Ca^{2+} transient. Note that Ca^{2+} oscillations were observed during the interval between the Ca^{2+} transients (arrows). (**B**) Time course of the changes in fluorescence before addition (**a–c**) and 10 min after the addition of 1 µM SEA0400 (**d–f**). Time course of the fluorescence intensity quantified in the whole field of view (**a,d**), cell 1 (**b,e**), and cell 2 (**c,f**) as shown in panel (**A**)/(**c**).

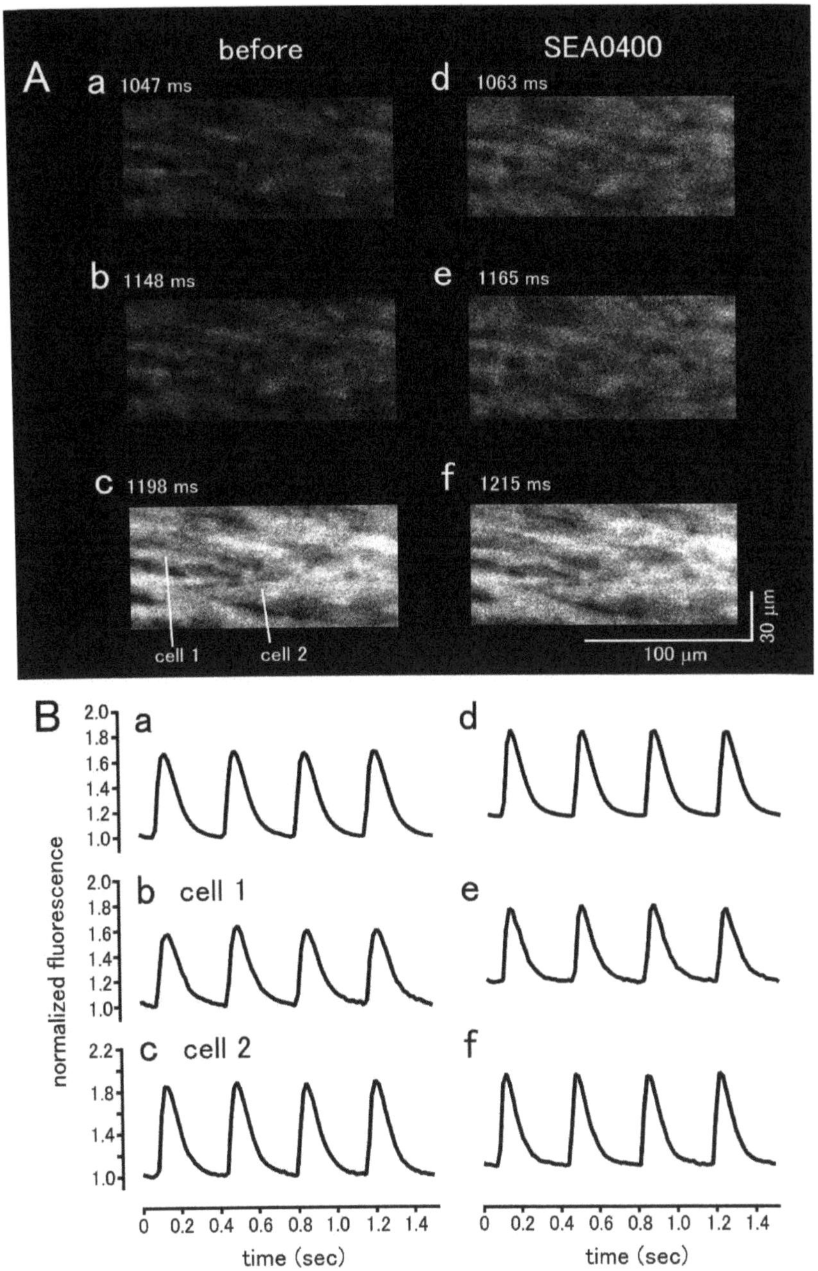

Figure 6. Effect of SEA0400 on spontaneously occurring Ca^{2+} transients in the guinea pig sinus node tissue. (**A**) Typical x-y images of the myocardium loaded with Cal-590 before (**a–c**) and 10 min after the addition of 1μM SEA0400 (**d–f**). Panels c and f are the images at the peak of the Ca^{2+} transient. Note that no Ca^{2+} oscillation was observed during the interval between the Ca^{2+} transients (**a**,**b**,**d**,**e**). (**B**) Time course of the changes in fluorescence before addition (**a–c**) and 10 min after the addition of 1 μM SEA0400 (**d–f**). Time course of the fluorescence intensity quantified in the whole field of view (**a**,**d**), cell 1 (**b**,**e**), and cell 2 (**c**,**f**) as shown in panel (**A**)/(**c**).

4. Discussion

In the present study, we intended to clarify the involvement of intracellular Ca^{2+}-mediated mechanisms in the pacemaker depolarization of the mouse and guinea pig sinus node. We applied standard microelectrode techniques on tissue preparations and used selective pharmacological agents. Compared to isolated cardiomyocytes, myocardial tissue preparations may have disadvantages such as less efficient delivery of oxygen and pharmacological agents to the cells of interest. On the other hand, they have advantages such as absence of cell isolation damage and cell selection bias. Above all, tissue preparations enable the observation of the sinus node as a whole working as a functional syncytium.

First of all, we examined whether or not intracellular Ca^{2+}-mediated mechanisms are involved in the pacemaking of the sinus node. Treatment of the sinus node tissue preparations with either BAPTA or ryanodine, agents that interfere with intracellular Ca^{2+}, decreased the firing rate and the slope of the pacemaker depolarization both in the mouse and guinea pig. These results indicated that Ca^{2+} released from the SR through the ryanodine receptor contributes to the formation of the pacemaker depolarization of the sinus node in both species. As both BAPTA and ryanodine may have a broad effect on intracellular Ca^{2+}-mediated events, their effects may also reflect changes other than the instantaneous reduction of intracellular Ca^{2+} [4]. Thus, we intended to clarify the transporters that convert intracellular Ca^{2+} into the pacemaker depolarization.

The Na^+/Ca^{2+} exchanger (NCX) is the major pathway for transsarcolemal Ca^{2+} extrusion from the cytoplasm; it pumps out one Ca^{2+} ion in exchange for three Na^+ ions (forward mode NCX). It has been postulated that the net inward current that occurs in this process contributes to the pacemaker depolarization of the sinus node. To examine this hypothesis in sinus node tissue preparations, we used SEA0400, a potent and selective inhibitor of NCX. In voltage-clamped guinea pig ventricular myocytes, 1 μM SEA0400 inhibited the NCX current by more than 80%, with no effect on the sodium current, L-type calcium current, delayed rectifier potassium current, and the inwardly rectifying potassium current [12]. The effectiveness of SEA0400 on tissue preparations was also confirmed in the mouse and guinea pig myocardium [13–15].

In the mouse sinus node, the slope of the pacemaker depolarization, as well as the firing rate of action potentials and Ca^{2+} transients, was decreased by SEA0400. Similar results were obtained with low Na^+ extracellular solution. Thus, the NCX current activated by the Ca^{2+} released from the SR contributes to the pacemaker depolarization in mouse sinus node. These results were consistent with the observation that the firing frequency of the sinus node was lower in atrial-specific NCX knock out mice than in wild type mice [16]. In contrast, in the guinea pig sinus node, treatment with SEA0400, as well as low Na^+ extracellular solution, affected neither the firing rate nor the slope of the pacemaker depolarization. SEA0400 caused a small increase in the basal Ca^{2+} concentration. These results indicated that NCX contributes to transarcolemmal Ca^{2+} efflux but does not play an essential role in the pacemaker depolarization of the guinea pig sinus node. This lack of SEA0400 effect in the guinea pig sinus node was not due to its lower firing rate because SEA0400 was ineffective in the guinea pig right atria even after its beating rate was increased by noradrenaline to a level similar to that of the mouse, and effective in the mouse right atria even after its beating rate was decreased by carbachol to a level similar to that of the guinea pig.

The effects of SEA0400 and low Na^+ on the maximum rate of rise and the peak potential could be explained by their effect on the pacemaker depolarization. SEA0400 and low Na^+ markedly reduced the slope of pacemaker depolarization in the mouse sinus node. This means that the membrane potential is kept in a partially depolarized potential range close to the threshold potential for a longer time, which may cause increased inactivation of the L-type Ca^{2+} channel leading to a reduced maximum rate of rise and peak potential. In contrast, in the guinea pig sinus node, in which the slope of the pacemaker depolarization was unchanged, neither the maximum rate of rise nor the peak potential was affected.

The pacemaker depolarization is considered to be formed by several inward currents such as the hyperpolarization-activated current (I_f), L-type and T-type Ca^{2+} currents, and the sustained inward current (I_{st}) [1–5]. In addition, the deactivation of I_K is dominant in the early phase of pacemaker depolarization in the guinea pig sinus node [4,6]. The NCX current in the sinus node was detected both in the mouse and guinea pig; the current density was not different between the two species under voltage-clamp conditions [17,18]. Since the NCX currents observed were relatively small (<5 pA/pF), they may possibly be masked by other membrane currents. As the NCX current density in the intact sinus node is affected by the intracellular supply of Ca^{2+} to the NCX, it is possible that some difference in intracellular Ca^{2+} handling properties causes a difference in NCX current density in the spontaneously firing sinus node. In fact, spontaneous Ca^{2+} oscillations were observed during the interval between Ca^{2+} transients in sinus node tissue preparations, which was more pronounced in the mouse than in the guinea pig. This appears to be consistent with the observation in isolated sinus node cardiomyocytes that the number of Ca^{2+} sparks occurring during the pacemaker depolarization was larger in the mouse than in the guinea pig [9,19].

Concerning the sinus node of other animal species, it was reported that SEA0400 had no effect on the heart rate in the isolated rabbit heart [20] and in anesthetized dogs [21]. To the best of our knowledge, there is no information on the effect of SEA0400 on the human sinus node. Concerning ectopic pacemakers, we observed in the guinea pig pulmonary vein that both SEA0400 and ryanodine inhibited the spontaneous electrical activity [22]. Ca^{2+} sparks occurred during the interval between Ca^{2+} transients, and the frequency of Ca^{2+} spark firing increased abruptly just before the onset of Ca^{2+} transients [23]. These results indicated that the NCX current activated by Ca^{2+} released from the SR was involved in the pacemaker depolarization of the pulmonary vein myocardium. Thus, the contribution of NCX to myocardial pacemaking varies among animal species and type of myocardium.

To obtain further information on the Ca^{2+}-mediated pacemaking mechanisms of the sinus node, we examined the effect of niflumic acid, a blocker of the Ca^{2+}-activated Cl^- channel, which was identified in sinus node cells [24,25]. As the estimated reversal potential for Cl^- ions in cardiomyocytes was about −30 mV [26], activation of the Ca^{2+}-activated Cl^- channel could generate a depolarizing current during the pacemaker depolarization. Niflumic acid reduced the firing rate and the slope of pacemaker depolarization in the sinus node of both mouse and guinea pig. These results indicated that the inward current through the Cl^- channel activated by Ca^{2+} released from the SR contributes to the pacemaker depolarization of the sinus node both in the mouse and guinea pig.

The mouse has a high heart rate (400 bpm>), and its sinus node has a steeper slope of pacemaker depolarization among experimental animal species. The present study showed that the Ca^{2+}-activated Cl^- current (I_{ClCa}) was involved in the pacemaker depolarization of both mouse and guinea pig, but the NCX current contributed to cardiac pacemaking only in the mouse. Using similar experimental techniques, we previously showed that the contribution of the T-type Ca^{2+} channel to cardiac pacemaking differs among animal species; the effects of *R*-(−)-efonidipine, which is a selective T-type Ca^{2+} channel inhibitor, was prominent in the mouse, small but significant in the guinea pig, and was not observed in the rabbit [2]. Voltage gated Na^+ channels, which are considered not to play a role in pacemaker activity in the majority of mammalian species including the guinea pig [10], was reported to play a certain role in the pacemaking of the mouse sinus node [27]. The present and earlier studies revealed that the guinea pig sinus node depends on I_{CaL}, I_{st}, I_f, and I_{ClCa}, but not I_{NCX} for its pacemaking [4,5,11]. In the mouse sinus node, additional currents including the I_{CaT}, I_{NCX}, and I_{Na} also contribute to pacemaking [2,7,16,27]. Thus, the mouse sinus node appears to have a greater variety of ionic mechanisms to maintain its high firing rate.

5. Conclusions

Intracellular Ca^{2+}-mediated mechanisms were involved in the pacemaker depolarization of the sinus node both in the mouse and guinea pig. The NCX current was involved in the mouse but not in the guinea pig. This difference could partly explain the higher firing rate of the mouse sinus node.

Author Contributions: Conceptualization, I.N. and H.T.; methodology, I.N. and H.T.; software, S.H.; formal analysis, I.N. and R.O.; investigation, R.O., K.J. and A.F.; resources, H.T.; data curation, I.N.; writing—original draft preparation, I.N.; writing—review and editing, H.T.; visualization, I.N.; supervision, H.T.; project administration, I.N. and H.T.; funding acquisition, I.N., S.H. and H.T. All authors have read and agreed to the published version of the manuscript.

Funding: This research was funded by JSPS KAKENHI, Grant Numbers 20K07299, 20K16013 and 20K07091.

Institutional Review Board Statement: The study was approved by the Ethics Committee of Toho University (21-55-362; March 2021) and conducted in accordance with the "Guiding Principles for the Care and Use of Laboratory Animals" approved by The Japanese Pharmacological Society.

Informed Consent Statement: Not applicable.

Data Availability Statement: Not applicable.

Conflicts of Interest: The authors declare no conflict of interest.

References

1. DiFrancesco, D. The role of the funny current in pacemaker activity. *Circ. Res.* **2010**, *106*, 434–446. [CrossRef] [PubMed]
2. Tanaka, H.; Komikado, C.; Namekata, I.; Nakamura, H.; Suzuki, M.; Tsuneoka, Y.; Shigenobu, K.; Takahara, A. Species difference in the contribution of T-type calcium current to cardiac pacemaking as revealed by *R*(−)-efonidipine. *J. Pharmacol. Sci.* **2008**, *107*, 99–102. [CrossRef] [PubMed]
3. Tanaka, H.; Namekata, I.; Ogawa, T.; Tsuneoka, Y.; Komikado, C.; Takahara, A.; Iida-Tanaka, N.; Izumi-Nakaseko, H.; Tsuru, H.; Adachi-Akahane, S. Effects of *S*(+)-efonidipine on the rabbit sinus node action potential and calcium channel subunits Ca_V 1.2, Ca_V 1.3 and Ca_V 3.1. *Eur. J. Pharmacol.* **2010**, *649*, 263–267. [CrossRef] [PubMed]
4. Himeno, Y.; Toyoda, F.; Satoh, H.; Amano, A.; Cha, C.Y.; Matsuura, H.; Noma, A. Minor contribution of cytosolic Ca^{2+} transients to the pacemaker rhythm in guinea pig sinoatrial node cells. *Am. J. Physiol. Heart Circ. Physiol.* **2011**, *300*, 251–261. [CrossRef] [PubMed]
5. Toyoda, F.; Ding, W.G.; Matsuura, H. Heterogeneous functional expression of the sustained inward Na^+ current in guinea pig sinoatrial node cells. *Pflug. Arch.* **2018**, *470*, 481–490. [CrossRef]
6. Satoh, H. Sino-atrial nodal cells of mammalian hearts: Ionic currents and gene expression of pacemaker ionic channels. *J. Smooth Muscle Res.* **2003**, *39*, 175–193. [CrossRef]
7. Ono, K.; Iijima, T. Cardiac T-type Ca^{2+} channels in the heart. *J. Mol. Cell. Cardiol.* **2010**, *48*, 65–70. [CrossRef]
8. Maltsev, V.A.; Vinogradova, T.M.; Lakatta, E.G. The emergence of a general theory of the initiation and strength of the heartbeat. *J. Pharmacol. Sci.* **2006**, *100*, 338–369. [CrossRef]
9. Sirenko, S.G.; Yang, D.; Maltseva, L.A.; Kim, M.S.; Lakatta, E.G.; Maltsev, V.A. Spontaneous, local diastolic subsarcolemmal calcium releases in single, isolated guinea-pig sinoatrial nodal cells. *PLoS ONE* **2017**, *12*, e0185222. [CrossRef]
10. Sanders, L.; Rakovic, S.; Lowe, M.; Mattick, P.A.; Terrar, D.A. Fundamental importance of Na^+-Ca^{2+} exchange for the pacemaking mechanism in guinea-pig sino-atrial node. *J. Physiol.* **2006**, *571*, 639–649. [CrossRef]
11. Mangoni, M.E.; Nargeot, J. Properties of the hyperpolarization-activated current (I_f) in isolated mouse sino-atrial cells. *Cardiovasc. Res.* **2001**, *52*, 51–64. [CrossRef]
12. Tanaka, H.; Nishimaru, K.; Aikawa, T.; Hirayama, W.; Tanaka, Y.; Shigenobu, K. Effect of SEA0400, a novel inhibitor of sodium-calcium exchanger, on myocardial ionic currents. *Br. J. Pharmacol.* **2002**, *135*, 1096–1100. [CrossRef] [PubMed]
13. Tanaka, H.; Namekata, I.; Takeda, K.; Kazama, A.; Shimizu, Y.; Moriwaki, R.; Hirayama, W.; Sato, A.; Kawanishi, T.; Shigenobu, K. Unique excitation-contraction characteristics of mouse myocardium as revealed by SEA0400, a specific inhibitor of Na^+-Ca^{2+} exchanger. *Naunyn Schmiedebergs Arch. Pharmacol.* **2005**, *371*, 526–534. [CrossRef] [PubMed]
14. Tanaka, H.; Shimada, H.; Namekata, I.; Kawanishi, T.; Iida-Tanaka, N.; Shigenobu, K. Involvement of the Na^+/Ca^{2+} exchanger in ouabain-induced inotropy and arrhythmogenesis in guinea-pig myocardium as revealed by SEA0400. *J. Pharmacol. Sci.* **2007**, *103*, 241–246. [CrossRef]
15. Namekata, I.; Nakamura, H.; Shimada, H.; Tanaka, H.; Shigenobu, K. Cardioprotection without cardiosuppression by SEA0400, a novel inhibitor of Na^+-Ca^{2+} exchanger, during ischemia and reperfusion in guinea-pig myocardium. *Life Sci.* **2005**, *77*, 312–324. [CrossRef]

16. Torrente, A.G.; Zhang, R.; Zaini, A.; Giani, J.F.; Kang, J.; Lamp, S.T.; Philipson, K.D.; Goldhaber, J.I. Burst pacemaker activity of the sinoatrial node in sodium-calcium exchanger knockout mice. *Proc. Natl. Acad. Sci. USA* **2015**, *112*, 9769–9774. [CrossRef]
17. Gao, Z.; Rasmussen, T.P.; Li, Y.; Kutschke, W.; Koval, O.M.; Wu, Y.; Wu, Y.; Hall, D.D.; Joiner, M.L.; Wu, X.Q.; et al. Genetic inhibition of Na^+-Ca^{2+} exchanger current disables fight or flight sinoatrial node activity without affecting resting heart rate. *Circ. Res.* **2013**, *112*, 309–317. [CrossRef]
18. Kojima, A.; Ito, Y.; Kitagawa, H.; Matsuura, H.; Nosaka, S. Direct negative chronotropic action of desflurane on sinoatrial node pacemaker activity in the guinea pig heart. *Anesthesiology* **2014**, *120*, 1400–1413. [CrossRef]
19. Chen, B.; Wu, Y.; Mohler, P.J.; Anderson, M.E.; Song, L.S. Local control of Ca^{2+}-induced Ca^{2+} release in mouse sinoatrial node cells. *J. Mol. Cell. Cardiol.* **2009**, *47*, 706–715. [CrossRef]
20. Magee, W.P.; Deshmukh, G.; Deninno, M.P.; Sutt, J.C.; Chapman, J.G.; Tracey, W.R. Differing cardioprotective efficacy of the Na^+/Ca^{2+} exchanger inhibitors SEA0400 and KB-R7943. *Am. J. Physiol. Heart Circ. Physiol.* **2003**, *284*, 903–910. [CrossRef]
21. Takahashi, T.; Takahashi, K.; Onishi, M.; Suzuki, T.; Tanaka, Y.; Ota, T.; Yoshida, S.; Nakaike, S.; Matsuda, T.; Baba, A. Effects of SEA0400, a novel inhibitor of the Na^+/Ca^{2+} exchanger, on myocardial stunning in anesthetized dogs. *Eur. J. Pharmacol.* **2004**, *505*, 163–168. [CrossRef] [PubMed]
22. Namekata, I.; Tsuneoka, Y.; Takahara, A.; Shimada, H.; Sugimoto, T.; Takeda, K.; Nagaharu, M.; Shigenobu, K.; Kawanishi, T.; Tanaka, H. Involvement of the Na^+/Ca^{2+} exchanger in the automaticity of guinea-pig pulmonary vein myocardium as revealed by SEA0400. *J. Pharmacol. Sci.* **2009**, *110*, 111–116. [CrossRef] [PubMed]
23. Namekata, I.; Tanaka, Y.; Ohmori, T.; Tsuneoka, Y.; Hamaguchi, S.; Tanaka, H.; Tanaka, H. Cell morphology and early-phase Ca^{2+} transients of guinea-pig pulmonary vein cardiomyocytes compared with atrial and ventricular cardiomyocytes. *Bioimages* **2019**, *27*, 1–12.
24. Verkerk, A.O.; Wilders, R.; Zegers, J.G.; van Borren, M.M.; Ravesloot, J.H.; Verheijck, E.E. Ca^{2+}-activated Cl^- current in rabbit sinoatrial node cells. *J. Physiol.* **2002**, *540*, 105–117. [CrossRef] [PubMed]
25. Mao, Z.; Wang, Y.; Peng, H.; He, F.; Zhu, L.; Huang, H.; Huang, X.; Lu, X.; Tan, X. A newly identified missense mutation in CLCA2 is associated with autosomal dominant cardiac conduction block. *Gene* **2019**, *714*, 143990. [CrossRef]
26. Turner, D.; Kang, C.; Mesirca, P.; Hong, J.; Mangoni, M.E.; Glukhov, A.V.; Sah, R. Electrophysiological and Molecular Mechanisms of Sinoatrial node mechanosensitivity. *Front. Cardiovasc. Med.* **2021**, *8*, 662410. [CrossRef]
27. Lei, M.; Jones, S.A.; Liu, J.; Lancaster, M.K.; Fung, S.S.; Dobrzynski, H.; Camelliti, P.; Maier, S.K.; Noble, D.; Boyett, M.R. Requirement of neuronal- and cardiac-type sodium channels for murine sinoatrial node pacemaking. *J. Physiol.* **2004**, *559*, 835–843. [CrossRef]

MDPI
St. Alban-Anlage 66
4052 Basel
Switzerland
Tel. +41 61 683 77 34
Fax +41 61 302 89 18
www.mdpi.com

Biomolecules Editorial Office
E-mail: biomolecules@mdpi.com
www.mdpi.com/journal/biomolecules

www.ingramcontent.com/pod-product-compliance
Lightning Source LLC
Chambersburg PA
CBHW061604210926
51850CB00009B/2213